Lecture Notes in Mathematics 785

Editors:
A. Dold, Heidelberg
B. Eckmann, Zürich

Springer
*Berlin*
*Heidelberg*
*New York*
*Barcelona*
*Budapest*
*Hong Kong*
*London*
*Milan*
*Paris*
*Santa Clara*
*Singapore*
*Tokyo*

Wolfgang M. Schmidt

# Diophantine Approximation

Springer

Author

Wolfgang M. Schmidt
Department of Mathematics
University of Colorado
Boulder, CO 80309, USA



1st edition 1980
2nd printing 1996 (with minor corrections)

Mathematics Subject Classification (1980): 10B16, 10E05, 10E15, 10E40,
10F05, 10F10, 10F20, 10F30, 10K15

Library of Congress Cataloging in Publication Data.
Schmidt, Wolfgang M., Diophantine approximation. (Lecture notes in mathematics ; 785)
Bibliography: p. Includes index. 1. Algebraic number theory. 2.Approximation, Diophantine.
I. Title. II. Series: Lecture notes in mathematics (Berlin) ; 785. QA3.L28 no. 785 [QA247]
510s [512'.74] 80-11695
ISBN 0-387-09762-7

ISBN 3-540-09762-7 Springer-Verlag Berlin Heidelberg New York



Printed in Germany

SPIN: 10530138 46/2142-543210 - Printed on acid-free paper

## Preface

In spring 1970 I gave a course in Diophantine Approximation at the University of Colorado, which culminated in simultaneous approximation to algebraic numbers. A limited supply of mimeographed Lecture Notes was soon gone. The completion of these new Notes was greatly delayed by my decision to add further material.

The present chapter on simultaneous approximations to algebraic numbers is much more general than the one in the original Notes. This generality is necessary to supply a basis for the subsequent chapter on norm form equations. There is a new last chapter on approximation by algebraic numbers. I wish to thank all those, in particular Professor C.L. Siegel, who have pointed out a number of mistakes in the original Notes. I hope that not too many new mistakes have crept into these new Notes.

The present Notes contain only a small part of the theory of Diophantine Approximation. The main emphasis is on approximation to algebraic numbers. But even here not everything is included. I follow the approach which was initiated by Thue in 1908, and further developed by Siegel and by Roth, but I do not include the effective results due to Baker. Not included is approximation in $p$-adic fields, for which see e.g. Schlickewei [1976, 1977], or approximation in power series fields, for which see e.g., Osgood [1977] and Ratliff [1978]. Totally missing are Pisot-Vijayaraghavan Numbers, inhomogeneous approximation and uniform distribution. For these see e.g. Cassels [1957] and Kuipers and Niederreiter [1974]. Also excluded are Weyl Sums, nonlinear approximation and diophantine inequalities involving forms in many variables.

My pace is in general very leisurely and slow. This will be especially apparent when comparing Baker's [1975] chapter on approximation to algebraic numbers with my two separate chapters, one dealing with Roth's Theorem on approximation to a single algebraic number, the other with simultaneous approximation to algebraic numbers.

Possible sequences are chapters

I, II, III, for a reader who is interested in game and measure theoretic results, or

I, II, V, for a reader who wants to study Roth's Theorem , or

I, II, IV, V, VI, VII (§ 11, 12), VIII (§ 7-10), for a general theory of simultaneous approximation to algebraic numbers, or

I, II, IV, V, VI, VII, if the goal is norm form equations, or

I, II, VIII (§ 1-6, §11), if the emphasis is on approximation by algebraic numbers.

December 1979 W.M. Schmidt

# Notation

A real number $\xi$ may uniquely be written as

$$\xi = [\xi] + \{\xi\} \quad ,$$

where $[\xi]$, the <u>integer part of</u> $\xi$, is an integer, and where $\{\xi\}$, the <u>fractional part of</u> $\xi$, satisfies $0 \leq \{\xi\} < 1$.

$\|\xi\| = \min(\{\xi\}, 1 - \{\xi\})$ is the distance from $\xi$ to the nearest integer,

$U$ denotes the unit interval $0 \leq \xi < 1$.

$\mathbb{R}^n$ denotes the $n$-dimensional real space,

$E^n$ denotes Euclidean $n$-space.

$\underline{\underline{x}}, \underline{\underline{y}}, \ldots$ will denote vectors; so $\underline{\underline{x}} = (x_1, \ldots, x_n) \in \mathbb{R}^n$, or

$\underline{\underline{x}} = (x_1, \ldots, x_m) \in \mathbb{R}^m$, etc.

Addition and multiplication of vectors by scalars is obvious.

$\underline{\underline{e}}_1, \ldots, \underline{\underline{e}}_n$ will denote basis vectors.

$\lambda K$, where $\lambda > 0$ and where $K$ is in $\mathbb{R}^n$, is the set of elements $\lambda\underline{\underline{x}}$ with $\underline{\underline{x}} \in K$.

$\delta_{ij}$ is the Kronecker Symbol.

$X, Y, \ldots$, in general will be variables, while $x, y, \ldots$ will be real, usually rational integers. But this rule is sometimes hard to follow: In chapter IV, the symbols $X, Y, \ldots$ will also be used to denote coordinates in compound spaces.

$\boxed{\underline{\underline{x}}} = \max(|x_1|, \ldots, |x_n|)$ if $\underline{\underline{x}} = (x_1, \ldots, x_n)$. However $\boxed{\underline{\underline{\beta}}}$, where $\underline{\underline{\beta}} = (\beta_1, \ldots, \beta_n)$ has coordinates in an algebraic number field $K$, is given by $\boxed{\underline{\underline{\beta}}} = \max(|\beta_1^{(1)}|, \ldots, |\beta_n^{(1)}|, \ldots, |\beta_1^{(k)}|, \ldots, |\beta_n^{(k)}|)$, if $\beta^{(1)} = \beta, \beta^{(2)}, \ldots, \beta^{(k)}$ are the conjugates of an elements $\beta$. (But, on p. 173, $\boxed{\gamma}$ for a single element $\gamma$ has a different meaning.)

$\lceil P \rceil$ is the maximum absolute value of the coefficients of a polynomial $P$,

$\mathbb{Q}$ is the field of rationals,

$\mathbb{R}$ is the field of reals,

$\mathbb{C}$ is the field of complex numbers.

$[L : K]$ is the degree of a field extension $L$ over $K$.

$\{a,b,\dots,w\}$ denotes the set consisting of $a,b,\dots,w$, and

$\sim$ denotes a set theoretic difference.

$\ll$ is the Vinogradov symbol. Thus e.g. $f(\underline{\underline{x}}) \ll g(\underline{\underline{x}})$ means that $|f(\underline{\underline{x}})| \leq c|g(\underline{\underline{x}})|$ with a constant $c$. Often this "implied" constant $c$ may depend on extra parameters, such as the dimension, etc.

$\gg\ll$, in the context $f \gg\ll g$, means that both $f \ll g$ and $g \ll f$.

$o$, the "little o", in the context $f(n) = o(g(n))$, means that $f(n)/g(n)$ tends to $0$ as $n \to \infty$.

g.c.d. denotes the greatest common divisor of integers.

Starred Theorems, such as Theorem 6A*, are not proved in these Notes.

Table of Contents

## I. Approximation to Irrational Numbers by Rationals.

References: Dirichlet (1842), Hurwitz (1891), Perron (1954), Cassels (1957).

### §1. Dirichlet's Theorem.

Given a real number $\alpha$, let $[\alpha]$, the *integer part* of $\alpha$, denote the greatest integer $\leq \alpha$, and let $\{\alpha\} = \alpha - [\alpha]$. Then $\{\alpha\}$ is the *fractional part* of $\alpha$, and satisfies $0 \leq \{\alpha\} < 1$. Also, let $\|\alpha\|$ denote the distance from $\alpha$ to the nearest integer. Then always $0 \leq \|\alpha\| \leq \frac{1}{2}$.

THEOREM 1A. (Dirichlet (1842)). *Let $\alpha$ and $Q$ be real numbers with $Q > 1$. Then there exist integers $p, q$ such that $1 \leq q < Q$ and $|\alpha q - p| \leq \frac{1}{Q}$.*

*Proof*. First assume that $Q$ is an integer. Consider the following $Q + 1$ numbers:

$$0, 1, \{\alpha\}, \{2\alpha\}, \ldots, \{(Q-1)\alpha\} .$$

They all lie in the unit interval $0 \leq x \leq 1$. We divide the unit interval into $Q$ subintervals

$$\frac{u}{Q} \leq x < \frac{u+1}{Q} \qquad (u = 0, 1, \ldots, Q-1) ,$$

but with $<$ replaced by $\leq$ if $u = Q-1$. At least one such subinterval contains two (or more) of the $Q+1$ numbers above. Hence there are integers $r_1, r_2, s_1, s_2$ with $0 \leq r_i < Q$ $(i = 1,2)$ and $r_1 \neq r_2$ such that

$$|(r_1\alpha - s_1) - (r_2\alpha - s_2)| \leq \frac{1}{Q} .$$

If, say, $r_1 > r_2$ , put $q = r_1 - r_2$ , $p = s_1 - s_2$ . Then $1 \le q < Q$ and $|q\alpha - p| \le \frac{1}{Q}$ , proving the theorem when $Q$ is an integer.

Next, suppose $Q$ is <u>not</u> an integer. Apply what has already been proved to $Q' = [Q]+1$ . Then $1 \le q < Q'$ implies $1 \le q \le [Q]$ , whence $1 \le q < Q$ , and the theorem is true for $Q$ .

<u>Remark</u>. The two inequalities in Dirichlet's Theorem yield

$$|\alpha - \frac{p}{q}| \le \frac{1}{Qq} < \frac{1}{q^2} .$$

<u>COROLLARY 1B</u>. <u>Suppose that $\alpha$ is irrational. Then there exist infinitely many pairs</u> $p,q$ <u>of relatively prime integers with</u>

$$|\alpha - \frac{p}{q}| < \frac{1}{q^2} . \tag{1.1}$$

<u>Proof</u>. Dirichlet's Theorem obviously remains true if we ask for <u>relatively prime</u> integers $p,q$ satisfying $1 \le q < Q$ and $|\alpha q - p| \le \frac{1}{Q}$ . Since $\alpha$ is irrational, $\alpha q - p$ is never zero, and hence for any given $p,q$ , the inequality $|\alpha q - p| \le \frac{1}{Q}$ can only be satisfied for $Q \le Q_0(p,q)$. Hence as $Q \to \infty$ , there will be infinitely many <u>distinct</u> pairs $p,q$ of relatively prime integers occuring in Dirichlet's Theorem.

<u>Remark</u>. This corollary is <u>not</u> true if $\alpha$ is rational. For suppose that $\alpha = \frac{u}{v}$ . If $\alpha \ne \frac{p}{q}$ , then $|\alpha - \frac{p}{q}| = |\frac{u}{v} - \frac{p}{q}| = |\frac{qu - pv}{vq}| \ge \frac{1}{vq}$ , and therefore (1.1) can be satisfied by only <u>finitely</u> many pairs $p,q$ of relatively prime integers.

## <u>§2. Farey Series.</u>

<u>Definition</u>. The <u>Farey series</u> $\mathfrak{F}_n$ <u>of order</u> $n(\ge 1)$ is the sequence of rationals in their lowest terms between 0 and 1 with

denominators $\leq n$ , written in ascending order. For example,

$$\mathcal{F}_5: 0, \frac{1}{5}, \frac{1}{4}, \frac{1}{3}, \frac{2}{5}, \frac{1}{2}, \frac{3}{5}, \frac{2}{3}, \frac{3}{4}, \frac{4}{5}, 1 .$$

THEOREM 2A. *If* $\frac{h}{k}, \frac{h'}{k'}$ *are successive terms in* $\mathcal{F}_n$ , *then* $h'k - hk' = 1$ .

We need

LEMMA 2B. *Suppose that* $\underline{x} = (x_1, x_2)$ *and* $\underline{y} = (y_1, y_2)$ *are integer points in the plane, with* $\underline{0} = (0,0), \underline{x}, \underline{y}$ *not on a line. Suppose further that the closed triangle with vertices* $\underline{0}, \underline{x}, \underline{y}$ *contains no integer points but its vertices. Then*

$$x_1 y_2 - x_2 y_1 = \pm 1 .$$

*Proof of the Lemma.* Let $\mathcal{T}$ be the triangle mentioned above, and let $\mathcal{P}$ be the closed parallelogram with vertices $\underline{0}, \underline{x}, \underline{y}$ and $\underline{x}+\underline{y}$ . Then $\mathcal{P}$ contains no integer points but its vertices: for suppose that $\underline{z}$ is an integer point in $\mathcal{P}$ , $\underline{z} \notin \mathcal{T}$ . Then $\underline{x} + \underline{y} - \underline{z} \in \mathcal{T}$ , hence $\underline{x} + \underline{y} - \underline{z} = \underline{0}, \underline{x}$ or $\underline{y}$ , and therefore $\underline{z} = \underline{x} + \underline{y}, \underline{y}$ or $\underline{x}$ .

If $\underline{p}$ is any integer point, we may write $\underline{p} = \lambda\underline{x} + \mu\underline{y}$ with *real* coefficients $\lambda, \mu$ since $\underline{0}, \underline{x}, \underline{y}$ are not collinear. Then $\underline{p} = \underline{p}' + \underline{p}''$ , where

$$\underline{p}' = [\lambda]\underline{x} + [\mu]\underline{y} \text{ and } \underline{p}'' = \{\lambda\}\underline{x} + \{\mu\}\underline{y} .$$

Both $\underline{p}$ and $\underline{p}'$ are integer points, hence so is $\underline{p}''$ . Also $\underline{p}'' \in \mathcal{P}$ . Since $\underline{p}'' \neq \underline{x}, \underline{y}$ and $\underline{x} + \underline{y}$ , we have $\underline{p}'' = \underline{0}$ . Therefore $\underline{p} = \lambda\underline{x} + \mu\underline{y}$ with *integer* coefficients $\lambda, \mu$ .

In particular,

$$(1,0) = \lambda\underline{\underline{x}} + \mu\underline{\underline{y}} = (\lambda x_1 + \mu y_1, \lambda x_2 + \mu y_2) \quad ,$$

$$(0,1) = \lambda'\underline{\underline{x}} + \mu'\underline{\underline{y}} = (\lambda' x_1 + \mu' y_1, \lambda' x_2 + \mu' y_2)$$

for certain integers $\lambda, \mu, \lambda', \mu'$. It follows that

$$1 = \begin{vmatrix} 1 & 0 \\ 0 & 1 \end{vmatrix} = \begin{vmatrix} \lambda & \mu \\ \lambda' & \mu' \end{vmatrix} \cdot \begin{vmatrix} x_1 & x_2 \\ y_1 & y_2 \end{vmatrix} \quad ,$$

whence

$$\begin{vmatrix} x_1 & x_2 \\ y_1 & y_2 \end{vmatrix} = \pm 1$$

as claimed.

<u>Proof of the Theorem.</u> Put $\underline{\underline{x}} = (h,k)$, $\underline{\underline{y}} = (h',k')$. Then $\underline{\underline{0}}, \underline{\underline{x}}, \underline{\underline{y}}$ are not collinear since $\gcd(h,k) = \gcd(h',k') = 1$ and $\underline{\underline{x}} \neq \underline{\underline{y}}$. Let $\mathcal{T}$ denote the closed triangle with vertices $\underline{\underline{0}}, \underline{\underline{x}}, \underline{\underline{y}}$. Then there is no integer point in $\mathcal{T}$ besides $\underline{\underline{0}}, \underline{\underline{x}}, \underline{\underline{y}}$. For if there were such a point $(h'', k'')$, then there also would be a point with $\gcd(h'', k'') = 1$. Then $(h'', k'') = \lambda(h,k) + \mu(h', k')$ with $\lambda \geqq 0$, $\mu \geqq 0$, $0 < \lambda + \mu \leqq 1$ and $(\lambda, \mu)$ not equal to $(1,0)$ or to $(0,1)$. This implies that $k'' \leqq \lambda n + \mu n \leq n$. We have $\lambda > 0$, $\mu > 0$ (since $\gcd(h,k) = \gcd(h',k') = 1$), whence $\frac{h}{k} < \frac{h''}{k''} < \frac{h'}{k'}$. Thus $\frac{h''}{k''}$ would belong to $\mathcal{F}_n$, contradicting the supposition that $\frac{h}{k}$ and $\frac{h'}{k'}$ are consecutive elements of $\mathcal{F}_n$. The hypotheses of Lemma 2B are now satisfied, and we conclude that $h'k - hk' = \pm 1$. Since $\frac{h}{k} < \frac{h'}{k'}$, we have $h'k - hk' = 1$.

<u>COROLLARY 2C. *If* $\frac{h}{k}, \frac{h''}{k''}, \frac{h'}{k'}$ *are consecutive elements of* $\mathcal{F}_n$,</u>

<u>*then*</u>

$$\frac{h''}{k''} = \frac{h+h'}{k+k'} \quad .$$

Proof. By the theorem, $h''k - hk'' = 1$ and $h'k'' - h''k' = 1$, so that $h''(k+k') - k''(h+h') = 0$.

LEMMA 2D. *Suppose that $\frac{h}{k}$, $\frac{h'}{k'}$ are successive terms in the Farey series $\mathcal{F}_n$, and put $h'' = h + h'$, $k'' = k + k'$. (Note that $\frac{h''}{k''}$ does NOT belong to $\mathcal{F}_n$). Then for every $\alpha$ in $\frac{h}{k} \leqq \alpha \leqq \frac{h'}{k'}$, at least one of the following three inequalities holds*:

$$(2.1) \qquad \left|\alpha - \frac{h}{k}\right| < \frac{1}{\sqrt{5}k^2} \quad , \quad \left|\alpha - \frac{h''}{k''}\right| < \frac{1}{\sqrt{5}k''^2} \quad , \quad \left|\alpha - \frac{h'}{k'}\right| < \frac{1}{\sqrt{5}k'^2} \quad .$$

Proof. We may assume that $\alpha \geqq \frac{h''}{k''}$. Namely, otherwise replace $\alpha$ by $1 - \alpha$, $\frac{h}{k}$ by $1 - \frac{h'}{k'}$, etc. If none of the inequalities above hold, then

$$\alpha - \frac{h}{k} \geq \frac{1}{\sqrt{5}k^2} \quad , \quad \alpha - \frac{h''}{k''} \geq \frac{1}{\sqrt{5}k''^2} \quad , \quad \frac{h'}{k'} - \alpha \geq \frac{1}{\sqrt{5}k'^2} \quad .$$

Adding the first and third inequalities, we obtain

$$\frac{h'}{k'} - \frac{h}{k} = \frac{1}{kk'} \geq \frac{1}{\sqrt{5}}\left(\frac{1}{k^2} + \frac{1}{k'^2}\right) ;$$

adding the second and third inequalities, we obtain

$$\frac{h'}{k'} - \frac{h''}{k''} = \frac{1}{k'k''} \geq \frac{1}{\sqrt{5}}\left(\frac{1}{k'^2} + \frac{1}{k''^2}\right) \quad .$$

Then $\sqrt{5}kk' \geq k^2 + k'^2$ and $\sqrt{5}k'k'' \geq k'^2 + k''^2$, so that $\sqrt{5}\, k'(k+k'') \geq k^2 + 2k'^2 + k''^2$, and therefore $\sqrt{5}\, k'(2k+k') \geq 2k^2 + 3k'^2 + 2kk'$. It follows that

$$0 \geq \frac{1}{2}((\sqrt{5} - 1)k' - 2k)^2 \quad .$$

But this is impossible, since $k$ and $k'$ are nonzero integers.

LEMMA 2E. *Suppose* $\alpha$ *is a real quadratic irrational which is a root of a non-zero polynomial*

$$P(X) = aX^2 + bX + c$$

*with rational integer coefficients and discriminant* $D = b^2 - 4ac$. *Then for* $A > \sqrt{D}$, *the inequality*

$$|\alpha - \frac{p}{q}| < \frac{1}{Aq^2} \tag{2.2}$$

*has only finitely many solutions.*

Proof. Write $P(X) = a(X-\alpha)(X-\alpha')$, so that $D = a^2(\alpha-\alpha')^2$. Given $p/q$ with (2.2) we have

$$\frac{1}{q^2} \leq |P(\frac{p}{q})| = |\alpha - \frac{p}{q}|\,|a(\alpha' - \frac{p}{q})| < \frac{1}{Aq^2}|a(\alpha' - \alpha + \alpha - \frac{p}{q})| < \frac{\sqrt{D}}{Aq^2} + \frac{|a|}{A^2q^4},$$

which clearly is impossible if $A > \sqrt{D}$ and if $q$ is large.

THEOREM 2F. (Hurwitz (1891)).

(i) *For every irrational number* $\alpha$ *there are infinitely many distinct rationals* $\frac{p}{q}$ *with*

$$|\alpha - \frac{p}{q}| < \frac{1}{\sqrt{5}\,q^2}. \tag{2.3}$$

(ii) *This would be wrong if* $\sqrt{5}$ *were replaced by a constant* $A > \sqrt{5}$.

Proof. We may suppose that $0 < \alpha < 1$. If $\frac{h}{k}$ and $\frac{h'}{k'}$ are the successive terms in the Farey series $\mathfrak{F}_n$ with $\frac{h}{k} < \alpha < \frac{h'}{k'}$, then according to Lemma 2D, at least one of $\frac{h}{k}$ , $\frac{h'}{k'}$ , and $\frac{h''}{k''}$ with $h'' = h + h'$ , $k'' = k + k'$ satisfies (2.3). Then by the construction of the Farey Series, $\left| \frac{h}{k} - \frac{h'}{k'} \right| \leq 1/n$ and therefore $\left|\alpha - \frac{p}{q}\right| \leq 1/n$ . Since $\alpha$ is irrational, this can for fixed $\frac{p}{q}$ only be true for small values of $n$ . Hence as $n$ goes to infinity, we obtain infinitely many $\frac{p}{q}$ with (2.3).

Assertion (ii) follows from the case $\alpha = \frac{1}{2}(\sqrt{5} - 1)$ and $P(X) = X^2 + X - 1$ of Lemma 2E, since $D = 5$ in this case.

## §3. Continued Fractions: Algebraic Theory.

Let $a_0, a_1, a_2, \ldots$ be variables. We define polynomials $p_0, q_0, p_1, q_1, p_2, q_2, \ldots,$ with $p_n$ and $q_n$ being polynomials in $a_0, a_1, \ldots, a_n$ , as follows:

First, we define $p_0 = a_0$ and $q_0 = 1$ . Next, suppose that $p_0, q_0, \ldots, p_{n-1}, q_{n-1}$ have already been defined. Using the abbreviations

$$p'_k = p_k(a_1, a_2, \ldots, a_{k+1}) \text{ and } q'_k = q_k(a_1, a_2, \ldots, a_{k+1}) ,$$

we define

$$p_n = a_0 p'_{n-1} + q'_{n-1} \text{ and } q_n = p'_{n-1} .$$

$\frac{p_n}{q_n}$ is a rational function of $a_0, a_1, \ldots, a_n$ , and we write

$$\frac{p_n}{q_n} = [a_0, a_1, \ldots, a_n] .$$

In particular, $[a_0] = \frac{p_0}{q_0} = a_0$ . Unfortunately, if $a_0$ is not an integer, this notation conflicts with the greatest integer notation introduced in §1. However, the meaning of $[a_0]$ will be clear from the context.

For $n > 0$ , we have

$$[a_0,a_1,\dots,a_n] = \frac{p_n}{q_n} = \frac{a_0p'_{n-1} + q'_{n-1}}{p'_{n-1}} = a_0 + \frac{1}{p'_{n-1}/q'_{n-1}} = a_0 + \frac{1}{[a_1,\dots,a_n]} \ .$$

Applying this repeatedly, we obtain

$$[a_0,a_1,\dots,a_n] = a_0 + \cfrac{1}{a_1 + \cfrac{1}{a_2 + \cfrac{}{\ddots a_{n-1} + \cfrac{1}{a_n}}}} \ .$$

A rational function of this type is called a <u>continued fraction</u>.

<u>LEMMA 3A</u>. <u>For</u> $n \geq 2$ , <u>we have</u>

$$p_n = a_n p_{n-1} + p_{n-2} \ ,$$
$$q_n = a_n q_{n-1} + q_{n-2} \ .$$

<u>Proof</u>. For $n = 2$ , the assertion is verified directly. Assume, then that $n > 2$ , and that the assertion is true for $n - 1$ . Then

$$p'_{n-1} = a_n p'_{n-2} + p'_{n-3} \ ,$$
$$q'_{n-1} = a_n q'_{n-2} + q'_{n-3} \ .$$

We obtain

$$\begin{aligned} p_n &= a_0p'_{n-1} + q'_{n-1} = a_0(a_np'_{n-2} + p'_{n-3}) + a_nq'_{n-2} + q'_{n-3} \\ &= a_n(a_0p'_{n-2} + q'_{n-2}) + a_0p'_{n-3} + q'_{n-3} \\ &= a_np_{n-1} + p_{n-2} \ , \end{aligned}$$

and

$$q_n = p'_{n-1} = a_n p'_{n-2} + p'_{n-3}$$
$$= a_n q_{n-1} + q_{n-2} \quad ,$$

which proves the lemma.

For convenience, we also define $p_{-2} = 0$, $p_{-1} = 1$, $q_{-2} = 1$, $q_{-1} = 0$. It is easily checked that Lemma 3A then holds for all $n \geq 0$.

LEMMA 3B. *For each* $k$ *with* $1 \leq k \leq n$, *let*

$$r_k = [a_k, a_{k+1}, \ldots, a_n] \ .$$

*Then*

$$[a_0, a_1, \ldots, a_k, \ldots, a_n] = [a_0, a_1, \ldots, a_{k-1}, [a_k, \ldots, a_n]]$$

$$= \frac{p_{k-1} r_k + p_{k-2}}{q_{k-1} r_k + q_{k-2}} \ .$$

*Proof*. The second equality follows from Lemma 3A. The first equality may be proved by induction on $k$. The lemma is true for $k = 1$, since

$$[a_0, a_1, \ldots, a_n] = a_0 + \frac{1}{[a_1, \ldots, a_n]} = [a_0, [a_1, \ldots, a_n]] \ .$$

Now suppose that the lemma holds for $k - 1$, where $1 < k \leq n$. Then

$$[a_0, a_1, \ldots, a_n] = a_0 + \frac{1}{[a_1, \ldots, a_k, \ldots, a_n]} = a_0 + \frac{1}{[a_1, \ldots, a_{k-1}, [a_k, \ldots, a_n]]}$$

$$= [a_0, a_1, \ldots, a_{k-1}, [a_k, \ldots, a_n]] \quad ,$$

which proves the lemma.

LEMMA 3C. *For* $n \geq -1$, *we have*

$$q_n p_{n-1} - p_n q_{n-1} = (-1)^n \quad .$$

Proof. We proceed by induction on $n$. Now $q_{-1}p_{-2} - p_{-1}q_{-2} = 0^2 - (-1)^2 = (-1)^{-1}$, so the lemma is true for $n = -1$. Assuming that the lemma is true for $n-1$, we have

$$\begin{aligned} q_n p_{n-1} - p_n q_{n-1} &= (a_n q_{n-1} + q_{n-2})p_{n-1} - (a_n p_{n-1} + p_{n-2})q_{n-1} \\ &= -(q_{n-1}p_{n-2} - p_{n-1}q_{n-2}) = -(-1)^{n-1} = (-1)^n \quad , \end{aligned}$$

and the lemma is true for $n$.

LEMMA 3D. For $n \geq 0$, $q_n p_{n-2} - p_n q_{n-2} = (-1)^{n-1} a_n$ .

Proof. Using Lemma 3A and 3C, we have

$$\begin{aligned} q_n p_{n-2} - p_n q_{n-2} &= (a_n q_{n-1} + q_{n-2})p_{n-2} - (a_n p_{n-1} + p_{n-2})q_{n-2} \\ &= (q_{n-1}p_{n-2} - p_{n-1}q_{n-2})a_n = (-1)^{n-1} a_n \quad . \end{aligned}$$

LEMMA 3E. Put $\alpha = [a_0, a_1, \ldots, a_{n+1}]$ . Then

$$q_n \alpha - p_n = \frac{(-1)^n}{a_{n+1}q_n + q_{n-1}} \quad .$$

Proof. $q_n \alpha - p_n = q_n \dfrac{p_{n+1}}{q_{n+1}} - p_n = \dfrac{-(q_{n+1}p_n - p_{n+1}q_n)}{q_{n+1}} = \dfrac{(-1)^n}{a_{n+1}q_n + q_{n-1}}$

by Lemmas 3A and 3C.

LEMMA 3F. For $n \geq 1$, we have

$$\frac{q_n}{q_{n-1}} = [a_n, a_{n-1}, \ldots, a_1] \quad .$$

Proof. We use induction on $n$ . $\frac{q_1}{q_0} = \frac{a_1}{1} = [a_1]$ , so the lemma is true for $n = 1$ . Now suppose that the lemma is true for $n-1$ in place of $n$ . Then

$$\frac{q_n}{q_{n-1}} = \frac{a_n q_{n-1} + q_{n-2}}{q_{n-1}} = a_n + \frac{1}{q_{n-1}/q_{n-2}} = a_n + \frac{1}{[a_{n-1},\ldots,a_1]}$$

$$= [a_n, a_{n-1}, \ldots, a_1] ,$$

and the lemma is proved.

## §4. Simple continued fractions.

LEMMA 4A. *Suppose that* $a_0, a_1, a_2, \ldots,$ *are real numbers with* $a_1, a_2, \ldots$ *positive. Then*

$$\frac{p_0}{q_0} < \frac{p_2}{q_2} < \frac{p_4}{q_4} < \ldots ; \ldots < \frac{p_5}{q_5} < \frac{p_3}{q_3} < \frac{p_1}{q_1} .$$

Proof. By Lemma 3D,

$$\frac{p_{n-2}}{q_{n-2}} - \frac{p_n}{q_n} = \frac{(-1)^{n-1} a_n}{q_{n-2} q_n} .$$

Applying this with $n \geq 2$ , $n$ even, we get $\frac{p_{n-2}}{q_{n-2}} - \frac{p_n}{q_n} < 0$ ; applying this with $n \geq 3$ , $n$ odd, we get $\frac{p_{n-2}}{q_{n-2}} - \frac{p_n}{q_n} > 0$ .

It remains to show that $\frac{p_n}{q_n} < \frac{p_m}{q_m}$ if $n$ is even and $m$ is odd. Suppose, say, that $n < m$ . Then since $\frac{p_n}{q_n} \leq \frac{p_{m-1}}{q_{m-1}}$ , it will suffice to show that $\frac{p_{m-1}}{q_{m-1}} < \frac{p_m}{q_m}$ . But the last inequality is true since $q_m p_{m-1} - p_m q_{m-1} = (-1)^m < 0$ . (Recall that $m$ is odd.) The case

$n > m$ is proved similarly.

LEMMA 4B. Suppose that $a_0, a_1 \geq 1, \ldots, a_n \geq 1$ are integers. Then $[a_0, a_1, \ldots, a_n]$ is rational. Conversely, given a rational number $\frac{u}{v}$, there will be an $n$ and integers $a_0, a_1, \ldots, a_n$ as above with

$$\frac{u}{v} = [a_0, a_1, \ldots, a_n] \ . \tag{4.1}$$

Moreover, $a_0 \geq 1$ if $\frac{u}{v} \geq 1$.

Proof. Only the second part needs a proof. Without loss of generality, we assume that $v > 0$ and $(u,v) = 1$. The proof is by induction on $v$. If $v = 1$, $\frac{u}{v}$ is an integer and we have $\frac{u}{v} = [a_0]$ with $a_0 = \frac{u}{v}$. Suppose, then, that $v > 1$ (so that $\frac{u}{v}$ is not an integer). By the Division Algorithm, there exist integers $q, r$ with $u = vq + r$ and $1 \leq r < v$. By induction, we may write $\frac{v}{r}$ as a continued fraction, say

$$\frac{v}{r} = [a_1, \ldots, a_n] \ .$$

Since $\frac{v}{r} > 1$, we have $a_1 \geq 1, \ldots, a_n \geq 1$. Therefore

$$\frac{u}{v} = q + \frac{1}{v/r} = q + \frac{1}{[a_1, \ldots, a_n]} = [q, a_1, \ldots, a_n] \ ,$$

and (4.1) is established with $a_0 = q$. Clearly $a_0 = q \geq 1$ if $\frac{u}{v} \geq 1$, and the proof of the lemma is complete.

Definitions. If $a_0, a_1 \geq 1, \ldots, a_n \geq 1$ are integers and if $r = [a_0, a_1, \ldots, a_n]$, then this is an expansion of $r$ into a finite simple continued fraction. $\frac{p_i}{q_i}$ is the $i^{th}$ convergent to $\alpha$, $a_i$ is the $i^{th}$ partial quotient of $\alpha$, and $\alpha_i = [a_i, a_{i+1}, \ldots, a_n]$ is the

$i^{th}$ complete quotient of $\alpha$ . Note that $\frac{p_i}{q_i}$ is in reduced terms by Lemma 3C.

LEMMA 4C. (i) Suppose that $r$ is an integer. Then there are precisely two expansions of $r$ into a simple continued fraction as above, namely $r = [r]$ and $r = [r-1,1]$ . (ii) For $r$ rational but not integral, $r$ has precisely two simple continued fraction expansions: one is of the form $[a_0,a_1,\dots,a_{n-1},a_n]$ with $a_n \geq 2$ , and the other is $[a_0,a_1,\dots,a_{n-1},a_n-1,1]$.

Proof. In either case, $r$ is rational, and so by Lemma 4B, $r$ has an expansion as a simple continued fraction

$$r = [a_0,a_1,\dots,a_n] ,$$

where $a_0, a_1 \geq 1,\dots,a_n \geq 1$ are integers.

Suppose that $r$ is an integer. If $n = 0$ , then $r = [a_0] = a_0$ , so $r = [r]$. If $n > 0$ , then

$$r = a_0 + \frac{1}{[a_1,\dots,a_n]}$$

and $[a_1,\dots,a_n] \geq 1$ . Since $r - a_0$ is an integer, $[a_1,\dots,a_n] = 1$ . Because $a_1 \geq 1$ , it follows that $n = 1$ and $a_1 = 1$ . Then $a_0 = r-1$, and $r = [r-1,1]$. This proves (i).

Now suppose that $r = \frac{u}{v}$ with $(u,v) = 1$ and $v > 0$ . The case when $v = 1$ is treated in part (i) of the lemma. To prove the lemma in general we use induction on $v$ . If $v > 1$ , then $a_0$ in the continued fraction expansion of $\frac{u}{v}$ is uniquely determined: it is the integer part of $\frac{u}{v}$ . We have

$$\frac{u}{v} = a_0 + \frac{u_1}{v} = a_0 + \frac{1}{\alpha_1}$$

where $\alpha_1 = \frac{v}{u_1} > 1$ has a denominator less than $v$. So by induction, $\alpha_1$ has two expansions into a simple continued fraction, of the form $[a_1,\ldots,a_{n-1},a_n]$ with $a_n \geqq 2$ and $[a_1,\ldots,a_{n-1},a_n-1,1]$. The lemma follows.

LEMMA 4D. Suppose that $a_0, a_1 \geq 1$, $a_2 \geq 1, \ldots$ are integers. Then $\lim_{n\to\infty} \frac{p_n}{q_n}$ exists and is irrational. Conversely, if $\alpha$ is irrational, then there exist unique integers $a_0, a_1, a_2, \ldots$ as above with $\alpha = \lim_{n\to\infty} \frac{p_n}{q_n}$.

Notation. $\lim_{n\to\infty} [a_0,a_1,\ldots,a_n] = [a_0,a_1,a_2,\ldots]$. If $\alpha = [a_0,a_1,a_2,\ldots]$, then this is an expansion of $\alpha$ into a simple continued fraction. We retain the terminology used above for finite simple continued fractions: namely, $\frac{p_i}{q_i}$ is the $i^{th}$ convergent to $\alpha$, $a_i$ is the $i^{th}$ partial quotient of $\alpha$, $\alpha_i = [a_i,a_{i+1},\ldots]$ is the $i^{th}$ complete quotient of $\alpha$.

Proof of the lemma. (First part) By Lemma 3C,

$$\frac{p_{n-1}}{q_{n-1}} - \frac{p_n}{q_n} = \frac{(-1)^n}{q_{n-1}q_n} .$$

Since $\frac{p_0}{q_0} < \frac{p_2}{q_2} < \ldots$, and since these are bounded above by $\frac{p_1}{q_1}$, $\lim_{\substack{n\to\infty \\ n \text{ even}}} \frac{p_n}{q_n}$ exists. Similarly, $\lim_{\substack{n\to\infty \\ n \text{ odd}}} \frac{p_n}{q_n}$ exists. But these two limits are equal, since $\lim_{n\to\infty} \frac{(-1)^n}{q_{n-1}q_n} = 0$. (Here we use that $q_n = a_n q_{n-1} + q_{n-2}$, hence $q_n \geq q_{n-1} + 1$ if $n \geq 1$, and therefore $q_n \to \infty$ as $n \to \infty$.)

Let $\alpha = \lim_{n\to\infty} \frac{p_n}{q_n}$. Since $\alpha$ lies strictly between $\frac{p_n}{q_n}$ and $\frac{p_{n+1}}{q_{n+1}}$, we have

$$|\alpha - \frac{p_n}{q_n}| < \frac{1}{q_n q_{n+1}} < \frac{1}{q_n^2}$$

by Lemma 3C. Now $p_n$ and $q_n$ are relatively prime for each $n$ by Lemma 3C, so that there are infinitely many "reduced fractions" $\frac{p}{q}$ with $|\alpha - \frac{p}{q}| < \frac{1}{q^2}$. By the remark at the end of §1, $\alpha$ must be irrational.

(Second part). We suppose that $\alpha$ is irrational. Let $a_0$ be the integer part of $\alpha$, and define $\alpha_1$ by $\alpha = a_0 + \frac{1}{\alpha_1}$. Since $\alpha$ is irrational, $\alpha_1$ is irrational; also $\alpha_1 > 1$. In general, having defined integers $a_0, a_1 \geq 1, \ldots, a_{k-1} \geq 1$ and irrationals $\alpha_1 > 1$, $\alpha_2 > 1, \ldots, \alpha_k > 1$, let $a_k$ be the integer part of $\alpha_k$ and define $\alpha_{k+1}$ by $\alpha_k = a_k + \frac{1}{\alpha_{k+1}}$. Then $a_k \geq 1$ and $\alpha_{k+1} > 1$; also, $\alpha_{k+1}$ is irrational since $\alpha_k$ is irrational. We now show that

$$\alpha = [a_0, a_1, a_2, \ldots] .$$

For each $n \geq 0$, we have $\alpha = [a_0, a_1, \ldots, a_n, \alpha_{n+1}]$. By Lemma 3E,

$$q_n \alpha - p_n = \frac{(-1)^n}{\alpha_{n+1} q_n + q_{n-1}} ,$$

whence

$$|\alpha - \frac{p_n}{q_n}| < \frac{1}{q_n^2} .$$

This implies that $\lim_{n \to \infty} \frac{p_n}{q_n} = \alpha$.

It remains to prove that the integers $a_0, a_1 \geq 1, a_2 \geq 1, \ldots$ are unique. We have

$$\alpha = [a_0, a_1, a_2, \ldots] = a_0 + \frac{1}{[a_1, a_2, \ldots]} :$$

$0 \le \alpha - a_0 < 1$ , so $a_0$ is the integer part of $\alpha$ , and is therefore unique. It follows, then, that $\alpha_1 = [a_1, a_2, \ldots]$ is unique. Then $a_1$ is the integer part of $\alpha_1$ and is unique, etc.

## §5. Continued Fractions and Approximation to Irrationals by Rationals.

Recall that by a formula of the last section, every convergent $\frac{p}{q}$ of $\alpha$ satisfies

$$|\alpha - \frac{p}{q}| < \frac{1}{q^2} .$$

THEOREM 5A. (Vahlen (1895)). Let $\frac{p_{n-1}}{q_{n-1}}$ , $\frac{p_n}{q_n}$ be consecutive convergents to $\alpha$ . Then at least one of them satisfies

$$|\alpha - \frac{p}{q}| < \frac{1}{2q^2} .$$

Proof. The numbers $\alpha - \frac{p_n}{q_n}$ , $\alpha - \frac{p_{n-1}}{q_{n-1}}$ are of opposite sign, hence we have

$$|\alpha - \frac{p_n}{q_n}| + |\alpha - \frac{p_{n-1}}{q_{n-1}}| = |\frac{p_n}{q_n} - \frac{p_{n-1}}{q_{n-1}}| = \frac{1}{q_n q_{n-1}} < \frac{1}{2q_n^2} + \frac{1}{2q_{n-1}^2} ,$$

since $ab < \frac{1}{2}(a^2 + b^2)$ if $a \neq b$ . Hence, either

$$|\alpha - \frac{p_n}{q_n}| < \frac{1}{2q_n^2} \quad \text{or} \quad |\alpha - \frac{p_{n-1}}{q_{n-1}}| < \frac{1}{2q_{n-1}^2} .$$

THEOREM 5B. (E. Borel (1903a,b)). Let $\frac{p_{n-2}}{q_{n-2}}$ , $\frac{p_{n-1}}{q_{n-1}}, \frac{p_n}{q_n}$ be three consecutive convergents to $\alpha$ . Then at least one of them satisfies

$$|\alpha - \frac{p}{q}| < \frac{1}{\sqrt{5}\, q^2} .$$

Proof. Put $\alpha = [a_0, a_1, \ldots]$ , $\alpha_i = [a_i, a_{i+1}, \ldots]$ and

$\beta_i = \frac{q_{i-2}}{q_{i-1}}$ for $i = n-1, n, n+1$. We have

$$\alpha = [a_0, a_1, \ldots, a_n, \alpha_{n+1}] \quad ,$$

so

$$\alpha q_n - p_n = \frac{(-1)^n}{\alpha_{n+1} q_n + q_{n-1}}$$

by Lemma 3E. Thus

$$|\alpha - \frac{p_n}{q_n}| = \frac{1}{q_n(\alpha_{n+1} q_n + q_{n-1})} = \frac{1}{q_n^2(\alpha_{n+1} + \beta_{n+1})} \quad .$$

To complete the proof, it suffices to show that there cannot be three integers $i = n-1, n, n+1$ with

$$\alpha_i + \beta_i \le \sqrt{5} \quad . \tag{5.1}$$

Suppose that (5.1) were true for $i = n-1, n$. Now

$\alpha_{n-1} = a_{n-1} + \frac{1}{\alpha_n}$ and $\frac{1}{\beta_n} = \frac{q_{n-1}}{q_{n-2}} = a_{n-1} + \frac{q_{n-3}}{q_{n-2}} = a_{n-1} + \beta_{n-1}$, so

$$\frac{1}{\alpha_n} + \frac{1}{\beta_n} = \alpha_{n-1} + \beta_{n-1} \le \sqrt{5} \quad .$$

Therefore

$$1 = \alpha_n \left(\frac{1}{\alpha_n}\right) \le (\sqrt{5} - \beta_n)\left(\sqrt{5} - \frac{1}{\beta_n}\right) ,$$

which is equivalent to

$$\beta_n^2 - \sqrt{5}\,\beta_n + 1 \le 0 \quad .$$

It follows that $\beta_n \ge \frac{1}{2}(\sqrt{5} - 1)$; since $\beta_n$ is rational,

$$\beta_n > \frac{1}{2}(\sqrt{5} - 1) \quad .$$

If (5.1) also were true for $i = n, n+1$, then

$$\beta_{n+1} > \frac{1}{2}(\sqrt{5} - 1) \quad ,$$

and therefore

$$1 \le a_n = \frac{q_n}{q_{n-1}} - \frac{q_{n-2}}{q_{n-1}} = \frac{1}{\beta_{n+1}} - \beta_n < \frac{2}{\sqrt{5} - 1} - \frac{\sqrt{5} - 1}{2} = 1 \ ,$$

a contradiction. This completes the proof of the theorem.

THEOREM 5C (Legendre) *Suppose $p, q$ are relatively prime integers with $q > 0$ and*

$$|\alpha - \frac{p}{q}| < \frac{1}{2q^2} \ .$$

*Then $\frac{p}{q}$ is a convergent to $\alpha$.*

*Proof.* We assume that $\alpha \ne \frac{p}{q}$, else the theorem is trivially true. Then we may write

$$\alpha - \frac{p}{q} = \frac{\varepsilon \vartheta}{q^2} \quad ,$$

where $0 < \vartheta < \frac{1}{2}$ and $\varepsilon = \pm 1$. From Lemma 4C there is an expansion of $\frac{p}{q}$ into a simple continued fraction

$$\frac{p}{q} = [b_0, b_1, \ldots, b_{n-1}] \quad ,$$

where $n$ is chosen so that $(-1)^{n-1} = \varepsilon$.

Define $\omega$ by

$$\alpha = \frac{\omega p_{n-1} + p_{n-2}}{\omega q_{n-1} + q_{n-2}} \quad , \tag{5.2}$$

so that

$$\alpha = [b_0, b_1, \ldots, b_{n-1}, \omega] .$$

(Note that (5.2) is equivalent to $(\alpha q_{n-1} - p_{n-1})\omega = p_{n-2} - \alpha q_{n-2}$ . We may assume that $\alpha q_{n-1} - p_{n-1} \neq 0$ , else $\alpha = \frac{p_{n-1}}{q_{n-1}} = \frac{p}{q}$ .) Then

$$\frac{\varepsilon\vartheta}{q^2} = \alpha - \frac{p}{q} = \frac{1}{q_{n-1}} (\alpha q_{n-1} - p_{n-1}) = \frac{1}{q_{n-1}} \cdot \frac{(-1)^{n-1}}{\omega q_{n-1} + q_{n-2}}$$

from Lemma 3E, and therefore

$$\vartheta = \frac{q_{n-1}}{\omega q_{n-1} + q_{n-2}} .$$

Solving this for $\omega$ , we obtain

$$\omega = \frac{q_{n-1} - \vartheta q_{n-2}}{\vartheta q_{n-1}} = \frac{1}{\vartheta} - \frac{q_{n-2}}{q_{n-1}} :$$

it follows that $\omega > 2 - 1 = 1$ . Expand $\omega$ into a (finite or infinite) simple continued fraction:

$$\omega = [b_n, b_{n+1}, \ldots] .$$

Since $\omega > 1$ , each of the integers $b_j$ $(j = n, n+1, \ldots)$ is positive and therefore

$$\begin{aligned} \alpha &= [b_0, b_1, \ldots, b_{n-1}, [b_n, b_{n+1}, \ldots]] \\ &= [b_0, b_1, \ldots, b_{n-1}, b_n, b_{n+1}, \ldots] \end{aligned}$$

by Lemma 3B, passing to the limit if necessary. This is a simple continued fraction for $\alpha$ , and

$$\frac{p}{q} = \frac{p_{n-1}}{q_{n-1}} = [b_0, b_1, \ldots, b_{n-1}]$$

is a convergent to $\alpha$ , so the proof is finished.

LEMMA 5D. *Suppose* $\alpha$ *has a continued fraction expansion of the type*

$$\alpha = [a_0, a_1, \ldots, a_N, 1, 1, \ldots] .$$

*Then*

$$\lim_{n\to\infty} q_n^2 \left|\alpha - \frac{p_n}{q_n}\right| = \frac{1}{\sqrt{5}} .$$

*Proof*. Using the notation introduced in the proof of Theorem 5A,

$$\left|\alpha - \frac{p_n}{q_n}\right| = \frac{1}{q_n^2(\alpha_{n+1} + \beta_{n+1})} .$$

Here if n is sufficiently large,

$$\alpha_{n+1} = [1,1,1,\ldots] = \frac{1}{2}(\sqrt{5}+1),$$

and

$$\frac{1}{\beta_{n+1}} = \frac{q_n}{q_{n-1}} = [a_n, a_{n-1}, \ldots, a_N, \ldots, a_1, a_0]$$

$$= [\underbrace{1,1,\ldots,1}_{n-N}, a_N, \ldots, a_1, a_0] .$$

Since $[\underbrace{1,1,\ldots,1}_{n-N-1}]$ and $[\underbrace{1,1,\ldots,1}_{n-N}]$ are consecutive convergents to $1/\beta_{n+1}$ , the number $1/\beta_{n+1}$ lies between these convergents, and therefore $1/\beta_{n+1}$ approaches $[1,1,\ldots] = \frac{1}{2}(\sqrt{5} + 1)$ as $n \to \infty$ . Hence $\beta_{n+1} \to (\frac{1}{2}(\sqrt{5}+1))^{-1} = \frac{1}{2}(\sqrt{5} - 1)$ , and $\alpha_{n+1} + \beta_{n+1} \to \sqrt{5}$ .

Continued fractions may be used to give

Another proof of Theorem 2F. Namely, assertion (i) follows easily from Theorem 5B, while assertion (ii) is a consequence of Theorem 5C and Lemma 5D.

THEOREM 5E. (Lagrange (1770a). *Suppose that* $\alpha$ *is irrational, and let* $\frac{p_0}{q_0}, \frac{p_1}{q_1}, \ldots$ *be the convergents to* $\alpha$. *Then*

(i) $|\alpha q_0 - p_0| > |\alpha q_1 - p_1| > |\alpha q_2 - p_2| > \ldots$

(ii) *If* $n \geqq 1$ *and* $1 \leq q \leq q_n$, *and if* $(p,q) \neq (p_n, q_n)$, $(p,q) \neq (p_{n-1}, q_{n-1})$, *then* $|\alpha q - p| > |\alpha q_{n-1} - p_{n-1}|$.

Remark. It follows from (i) and (ii) that if $1 \leq q \leq q_n$, $(p,q) \neq (p_n, q_n)$ and $n \geq 1$, then $|\alpha q - p| > |\alpha q_n - p_n|$. This result is sometimes called the "Law of Best Approximation."

Proof of the theorem. From Lemma 3E,

$$|\alpha q_n - p_n| = \frac{1}{\alpha_{n+1} q_n + q_{n-1}} < \frac{1}{q_n + q_{n-1}},$$

and

$$|\alpha q_{n-1} - p_{n-1}| = \frac{1}{\alpha_n q_{n-1} + q_{n-2}} > \frac{1}{(a_n + 1) q_{n-1} + q_{n-2}} = \frac{1}{q_n + q_{n-1}},$$

which proves (i).

To prove (ii), we first define $\mu, \nu$ by the equations

$$\mu p_n + \nu p_{n-1} = p,$$

$$\mu q_n + \nu q_{n-1} = q.$$

The matrix determined by these two equations has determinant $\pm 1$, hence $\mu, \nu$ are integers.

If $\nu = 0$, then $(p,q) = \mu(p_n,q_n)$. But this is impossible since $0 < q \le q_n$ and $(p,q) \ne (p_n,q_n)$. If $\mu = 0$, then $(p,q) = \nu(p_{n-1},q_{n-1})$. Since $(p,q) \ne (p_{n-1},q_{n-1})$, we have $\nu \ge 2$, and therefore

$$|\alpha q - p| \ge 2|\alpha q_{n-1} - p_{n-1}| > |\alpha q_{n-1} - p_{n-1}| .$$

If both $\mu \ne 0$ and $\nu \ne 0$, then since $1 \le q \le q_n$, $\mu$ and $\nu$ are of opposite sign. Hence $\mu(\alpha q_n - p_n)$ and $\nu(\alpha q_{n-1} - p_{n-1})$ are of the same sign, and therefore

$$|\alpha q - p| = |\mu(\alpha q_n - p_n)| + |\nu(\alpha q_{n-1} - p_{n-1})| .$$

Thus $|\alpha q - p| > |\alpha q_{n-1} - p_{n-1}|$ since $\mu\nu \ne 0$, and (ii) is established.

Definition. An irrational number $\alpha$ is badly approximable if there is a constant $c = c(\alpha) > 0$ such that

$$|\alpha - \frac{p}{q}| > \frac{c}{q^2}$$

for every rational $\frac{p}{q}$. (In view of Theorem 2F, such a constant $c$ must satisfy $0 < c < \frac{1}{\sqrt{5}}$.)

THEOREM 5F. *$\alpha$ is badly approximable if and only if the partial quotients in its continued fraction expansion are bounded:*

$$|a_n| \le K(\alpha) .$$

Proof. In order to study the inequality

$$|\alpha - \frac{p}{q}| < \frac{c(\alpha)}{q^2}$$

where $0 < c(\alpha) < \frac{1}{2}$, we may restrict ourselves to convergents $\frac{p_n}{q_n}$ by

Theorem 5C. From the proof of Theorem 5B and from Lemma 3F, we have

$$|\alpha - \frac{p_n}{q_n}| = \frac{1}{q_n^2(\alpha_{n+1} + \beta_{n+1})} = \frac{1}{q_n^2\ [a_{n+1}, a_{n+2}, \ldots] + \frac{1}{[a_n, a_{n-1}, \ldots, a_1]}} ,$$

and hence

$$\frac{1}{q_n^2(a_{n+1} + 2)} \le |\alpha - \frac{p_n}{q_n}| \le \frac{1}{q_n^2 a_{n+1}} .$$

The theorem follows.

COROLLARY 5G. *There exist continuum many badly approximable numbers, and there exist continuum many numbers which are not badly approximable.*

## §6. Further results.

We state results without proofs. For proofs see Perron (1954), Cassels (1957), Hardy and Wright (1954), and the references given below.

*Definition.* An infinite continued fraction $[a_0, a_1, a_2, \ldots]$ is *periodic* if there are integers $k \ge 0$, $m \ge 1$ such that

$$a_{n+m} = a_n \quad \text{provided} \quad n \ge k .$$

In this case, we write the fraction as

$$[a_0, a_1, \ldots, a_{k-1}, \overline{a_k, a_{k+1}, \ldots, a_{k+m-1}}] .$$

THEOREM 6A*. *$\alpha$ has a periodic simple continued fraction expansion if and only if $\alpha$ is a quadratic irrational.*

The necessity was proved by Euler (1737); see also (1748). The sufficiency was proved by Lagrange (1770b). See also Lagrange's Oeuvres, II.

Definition. Two irrational numbers $\alpha$ and $\beta$ are equivalent if there exist integers $a,b,c,d$ with $ad - bc = \pm 1$ such that

$$\beta = \frac{a\alpha + b}{c\beta + d} \quad .$$

(It is not difficult to verify that this is an equivalence relation.)

THEOREM 6B*. (Serret, 1878)[†)] Let $\alpha = [a_0,a_1,a_2,\ldots]$ and $\beta = [b_0,b_1,b_2,\ldots]$ be irrational. Then $\alpha$ and $\beta$ are equivalent precisely if there are integers $u$ and $v$ such that $a_{u+n} = b_{v+n}$ $(n = 1,2,\ldots)$ .

THEOREM 6C* (Hurwitz, 1891). Suppose that $\alpha = [a_0,a_1,a_2,\ldots]$ with $a_n \geq 2$ for infinitely many $n$ . Then there are infinitely many distinct rationals $\frac{p}{q}$ with

$$|\alpha - \frac{p}{q}| < \frac{1}{\sqrt{8}\, q^2} \quad .$$

COROLLARY 6D*. The inequality

$$|\alpha - \frac{p}{q}| < \frac{1}{\sqrt{8}\, q^2}$$

has infinitely many rational solutions $\frac{p}{q}$ whenever $\alpha$ is not equivalent to $\frac{1}{2}(\sqrt{5}+1)$.

Definition. $\nu(\alpha) = \liminf_{q\to\infty} q\|q\alpha\|$ .

Remarks. Trivially, $\nu(\alpha) = 0$ whenever $\alpha$ is rational. It is clear that $\nu(\alpha) > 0$ if and only if $\alpha$ is badly approximable. Theorem 2F implies that $\nu(\alpha) \leq \frac{1}{\sqrt{5}}$ for every real number $\alpha$ . There are

---

[†)] This is the date of the German translation. I was unable to find the original date.

numbers $\alpha$ with $\nu(\alpha) = \frac{1}{\sqrt{5}}$ : $\alpha = \frac{1}{2}(\sqrt{5}+1)$, for example. Corollary 6D* implies that $\nu(\alpha) \le \frac{1}{\sqrt{8}}$ whenever $\alpha$ is not equivalent to $\frac{1}{2}(\sqrt{5}+1)$.

THEOREM 6E* (Markoff (1879). See also Hurwitz (1906)). *There exist numbers*

$$\mu_1 = \frac{1}{\sqrt{5}} > \mu_2 = \frac{1}{\sqrt{8}} > \mu_3 > \mu_4 > \dots$$

*with limit* $\frac{1}{3}$ *such that for every* $\mu_i$ *there are finitely many equivalence classes (as defined above Theorem 6B*) of numbers such that*

$$\nu(\alpha) = \mu_i$$

*precisely if* $\alpha$ *lies in such a class. Furthermore, if* $\mu > \frac{1}{3}$ *and* $\mu \ne \mu_i$ $(i = 1,2,3,\dots)$, *then there is no* $\alpha$ *with* $\nu(\alpha) = \mu$. *Finally, there are continuum many* $\alpha$ *with* $\nu(\alpha) = \frac{1}{3}$.

THEOREM 6F*. (Perron, 1921a). *There are no* $\alpha$ *with* $\frac{1}{\sqrt{13}} < \nu(\alpha) < \frac{1}{\sqrt{12}}$, *but there are* $\alpha$ *with* $\nu(\alpha) = \frac{1}{\sqrt{13}}$ *or* $\nu(\alpha) = \frac{1}{\sqrt{12}}$.

THEOREM 6G* (M. Hall. The theorem follows from his (1947) paper). *There is a constant* $c > 0$ *such that, for every number* $\nu$ *satisfying* $0 \le \nu < c$, *there is an* $\alpha$ *with* $\nu(\alpha) = \nu$.

THEOREM 6H* (Hightower, 1970). *There are countably many disjoint intervals* $I_1, I_2, \dots$ *contained in* $0 \le x \le \frac{1}{3}$ *such that*

(i) *There is no* $\alpha$ *with* $\nu(\alpha) \in \bigcup_{i=1}^{\infty} I_i$.

(ii) *If* $I_i, I_j$ *are any two distinct such intervals, there is an*

$\alpha$ with $\nu(\alpha)$ "between" $I_i$ and $I_j$ .

The above results can be formulated in terms of the Lagrange spectrum, which is defined as the set of numbers $\nu$ of the type $\nu = \nu(\alpha)$ for some $\alpha$ . For recent references on this area, see Cusick (1975).

Definition. We say that Dirichlet's Theorem can be improved for some particular $\alpha$ if there is a constant $c(\alpha) < 1$ such that for every $Q > Q_0(\alpha)$ , there exist integers $p,q$ with $1 \leq q < Q$ and $|\alpha q - p| < \frac{c(\alpha)}{Q}$ .

THEOREM 6I* (Davenport and Schmidt (1968)). Dirichlet's Theorem can be improved for $\alpha$ precisely if $\alpha$ is badly approximable. Namely, put

$$\gamma(\alpha) = \liminf \frac{1}{[a_{n+1},a_{n+2},\ldots][a_n,a_{n-1},\ldots,a_1]} .$$

Then one may pick $c(\alpha) > (1 + \gamma(\alpha))^{-1}$ , but one may not pick $c(\alpha) < (1 + \gamma(\alpha))^{-1}$ .

THEOREM 6J* (Segre (1945)). Suppose that $\alpha$ is irrational and $\tau \geq 0$ . Then there exist infinitely many rationals $\frac{h}{k}$ with

$$-\frac{1}{(1+4\tau)^{1/2}k^2} < \alpha - \frac{h}{k} < \frac{\tau}{(1+4\tau)^{1/2}k^2} .$$

Remark. If $\tau = 1$ , the theorem reduces to Hurwitz' Theorem. A proof of the theorem, using Farey series, may be found, e.g., in Niven (1962).

## II. Simultaneous Approximation

References: Dirichlet (1842), Minkowski (1896), Perron (1921b).

### §1. Dirichlet's Theorem on Simultaneous Approximation

THEOREM 1A. *Suppose that* $\alpha_1,\ldots,\alpha_n$ *are* $n$ *real numbers and that* $Q > 1$ *is an integer. Then there exist integers* $q, p_1, \ldots, p_n$ *with*

$$(1.1) \qquad 1 \le q < Q^n \quad \text{and} \quad |\alpha_i q - p_i| \le \frac{1}{Q} \qquad (1 \le i \le n) .$$

COROLLARY 1B. *Suppose that at least one of* $\alpha_1,\ldots,\alpha_n$ *is irrational. Then there are infinitely many n-tuples* $\frac{p_1}{q}, \ldots, \frac{p_n}{q}$ *with*

$$(1.2) \qquad \left|\alpha_i - \frac{p_i}{q}\right| < \frac{1}{q^{1+1/n}} \qquad (1 \le i \le n) .$$

*Deduction of the Corollary*. In Theorem 1A, we may require in addition that $\text{g.c.d.}(q,p_1,\ldots,p_n) = 1$. The inequalities of the theorem clearly imply (1.2).

If, say, $\alpha_1$ is irrational, then

$$|\alpha_1 q - p_1| \ne 0 .$$

Hence for *fixed* $q, p_1, \ldots, p_n$, (1.1) can hold only for $Q \le Q_0$. Hence as $Q \to \infty$, we get infinitely many different solutions.

THEOREM 1C. *Suppose that* $\alpha_1,\ldots,\alpha_n$ *and* $Q$ *are given as in Theorem 1A. Then there exist integers* $q_1,\ldots,q_n,p$ *with*

$$(1.3) \qquad 1 \le \max(|q_1|,\ldots,|q_n|) < Q^{1/n} \quad \text{and} \quad |a_1 q_1 + \ldots + \alpha_n q_n - p| \le \frac{1}{Q} .$$

COROLLARY 1D. *Suppose that* $1, \alpha_1, \ldots, \alpha_n$ *are linearly independent over the rationals. Then there are infinitely many coprime* $(n+1)$*-tuples*

$(q_1,\ldots,q_n,p)$ with

(1.4) $$q = \max(|q_1|,\ldots,|q_n|) > 0 \quad \text{and} \quad |\alpha_1 q_1+\ldots+\alpha_n q_n - p| < \frac{1}{q^n} .$$

Deduction of the Corollary. It is clear that (1.3) implies (1.4). By linear independence, we always have

$$|\alpha_1 q_1 + \ldots + \alpha_n q_n - p| \neq 0.$$

Hence for fixed $q_1,\ldots,q_n,p$, (1.3) can hold only for $Q \le Q_0$. Hence as $Q \to \infty$, we obtain infinitely many solutions.

THEOREM 1E. (Dirichlet (1842)) Suppose that $\alpha_{ij}$ $(1 \le i \le n,$ $1 \le j \le m)$ are $nm$ real numbers and that $Q > 1$ is an integer. Then there exist integers $q_1,\ldots,q_m,p_1,\ldots,p_n$ with

(1.5) $$1 \le \max(|q_1|,\ldots,|q_m|) < Q^{n/m} ,$$
$$|\alpha_{i1} q_1 + \ldots + \alpha_{im} q_m - p_i| \le \frac{1}{Q} \quad (1 \le i \le n) .$$

Remark. The hypothesis that "Q is an integer" is unnecessary. Our proof of Theorem 1E cannot immediately be adopted to this generalization, but see the remark following the proof of Theorem 2A. It is clear that Theorem 1E implies the two preceding theorems. We omit the obvious proof of the following

COROLLARY 1F. Suppose, for some $i$ in $1 \le i \le n$, that $\alpha_{i1},\ldots,\alpha_{im},1$ are linearly independent over the rationals. Or suppose, more generally, that

$$(\alpha_{11}q_1 + \ldots + \alpha_{1m}q_m , \ldots , \alpha_{n1}q_1 + \ldots + \alpha_{nm}q_m)$$

is never an integer point when $(q_1,\ldots,q_m)$ is a nonzero integer point. The

*there exist infinitely many coprime* $(m+n)$-*tuples* $(q_1,\ldots,q_m,p_1,\ldots,p_n)$ *with*

$$q = \max(|q_1|,\ldots,|q_m|) > 0 \quad \text{and} \quad |\alpha_{i1}q_1+\ldots+\alpha_{im}q_m - p_i| < \frac{1}{q^{m/n}} \quad (1 \leq i \leq n) .$$

Proof of Theorem 1E. Consider points

$$(\{\alpha_{11}x_1 + \ldots + \alpha_{1m}x_m\},\ldots,\{\alpha_{n1}x_1 + \ldots + \alpha_{nm}x_m\}) ,$$

where each $x_j$ is an integer satisfying

$$0 \leq x_j < Q^{n/m} \quad (1 \leq j \leq m) .$$

There are at least $Q^n$ such points, and each of these points lies in the closed unit cube $\bar{U}^n$ consisting of points $(t_1,\ldots,t_n)$ with $0 \leq t_k \leq 1$ $(k = 1,\ldots,n)$. The point $(1,1,\ldots,1)$ also lies in $\bar{U}^n$, so together there are at least $Q^n + 1$ points under consideration.

We divide $\bar{U}^n$ into $Q^n$ pairwise disjoint subcubes of side length $\frac{1}{Q}$. (Thus some of the cubes will contain some of their faces or edges and not others). Two of the points under consideration will be in the same subcube. These points are, say,

$$(\alpha_{11}x_1 + \ldots + \alpha_{1m}x_m - y_1,\ldots,\alpha_{n1}x_1 + \ldots + \alpha_{nm}x_m - y_n) ,$$

$$(\alpha_{11}x_1' + \ldots + \alpha_{1m}x_m' - y_1',\ldots,\alpha_{n1}x_1' + \ldots + \alpha_{nm}x_m' - y_n') .$$

Here, $(x_1,\ldots,x_m) \neq (x_1',\ldots,x_m')$ . Put $q_i = x_i - x_i'$ $(1 \leq i \leq m)$ and $p_j = y_j - y_j'$ $(1 \leq j \leq n)$. Then (1.5) is easily seen to be satisfied, and the proof is complete.

## §2. Theorems of Blichfeldt and Minkowski.

Notation. In this section and throughout the remainder of these notes, we make the following conventions: $E^n$ denotes real n-dimensional

Euclidean space, points of which are denoted by $\underline{\underline{x}} = (x_1,\ldots,x_n)$, $\underline{\underline{y}}$, etc.; $U$ is the closed-open unit interval $0 \le x < 1$, and $U^n$ is the unit cube of points $\underline{\underline{x}} = (x_1,\ldots,x_n)$ with each $x_k \in U$. $\bar{U}^n$ is the closure of $U^n$.

Suppose that $\underline{\underline{x}} = (x_1,\ldots,x_n) \in E^n$. We put

$$\lceil \underline{\underline{x}} \rceil = \max(|x_1|,\ldots,|x_n|) .$$

If each $x_k$ is an integer, we say that $\underline{\underline{x}}$ is an <u>integer point</u>.

Let $\mathfrak{S}$ be any subset of $E^n$. If $\underline{\underline{x}} \in E^n$, we denote by $\mathfrak{S} + \underline{\underline{x}}$ the set[†)] of all points $\underline{\underline{s}} + \underline{\underline{x}} = (s_1 + x_1,\ldots,s_n + x_n)$ with $\underline{\underline{s}} \in \mathfrak{S}$. If $\lambda$ is a real number, then $\lambda\mathfrak{S}$ denotes the set of all points $\lambda\underline{\underline{s}} = (\lambda s_1,\ldots,\lambda s_n)$ with $\underline{\underline{s}} \in \mathfrak{S}$.

<u>THEOREM 2A</u>. (Blichfeldt (1914)) <u>Let $\mathcal{P}$ be a discrete (i.e., without limit point) set of points in $E^n$, invariant under translation by integer points, and with precisely $N$ points in $U^n$. Let $\mathcal{R}$ be a measurable set of measure $\mu(\mathcal{R}) > 0$. Then there is a point $\underline{\underline{x}}$ in $U^n$ such that $\mathcal{R} + \underline{\underline{x}}$ contains at least $N\mu(\mathcal{R})$ points of $\mathcal{P}$. Moreoever, if $\mathcal{R}$ is compact, then there is a translation of $\mathcal{R}$ containing more than $N\mu(\mathcal{R})$ points of $\mathcal{P}$.</u>

<u>Proof</u>. Write $\nu(\mathfrak{S})$ for the number of points of $\mathcal{P}$ lying in a set $\mathfrak{S}$. Now $\mathcal{P}$ has $N$ points in $U^n$, say $\underline{\underline{p}}_1,\ldots,\underline{\underline{p}}_N$. Write $\nu_i(\mathfrak{S})$ for the number of points $\underline{\underline{p}}_i + \underline{\underline{g}}$ ($\underline{\underline{g}}$ an integer point) which lie in $\mathfrak{S}$ ($1 \le i \le n$). Since $\mathcal{P}$ is invariant under translation by integer points,

$$\nu(\mathfrak{S}) = \sum_{i=1}^{N} \nu_i(\mathfrak{S}) . \tag{2.1}$$

---

†) called the <u>translation</u> of $\mathfrak{S}$ by $\underline{\underline{x}}$.

For each $i$,

$$\nu_i(\mathfrak{R} + \underline{\underline{x}}) = \sum_{\underline{\underline{g}}} \chi(\underline{\underline{p}}_i + \underline{\underline{g}} - \underline{\underline{x}}) ,$$

where $\chi$ is the characteristic function of $\mathfrak{R}$ and the sum is over all integer points $\underline{\underline{g}} \in E^n$. We obtain

$$\int_{U^n} \nu_i(\mathfrak{R} + \underline{\underline{x}}) d\underline{\underline{x}} = \int_{U^n} \sum_{\underline{\underline{g}}} \chi(\underline{\underline{p}}_i + \underline{\underline{g}} - \underline{\underline{x}}) d\underline{\underline{x}}$$

$$= \int_{E^n} \chi(\underline{\underline{z}}) d\underline{\underline{z}}$$

$$= \mu(\mathfrak{R}) .$$

It follows from (2.1) that

$$\int_{U^n} \nu(\mathfrak{R} + \underline{\underline{x}}) d\underline{\underline{x}} = N\mu(\mathfrak{R}) ,$$

and therefore $\nu(\mathfrak{R} + \underline{\underline{x}}) \geq N\mu(\mathfrak{R})$ for some $\underline{\underline{x}} \in U^n$.

If $\mathfrak{R}$ is compact, there is nothing more to prove unless $N\mu(\mathfrak{R})$ is an integer, say $\nu_0$. Suppose, then, that this is the case. It follows from the first part of the proof that for every positive integer $k$, there is a point $\underline{\underline{x}}_k \in U^n$ such that $\nu(\mathfrak{S}_k) \geq \nu_0 + 1$, where

$$\mathfrak{S}_k = (1 + \frac{1}{k})\mathfrak{R} + \underline{\underline{x}}_k .$$

Since each $\underline{\underline{x}}_k$ belongs to the compact set $\bar{U}^n$, the sequence $(\underline{\underline{x}}_k)_{k=1}^{\infty}$ contains a convergent subsequence $(\underline{\underline{x}}_{k_j})_{j=1}^{\infty}$ : say $\underline{\underline{x}}_{k_j} \to \underline{\underline{x}}_0$.

We have $\nu(\mathfrak{S}_{k_j}) \geq \nu_0 + 1$ for each $j$ ; since all of the sets $\mathfrak{S}_{k_j}$ are uniformly bounded, there are $\nu_0 + 1$ points $\underline{\underline{u}}_1, \ldots, \underline{\underline{u}}_{\nu_0+1}$ which lie in infinitely many of the sets $\mathfrak{S}_{k_j}$. Then $\underline{\underline{u}}_1, \ldots, \underline{\underline{u}}_{\nu_0+1}$

also lie in $\mathfrak{R} + \underline{\underline{x}}_0$ , in view of the compactness of $\mathfrak{R}$ . Hence

$$\nu(\mathfrak{R} + \underline{\underline{x}}_0) \geq \nu_0 + 1 > N\mu(\mathfrak{R}) .$$

If $\underline{\underline{x}}_0$ does not lie in $U^n$ , we may replace it by a point in $U^n$ which is obtained from it by a translation by an integer point.

The proof is now complete.

<u>Remark</u>. Theorem 2A may be used to extend Theorem 1E to hold for any $Q > 1$ . Namely, consider points

$$(\alpha_{11}x_1 + \ldots + \alpha_{1m}x_m, \ldots, \alpha_{n1}x_1 + \ldots + \alpha_{nm}x_m) ,$$

where each $x_j$ is an integer satisfying

$$0 \leq x_j < Q^{n/m} \quad (1 \leq j \leq m) .$$

Let $\mathcal{P}$ be the set of such points, together with all of their translations by integer points. Note that the number $N$ of points in $\mathcal{P}$ lying in $U^n$ satisfies $N \geq Q^n$ . Let $\mathcal{R}$ be the cube $0 \leq t_k \leq \frac{1}{Q}$ $(1 \leq k \leq n)$ . $\mathcal{R}$ is compact and $\mu(\mathcal{R}) = 1/Q^n$ .

By Theorem 2A, there is a translation $\mathcal{R} + \underline{\underline{x}}$ with $\nu(\mathcal{R} + \underline{\underline{x}}) > N\mu(\mathcal{R}) \geq 1$ : i.e., $\nu(\mathcal{R} + \underline{\underline{x}}) \geq 2$ . The proof is now completed as before.

<u>THEOREM 2B</u>. (Minkowski's Convex Body Theorem. Minkowski (1896)). <u>Let</u> $\mathcal{R}$ <u>be a convex set in</u> $E^n$ , <u>symmetric about</u> $\underline{\underline{0}}$ , <u>bounded and with volume</u> $\mu(\mathcal{R})$ †) . <u>Assume either that</u> $\mu(\mathcal{R}) > 2^n$ <u>or that</u> $\mathcal{R}$ <u>is compact and</u> $\mu(\mathcal{R}) \geq 2^n$ . <u>Then</u> $\mathcal{R}$ <u>contains an integer point</u> $\neq 0$ .

By considering the cube of $\underline{\underline{x}}$ with $|x_i| < 1$ $(i = 1,\ldots,n)$ we see that Minkowski's Theorem is best possible.

---

†) It may be shown that such a set always does have a (Jordan)-volume. A reader who does not wish to use this fact may regard the existence of a volume as a further assumption on $\mathcal{R}$ .

Proof. Either $\mu(\frac{1}{2}\mathfrak{K}) > 1$ , or $\mathfrak{K}$ is compact and $\mu(\frac{1}{2}\mathfrak{K}) \geq 1$ . In either case, we apply Theorem 2A to $\frac{1}{2}\mathfrak{K}$ and $\mathcal{P}$ , where $\mathcal{P}$ is the set of all integer points in $E^n$ . Then there is a translation $\frac{1}{2}\mathfrak{K} + \underline{\underline{x}}$ which contains two distinct integer points, say $\underline{\underline{g}}_1$ and $\underline{\underline{g}}_2$ .

Thus $\underline{\underline{g}}_1 - \underline{\underline{x}} \in \frac{1}{2}\mathfrak{K}$ and $\underline{\underline{g}}_2 - \underline{\underline{x}} \in \frac{1}{2}\mathfrak{K}$ . By symmetry, $\underline{\underline{x}} - \underline{\underline{g}}_2 \in \frac{1}{2}\mathfrak{K}$ . Therefore, we may put $\underline{\underline{g}}_1 - \underline{\underline{x}} = \frac{1}{2}\underline{\underline{x}}_1$ and $\underline{\underline{x}} - \underline{\underline{g}}_2 = \frac{1}{2}\underline{\underline{x}}_2$ , where $\underline{\underline{x}}_1$ and $\underline{\underline{x}}_2$ lie in $\mathfrak{K}$ . Now $\mathfrak{K}$ is convex, so $\underline{\underline{g}}_1 - \underline{\underline{g}}_2 = \frac{1}{2}\underline{\underline{x}}_1 + \frac{1}{2}\underline{\underline{x}}_2 \in \mathfrak{K}$ . Hence $\underline{\underline{g}} = \underline{\underline{g}}_1 - \underline{\underline{g}}_2$ satisfies the conditions of the theorem.

Here is

Another proof of Minkowski's Theorem, due to Mordell (1934). Suppose that $\mu(\mathfrak{R}) > 2^n$ . For positive integers m let $\mathfrak{R}_m$ be the set of points in $\mathfrak{R}$ which have rational coordinates with denominator m . As $m \to \infty$ , the number of points in $\mathfrak{R}_m$ will be asymptotically equal to $\mu(\mathfrak{R})m^n$ , and hence for certain large m this number will be greater than $(2m)^n$ . There will then be two points $\left(\frac{a_1}{m},\ldots,\frac{a_n}{m}\right)$ and $\left(\frac{b_1}{m},\ldots,\frac{b_n}{m}\right)$ in $\mathfrak{R}_m$ having

(2.2) $$a_i \equiv b_i \pmod{2m} \qquad (i = 1,\ldots,n) \quad .$$

By the symmetry and convexity of $\mathfrak{R}$ , the point

$$\underline{\underline{g}} = \frac{1}{2}\left(\frac{a_1}{m},\ldots,\frac{a_n}{m}\right) - \frac{1}{2}\left(\frac{b_1}{m},\ldots,\frac{b_n}{m}\right)$$

is a non-zero point in $\mathfrak{R}$ , which is an integer point by (2.2).

The case when $\mu(\mathfrak{R}) = 2^n$ and $\mathfrak{R}$ is compact can easily be reduced to the case when $\mu(\mathfrak{R}) > 2^n$ .

THEOREM 2C. (Minkowski's Linear Forms Theorem) Suppose that $\beta_{ij}$ $(1 \leq i , j \leq n)$ are real numbers with determinant $\pm 1$ . Suppose that

$A_1,\dots,A_n$ *are positive with* $A_1 A_2 \dots A_n = 1$ . *Then there exists an integer point* $\underline{x} = (x_1,\dots,x_n) \neq \underline{0}$ *such that*

$$(2.3) \qquad |\beta_{i1}x_1 + \dots + \beta_{in}x_n| < A_i , \qquad (1 \le i \le n-1) ,$$

*and*

$$(2.4) \qquad |\beta_{n1}x_1 + \dots + \beta_{nn}x_n| \le A_n .$$

*Proof*. Write

$$L_i(\underline{x}) = \beta_{i1}x_1 + \dots + \beta_{in}x_n \qquad (1 \le i \le n) ,$$

and put

$$L_i'(\underline{x}) = \frac{1}{A_i} L_i(\underline{x}) \qquad (1 \le i \le n).$$

Then (2.3) and (2.4) may be rewritten as

$$|L_i'(\underline{x})| < 1 \qquad (1 \le i \le n-1)$$

and

$$|L_n'(\underline{x})| \le 1 .$$

The determinant of $L_1',\dots,L_n'$ is again equal to $\det(\beta_{ij}) = \pm 1$ , so that we may restrict ourselves to the case in which $A_1 = \dots = A_n = 1$ .

The basic idea of the proof is now as follows. The set $\mathcal{R}$ of all $\underline{x} \in E^n$ satisfying

$$|L_i(\underline{x})| \le 1 \qquad (1 \le i \le n)$$

is obtained from $\bar{U}^n$ by a linear transformation of determinant 1 . Hence $\mathcal{R}$ is a symmetric, closed parallelepiped of volume $2^n$ . By

Minkowski's Convex Body Theorem, there is an integer point $\underline{\underline{x}} \neq \underline{\underline{0}}$ in this parallelepiped.

To get strict inequality in the first n-1 inequalities we modify this argument as follows. For each $\varepsilon > 0$, the system of inequalities $|L_i(\underline{\underline{x}})| < 1$ $(1 \le i \le n-1)$, $|L_n(\underline{\underline{x}})| < 1+\varepsilon$ defines a symmetric parallelepiped $\Pi_\varepsilon$ of volume $2^n(1+\varepsilon) > 2^n$. By Minkowski's Convex Body Theorem, there is an integer point $\underline{\underline{x}}_\varepsilon \neq \underline{\underline{0}}$ in $\Pi_\varepsilon$.

Now take a sequence $(\varepsilon_k)_{k=1}^\infty$, $\varepsilon_k \to 0^+$. This sequence gives rise to a sequence $(\underline{\underline{x}}_{\varepsilon_k})_{k=1}^\infty$ of nonzero integer points, $\underline{\underline{x}}_{\varepsilon_k} \in \Pi_{\varepsilon_k}$. Since all of the sets $\Pi_{\varepsilon_k}$ are uniformly bounded, there is an integer point $\underline{\underline{x}} \neq \underline{\underline{0}}$ such that $\underline{\underline{x}} = \underline{\underline{x}}_{\varepsilon_k}$ for infinitely many $k$. Thus $\underline{\underline{x}} \in \Pi_{\varepsilon_k}$ for infinitely many $k$, whence $\underline{\underline{x}}$ satisfies (2.3) and (2.4).

<u>Another Proof of Theorem 1E</u>. Put $\ell = m + n$, and consider the following $\ell$ linear forms in $\underline{\underline{x}} = (x_1,\ldots,x_\ell)$:

$$L_i(\underline{\underline{x}}) = x_i, \qquad (1 \le i \le m)$$

$$L_{m+j}(\underline{\underline{x}}) = \alpha_{j1}x_1 + \ldots + \alpha_{jm}x_m - x_{m+j} \qquad (1 \le j \le n).$$

Their determinant clearly is $\pm 1$.

Let $Q > 1$ be given, $Q$ not necessarily an integer. By Lemma 2C, there is an integer point $\underline{\underline{x}} \neq \underline{\underline{0}}$ with

$$|L_i(\underline{\underline{x}})| < Q^{n/m}, \qquad (1 \le i \le m)$$

$$|L_{m+j}(\underline{\underline{x}})| \le Q^{-1} \qquad (1 \le j \le n).$$

Put $q_i = x_i$ $(1 \le i \le m)$ and $p_j = x_{m+j}$ $(1 \le j \le n)$. Then

$$q = \max(|q_1|,\ldots,|q_n|) < Q^{n/m} \quad ,$$

and

$$(2.5) \qquad |\alpha_{j1}q_1 + \ldots + \alpha_{jm}q_m - p_j| \le \frac{1}{Q} \qquad (1 \le j \le n).$$

It remains to be shown that $q \ge 1$ . If not, then all $q_i = 0$ , and therefore

$$|p_j| \le \frac{1}{Q} < 1 \qquad (1 \le i \le n)$$

from (2.5). But then all $p_j = 0$, whence $\underline{x} = \underline{0}$ , a contradiction.

## §3. Improvement of the Simultaneous Approximation Constants.

THEOREM 3A. (Minkowski (1910) did the case $m = 1$). *Consider linear forms*

$$L_i(\underline{x}) = \alpha_{i1}x_1 + \ldots + \alpha_{im}x_m \qquad (1 \le i \le n)$$

*where* $\underline{x} = (x_1,\ldots,x_m)$. *Put*

$$\mathcal{L}(\underline{x}) = (L_1(\underline{x}),\ldots,L_n(\underline{x})).$$

*Then there is an integer point* $(\underline{x},\underline{y}) = (x_1,\ldots,x_m,y_1,\ldots,y_n)$ *in* $E^{m+n}$ *with* $\underline{x} \ne \underline{0}$ *such that*

$$(3.1) \qquad \overline{|\mathcal{L}(\underline{x}) - \underline{y}|}^{n} < C_{m,n} \frac{1}{\overline{|\underline{x}|}^{m}} \quad ,$$

*where*

$$C_{m,n} = \frac{m^m n^n}{(m+n)^{m+n}} \cdot \frac{(m+n)!}{m!\,n!} \quad .$$

*Furthermore, suppose that whenever* $\underline{x} \ne \underline{0}$ *is an integer point in* $E^m$ , *then* $\mathcal{L}(\underline{x})$ *is* ***not*** *an integer point in* $E^n$ . *Then there exist infinitely* *many integer points* $(\underline{x},\underline{y})$ *with* $\underline{x} \neq \underline{0}$ *and with coprime components satisfying* (3.1) .

*Remark.* We make explicit the connection between Theorems 1E and 3A. The conclusion of Theorem 1E may be restated as follows: *There is an integer points* $(\underline{x},\underline{y}) = (x_1,\ldots,x_m,y_1,\ldots,y_n)$ *in* $E^{m+n}$ *with*

$$1 \le \overline{|\underline{x}|} < Q^{n/m} \quad \text{and} \quad \overline{|\mathfrak{L}(\underline{x}) - \underline{y}|} \le \frac{1}{Q} .$$

These inequalities yield

$$\overline{|\mathfrak{L}(\underline{x})-\underline{y}|}^n \le \frac{1}{Q^n} < \frac{1}{\overline{|\underline{x}|}^m} .$$

Theorem 3A is an improvement over Theorem 1E since $C_{m,n} < 1$. Namely $C_{m,n} = \binom{m+n}{m}\left(\frac{m}{m+n}\right)^m\left(\frac{n}{m+n}\right)^n$ is just one of the $m + n + 1$ terms in the binomial expansion of

$$1 = 1^{m+n} = \left(\frac{m}{m+n} + \frac{n}{m+n}\right)^{m+n} .$$

*LEMMA 3B. Suppose that* $m \ge 1$ , $n \ge 1$ *are integers and that* $t > 0$ . *Let* $K_{m,n}$ *be the set of points* $(\underline{x},\underline{y}) = (x_1,\ldots,x_m,y_1,\ldots,y_n)$ *in* $E^{m+n}$ *satisfying*

$$t^{-n}\overline{|\underline{x}|} + t^m\overline{|\underline{y}|} \le 1 . \tag{3.2}$$

*Then* $K_{m,n}$ *is compact, symmetric (with respect to* $\underline{0}$*) and convex, with volume*

$$V(m,n) = 2^{m+n} \frac{m!n!}{(m+n)!} .$$

*Proof.* Let us write $K_{m,n}(t)$ in place of $K_{m,n}$ for the moment. The transformation given by

$$x_i \mapsto t^n x_i \qquad (1 \le i \le m)$$
$$y_j \mapsto t^{-m} y_j \qquad (1 \le j \le n)$$

is linear and has determinant $(t^n)^m (t^{-m})^n = 1$ ; furthermore, this transformation maps $K_{m,n}(1)$ onto $K_{m,n}(t)$. Since a linear transformation preserves compactness, symmetry and convexity, it is enough to verify that $K_{m,n}$ has these properties when $t = 1$.

The compactness and symmetry of $K_{m,n}$ are obvious. To prove convexity, suppose that $(\underline{\underline{x}},\underline{y})$ and $(\underline{\underline{x}}',\underline{y}')$ belong to $K_{m,n}$ and that $\lambda \ge 0$ , $\mu \ge 0$ satisfy $\lambda + \mu = 1$. Then $\lambda(\underline{\underline{x}},\underline{y}) + \mu(\underline{\underline{x}}',\underline{y}') =$ $(\lambda\underline{\underline{x}} + \mu\underline{\underline{x}}' , \lambda\underline{y} + \mu\underline{y}')$ belongs to $K_{m,n}$ , since

$$\lceil \lambda\underline{\underline{x}} + \mu\underline{\underline{x}}' \rceil + \lceil \lambda\underline{y} + \mu\underline{y}' \rceil \le \lambda(\lceil \underline{\underline{x}} \rceil + \lceil \underline{y} \rceil) + \mu(\lceil x' \rceil + \lceil \underline{y}' \rceil) \le \lambda + \mu = 1.$$

It remains to compute $V(m,n)$ , the volume of $K_{m,n}$ . Let $K^*_{m,n}$ denote the set of points $(\underline{\underline{x}},\underline{y})$ in $K_{m,n}$ for which $x_i \ge 0$ $(1 \le i \le m)$, $y_j \ge 0$ $(1 \le j \le n)$, and $x_1 = \lceil \underline{\underline{x}} \rceil$ . By (3.2), $0 \le x_1 \le 1$ ; moreoever, for each fixed $x_1$ , $0 \le x_i \le x_1$ $(2 \le i \le m)$ and $0 \le y_j \le 1-x_1$ $(1 \le j \le n)$. If $V^*(m,n)$ denotes the volume of $K^*(m,n)$ , it follows that

$$V(m,n) = m2^{m+n}\, V^*(m,n)$$
$$= m2^{m+n} \int_0^1 x_1^{m-1}(1-x_1)^n dx_1 \ .$$

The value of the integral is $\frac{(m-1)!n!}{(m+n)!}$ (this follows by induction on $m$ or by the reader's knowledge of the Beta function), and the proof of the lemma is complete.

<u>Proof of Theorem 3A</u>. Suppose that $t > 0$ . Let $K_{m,n}$ be as in Lemma 3B, and let $\tilde{K}_{m,n}$ be the set of points $(\underline{\underline{x}},\underline{y}) = (x_1,\ldots,x_m,y_1,\ldots,y_n)$ in $E^{m+n}$ satisfying

$$t^{-n}\lceil \underline{\underline{x}} \rceil + t^m \lceil \mathcal{L}(\underline{\underline{x}}) - (\underline{y}) \rceil \le C ,$$

where $C = 2V(m,n)^{-1/(m+n)}$ . The linear transformation defined by

$$x_i \mapsto Cx_i \qquad (1 \le i \le m)$$

$$y_j \mapsto C(L_j(\underline{\underline{x}}) - y_j) \qquad (1 \le j \le n)$$

maps $K_{m,n}$ onto $\tilde{K}_{m,n}$ and has determinant $(-1)^n C^{m+n}$ , so that $\tilde{K}_{m,n}$ has volume $C^{m+n}V(m,n) = 2^{m+n}$ . Since compactness, symmetry and convexity are preserved under linear transformations, $\tilde{K}_{m,n}$ has these three properties in view of Lemma 3B. By Minkowski's Convex Body Theorem, it follows that $\tilde{K}_{m,n}$ contains an integer point $(\underline{\underline{x}},\underline{y}) \ne (\underline{\underline{0}},\underline{\underline{0}})$.

Now put $M = \lceil \underline{\underline{x}} \rceil$ , $L = \lceil \mathcal{L}(\underline{\underline{x}}) - \underline{y} \rceil$ : then

(3.3) $$t^{-n}M + t^m L \le C .$$

For a given integer point $(\underline{\underline{x}},\underline{y})$ , equality in (3.3) can hold for at most finitely many values of $t$ . Since the number of integer points in $E^{m+n}$ is countable, it follows that

(3.4) $$t^{-n}M + t^m L < C$$

for all but countably many $t$ . In the remainder of the proof, we consider only those $t$ which satisfy (3.4).

Recall the Arithmetic-Geometric Inequality: If $z_1, z_2, \ldots, z_\ell$ are $\ell$ nonnegative numbers, then

$$(z_1 z_2 \ldots z_\ell)^{1/\ell} \le \frac{z_1 + z_2 + \ldots + z_\ell}{\ell} .$$

We apply this with $\ell = m+n$ , $z_1 = \ldots = z_m = m^{-1}t^{-n}M$ , $z_{m+1} = \ldots = z_{m+n} = n^{-1}t^m L$ , to obtain

$$(m^{-1}t^{-n}M)^m(n^{-1}t^mL)^n \le (\frac{t^{-n}M + t^mL}{\ell})^\ell < (\frac{C}{\ell})^\ell ,$$

which yields

$$M^mL^n < \frac{m^m n^n}{(m+n)^{m+n}} C^{m+n} = C_{m,n} .$$

This inequality is equivalent to (3.1), provided $M \neq 0$. If we choose $t$ to satisfy (3.4) and $t \ge C^{1/m}$, then $M \neq 0$, since otherwise (3.4) implies that

$$\overline{|\mathcal{L}(\underline{\underline{0}}) - \underline{\underline{y}}|} = \overline{|\underline{\underline{y}}|} < Ct^{-m} \le 1 ,$$

hence $\underline{\underline{y}} = \underline{\underline{0}}$, a contradiction. This establishes the first assertion of the theorem.

To complete the proof of the theorem, suppose now that, whenever $\underline{\underline{x}} \neq \underline{\underline{0}}$ is an integer point in $E^m$, $\mathcal{L}(\underline{\underline{x}})$ is <u>not</u> an integer point in $E^n$. Further suppose that $t$ satisfies (3.4) and $t \ge C^{1/m}$. Now $M \neq 0$ by the preceding paragraph, hence $L \neq 0$. It follows that for fixed $(\underline{\underline{x}},\underline{\underline{y}})$, inequality (3.4) can hold only for $t \le t_0$. Hence as $t \to \infty$, there will be infinitely many distinct integer points $(\underline{\underline{x}},\underline{\underline{y}})$ with coprime components and with $\underline{\underline{x}} \neq \underline{\underline{0}}$ satisfying (3.1).

<u>Remarks</u>. Consider the special case of Theorem 3A in which $m = 1$. Then $C_{1,n} = (\frac{n}{n+1})^n$. It was shown by Blichfeldt (1914) that $C_{1,n}$ may be replaced by the still smaller constant

$$\gamma_n = \left(\frac{n}{n+1}\right)^n \left(1 + \left(\frac{n-1}{n+1}\right)^{n+3}\right)^{-1} .$$

When $m = 1$, the inequality (3.1) may be written as

$$\max(|\alpha_1 x - y_1|, \ldots, |\alpha_n x - y_n|) < \frac{n}{n+1} x^{-1/n} .$$

If $n = 1$ , the coefficient of $x^{-1/n}$ here is $\frac{1}{2}$. By Hurwitz' Theorem of Chapter I, a coefficient $1/\sqrt{5}$ is best possible. If $n = 2$ , the coefficient in question is $2/3$. The best possible coefficient is unknown. However, if $c_0$ denotes the infimum of admissible coefficients when $n = 2$, we have shown that $c_0 \le 2/3$, and it is known that

$$0.53 \approx \sqrt{2/7} \le c_0 \le 0.615 \quad .$$

See Cassels (1955a), Davenport (1952), Davenport & Mahler (1946), Mack (1977).

In the next section we are going to show that the best coefficient is positive for every $n,m$.

## §4. Badly Approximable Systems of Linear Forms.

Let $\alpha_{ij}$ $(1 \leq i \leq n \;,\; 1 \leq j \leq m)$ be $nm$ real numbers, and consider linear forms

$$L_i(\underline{\underline{x}}) = \alpha_{i1}x_1 + \dots + \alpha_{im}x_m \;, \qquad (1 \le i \le n)$$

where $\underline{\underline{x}} = (x_1,\dots,x_m)$. Put

$$\mathcal{L}(\underline{\underline{x}}) = (L_1(\underline{\underline{x}}),\dots,L_n(\underline{\underline{x}})).$$

Definition. $L_1,\dots,L_n$ is a badly approximable system of linear forms if there is a constant $\gamma = \gamma(L_1,\dots,L_n) > 0$ such that

$$\text{(4.1)} \qquad \overline{|\underline{\underline{x}}|}^{\,m}\ \overline{|\mathcal{L}(\underline{\underline{x}}) - \underline{\underline{y}}|}^{\,n} > \gamma$$

for every integer point $(\underline{\underline{x}},\underline{\underline{y}}) = (x_1,\dots,x_m,y_1,\dots,y_n)$ with $\underline{\underline{x}} \neq \underline{\underline{0}}$ . (In other words, these are linear forms for which $C_{m,n}$ in Theorem 3A cannot be replaced by arbitrarily small constants.)

Remark. Consider, for the moment, the special case $m = n = 1$ . Then

(4.1) reduces to $|x| \cdot |\alpha_{11}x-y| > \gamma$ , or

$$|\alpha_{11} - \frac{y}{x}| > \frac{\gamma}{x^2} .$$

Thus, the linear form $L(x) = \alpha x$ is badly approximably precisely if the coefficient $\alpha$ is badly approximably as defined in §5 of Chapter I.

In the special case when $n = 1$ , a single linear form $L(\underline{\underline{x}}) = \alpha_1 x_1 + \ldots + \alpha_m x_m$ is badly approximable if

$$|\alpha_1 q_1 + \ldots + \alpha_m q_m - p| > \frac{\gamma_1}{q^m}$$

for every integer point $(q_1,\ldots,q_m,p)$ with $q = \max(|q_1|,\ldots,|q_m|) > 0$. The other extreme is when $m = 1$ and we have linear forms $L_1(x) = \alpha_1 x,\ldots,L_n(x) = \alpha_n x$ , and $\max(|\alpha_1 q-p_1|,\ldots,|\alpha_n q-p_n|) > \gamma_2/q^{1/n}$ for integers $q > 0$ , $p_1,\ldots,p_n$ , or

$$\max(|\alpha_1 - \frac{p_1}{q}|,\ldots,|\alpha_n - \frac{p_n}{q}|) > \frac{\gamma_2}{q^{1+(1/n)}} .$$

In this case we simply call $(\alpha_1,\ldots,\alpha_n)$ a <u>badly approximable</u> n-tuple.

It will be shown in Theorem 5B of Chapter IV (Khintchine's transference principle) that the n-tuple $(\alpha_1,\ldots,\alpha_n)$ is badly approximable precisely if the linear form $\alpha_1 x_1 + \ldots + \alpha_n x_n$ is badly approximable. More generally, it will be shown that the system of linear forms $L_1(\underline{\underline{x}}),\ldots,L_n(\underline{\underline{x}})$ is badly approximably if and only if the "dual" system $L_1^*(\underline{\underline{y}}),\ldots,L_m^*(\underline{\underline{y}})$ is badly approximable, where

$$L_j^*(\underline{\underline{y}}) = \alpha_{1j}y_1 + \ldots + \alpha_{nj}y_n \qquad (1 \leq j \leq m).$$

<u>THEOREM 4A</u>. <u>Suppose</u> $1,\alpha_1,\ldots,\alpha_m$ <u>is a basis of a real algebraic number field of degree</u> $m + 1$. <u>Then the linear form</u> $\alpha_1 x_1 + \ldots + \alpha_m x_m$ <u>is badly approximable</u>.

Proof. We may restrict ourselves to integers $q_1,\ldots,q_m,p$ with $|\alpha_1 q_1 + \ldots + \alpha_m q_m - p| < 1$, and then†) $|p| \ll q = \max(|q_1|,\ldots,|q_m|)$. Each conjugate $\alpha_1^{(i)} q_1 + \ldots + \alpha_m^{(i)} q_m - p$ has an absolute value $\ll q$, and therefore the norm satisfies†)

$$\mathfrak{N}(\alpha_1 q_1 + \ldots + \alpha_m q_m - p) \ll q^m |\alpha_1 q_1 + \ldots + \alpha_m q_m - p|. \tag{4.2}$$

On the other hand, there is a rational integer $h > 0$ such that $h\alpha_1,\ldots,h\alpha_n$ are algebraic integers. Then $|\mathfrak{N}(h\alpha_1 q_1 + \ldots + h\alpha_m q_m - hp)| \geqq 1$ and

$$|\mathfrak{N}(\alpha_1 q_1 + \ldots + \alpha_m q - p)| \geqq h^{-m-1} \gg 1. \tag{4.3}$$

Combining (4.2) and (4.3) we obtain the theorem.

In general we have

THEOREM 4B. (Perron (1921b)) *For every pair of integers* $m \geq 1$, $n \geq 1$, *there exist algebraic numbers* $\alpha_{ij}$ $(1 \leq i \leq n,\ 1 \leq j \leq m)$ *such that the system of linear forms* $L_1,\ldots,L_n$ *given above is badly approximable*.

Proof. Put $\ell = m + n$, and let $\varphi_1,\ldots,\varphi_\ell$ be $\ell$ real conjugate algebraic integers. (The existence of such integers will be discussed below.) Writing $\underline{x} = (x_1,\ldots,x_m)$ and $\underline{y} = (y_1,\ldots,y_n)$, we define

$$M_k(\underline{x},\underline{y}) = \sum_{i=1}^{n} \varphi_k^{i-1} y_i + \sum_{j=1}^{m} \varphi_k^{n+j-1} x_j \qquad (1 \leq k \leq \ell).$$

If $(\underline{x},\underline{y}) \neq (\underline{0},\underline{0})$ is an integer point, then $M_k(\underline{x},\underline{y}) \neq 0$ for each $k$, since $1,\varphi_k,\varphi_k^2,\ldots,\varphi_k^{\ell-1}$ are linearly independent over the rationals. Furthermore, $M_1(\underline{x},\underline{y})$, $M_2(\underline{x},\underline{y}),\ldots,M_\ell(\underline{x},\underline{y})$ are conjugate algebraic integers, whence

†) The "Vinogradov symbol" $\ll$ is explained in the notation section. In the present context the implied constant depends on $n,m$

LEMMA 4C. For every $\ell$ there exists a real number field of degree $\ell$ which is normal (over the rationals).

Proof. Let $\varphi$ be the Euler $\varphi$- function and pick an integer $n$ for which $\varphi(n)$ is a multiple of $2\ell$. Let $\zeta$ be a primitive $n$-th root of unity. The field $\mathbb{Q}(\zeta)$ (obtained by adjoining $\zeta$ to the field $\mathbb{Q}$ of rationals) is of degree $\varphi(n)$ and normal, and its Galois group is abelian (Van der Waerden (1955) §55). The subfield $\mathbb{Q}(\zeta + (1/\zeta))$ is real, of degree $\frac{1}{2}\varphi(n)$ and normal, and its Galois group $G$ is abelian. The group $G$ contains a subgroup $H$ of order $(1/2\ell)\varphi(n)$. The fixed field of $H$ is real, normal and of degree $\ell$.

COROLLARY 4D. The linear forms of Theorem 4B can be constructed to have their coefficients in a field of degree $m + n$.

Proof. They can be constructed to have their coefficients in any real field $F$ which is normal and of degree $\ell = m + n$. We may write $F$ as $F = \mathbb{Q}(\varphi)$ where $\varphi$ is an algebraic integer, and in the proof of Theorem 4B let $\varphi_1 = \varphi, \varphi_2, \ldots, \varphi_\ell$ be the conjugates of $\varphi$.

Furtwängler (1927-28) gave a quantitative version of Theorem 4B†).

We close this chapter with a famous conjecture of Littlewood. We have

$$|x| \max(\|\alpha_1 x\|, \ldots, \|\alpha_n x\|)^n \geq |x| \; \|\alpha_1 x\| \ldots \|\alpha_n x\| \quad ,$$

with equality when $n = 1$. A badly approximably n-tuple has

$$|x| \max(\|\alpha_1 x\|, \ldots, \|\alpha_n x\|)^n > \gamma > 0 \; .$$

On the other hand, Littlewood conjectured that for $n \geq 2$, arbitrary

†) See also forthcoming work of T.W. Cusick.

$\alpha_1,\ldots,\alpha_n$ and arbitrary $\varepsilon > 0$, there is always an $x > 0$ with

$$x\|\alpha_1 x\| \ldots \|\alpha_n x\| < \varepsilon .$$

Put differently,

$$\liminf_{x\to\infty} x\|\alpha_1 x\| \ldots \|\alpha_n x\| = 0.$$

If this conjecture is true for some $n$, then it is true a fortiori for $n' > n$.

# III. Games and Measures

## First Part: Games

References: Cassels (1956), Schmidt (1966).

### §1. The $(\alpha,\beta)$-Game.

Suppose that $0 < \alpha < 1$, $0 < \beta < 1$. Consider the following game by players Black and White. First, Black picks a compact real interval $B_1$ of length $\ell(B_1)$. Next, White picks a compact interval $W_1 \subset B_1$ of length $\ell(W_1) = \alpha\ell(B_1)$. Then Black picks a compact interval $B_2 \subset W_1$ of length $\ell(B_2) = \beta\ell(W_1)$, etc. In this way, a nested sequence of compact intervals

$$B_1 \supset W_1 \supset B_2 \supset W_2 \supset \cdots$$

is generated, with lengths

$$\ell(B_k) = (\alpha\beta)^{k-1}\ell(B_1) \quad \text{and} \quad \ell(W_k) = (\alpha\beta)^{k-1}\alpha\ell(B_1) \qquad (k = 1,2,3,\ldots).$$

It is clear that $\bigcap_{k=1}^{\infty} B_k = \bigcap_{k=1}^{\infty} W_k$ consists of a single point.

In $E^n$, the same game is played with intervals replaced by closed balls and length replaced by radius. Let $S$ be a subset of $E^n$. We say that <u>White wins the game</u> if $\bigcap_{k=1}^{\infty} B_k \in S$. Furthermore, $S$ is an $(\alpha,\beta)$-<u>winning set</u> if White is able to win the game no matter how Black plays.

The game just described is a metrical variation of the Banach-Mazur game as discussed in, for example, J.C. Oxtoby (1957).

THEOREM 1A. *Suppose that* $2\beta < 1 + \alpha\beta$. *Then every* $(\alpha,\beta)$-*winning set has the power of the continuum.*

We need

LEMMA 1B. *Suppose that* $0 < \alpha < 1$, $0 < \beta < 1$ *and* $\gamma = 1+\alpha\beta-2\beta > 0$. *Let* $t$ *be an integer with* $(\alpha\beta)^t < \frac{1}{2}\gamma$, *and* $\underline{u}$ *a vector of length* 1. *Suppose a ball* $W_k$ *occurs in the* $(\alpha,\beta)$-*game*. *Then Black can play so that (no matter how White plays) every point* $\underline{x}$ *of* $W_{k+t}$ *satisfies*

$$(\underline{x} - \underline{w}_k)\underline{u} > \frac{1}{2}\gamma\,\rho_k \quad {}^{\dagger)}$$

*where* $\underline{w}_k$ *and* $\rho_k$ *denote respectively the center and the radius of* $W_k$.

*Proof*. Black can choose $B_{k+1} \subset W_k$ with center $\underline{b}_{k+1} = \underline{w}_k + (1-\beta)\rho_k\underline{u}$. Then

$$(\underline{b}_{k+1} - \underline{w}_k)\underline{u} = (1-\beta)\rho_k . \tag{1.1}$$

If now White picks any ball $W_{k+1} \subset B_{k+1}$ with center $\underline{w}_{k+1}$, then we must have

$$|\underline{w}_{k+1} - \underline{b}_{k+1}| \leq (\text{radius of } B_{k+1}) - (\text{radius of } W_{k+1}) = \beta\rho_k - \alpha\beta\rho_k .$$

By the Cauchy-Schwarz Inequality, it follows that

$$(\underline{w}_{k+1} - \underline{b}_{k+1})\underline{u} \geq -|\underline{w}_{k+1} - \underline{b}_{k+1}|\cdot|\underline{u}| \geq (\alpha\beta - \beta)\rho_k ,$$

and in conjunction with (1.1) we get

$$(\underline{w}_{k+1} - \underline{w}_k)\underline{u} \geq (1 + \alpha\beta - 2\beta)\rho_k = \gamma\rho_k .$$

Repetition of this argument gives

$$(\underline{w}_{k+j} - \underline{w}_{k+j-1})\underline{u} \geq \gamma\,\rho_{k+j-1} \qquad (j = 1,\dots,t) ,$$

where $\underline{w}_{k+j}$ and $\rho_{k+j}$ is the center and the radius of $W_{k+j} (j = 0,\dots,t)$. By adding these $t$ inequalities we obtain

---

†) $\underline{vu}$ denotes the inner product of vectors $\underline{v}$, $\underline{u}$.

(1.2) $$(\underline{\underline{w}}_{k+t} - \underline{\underline{w}}_k)\underline{\underline{u}} \geq \gamma\rho_k + \ldots + \gamma\rho_{k+t-1} \geq \gamma\rho_k \quad .$$

The ball $W_{k+t}$ has radius $(\alpha\beta)^t\rho_k < \frac{1}{2}\gamma\rho_k$, and hence every $\underline{\underline{x}} \in W_{k+t}$ has $|\underline{\underline{x}} - \underline{\underline{w}}_{k+t}| < \frac{1}{2}\gamma\rho_k$, which together with (1.2) yields

$$(\underline{\underline{x}} - \underline{\underline{w}}_k)\underline{\underline{u}} > \frac{1}{2}\gamma\rho_k \ .$$

<u>Proof of Theorem 1A</u>. Put $\underline{\underline{u}} = (0,\ldots,0,1)$, and given a ball $W$ with center $\underline{\underline{w}}$ let the <u>upper half</u> $W^+$ of $W$ consist of $\underline{\underline{x}} \in W$ having $(\underline{\underline{x}} - \underline{\underline{w}})\underline{\underline{u}} > 0$, and the <u>lower half</u> $W^-$ of $\underline{\underline{x}} \in W$ with $(\underline{\underline{x}} - \underline{\underline{w}})\underline{\underline{u}} < 0$.

Suppose White plays a winning strategy, and suppose $B_1$, $W_1$ are already given. Choose $t$ with $(\alpha\beta)^t < \frac{1}{2}\gamma$. By Lemma 1B, Black can play such that $W_{1+t}$ lies in the upper half $W_1^+$ of $W_1$. Or, if he chooses otherwise, he can play such that $W_{1+t} \subset W_1^-$. Let $W_{1+t}^+$ be the ball $W_{1+t}$ if Black uses his first strategy, and $W_{1+t}^-$ the ball $W_{1+t}$ if Black uses his second strategy. In the next $t$ moves Black can play such that $W_{1+2t}$ lies in the upper half of $W_{1+t}$ or in the lower half of $W_{1+t}$. In this way there are four possibilities for $W_{1+2t}$, which we denote by $W_{1+2t}^{++}$, $W_{1+2t}^{+-}$, $W_{1+2t}^{-+}$, $W_{1+2t}^{--}$. And so on. If $i_1, i_2, \ldots$ is a sequence of $+$ and $-$ signs, we get a sequence of balls $W_{1+t}^{i_1}$, $W_{1+2t}^{i_1 i_2}$, $W_{1+3t}^{i_1 i_2 i_3}$, ... whose intersection must lie in $S$. Since the number of sequences $i_1, i_2, \ldots$ has the power of the continuum, and since for distinct sequences we obtain distinct intersections, it follows that $S$ has the power of the continuum.

<u>LEMMA 1C</u>. <u>Suppose</u> $S$ <u>is</u> $(\alpha,\beta)$-<u>winning</u>. <u>Then it is</u> $(\alpha(\beta\alpha)^k,\beta)$-<u>winning for every positive integer</u> $k$.

<u>Proof</u>. Suppose in the $(\alpha,\beta)$-game, White not only chooses the balls $W_n$, but also the balls $B_n$ except those where $n \equiv 1 \pmod{k+1}$.

Thus Black can pick only the balls $B_1, B_{1+(k+1)}, B_{1+2(k+1)}, \ldots$ . The balls

$$B_1 , W_{k+1} , B_{1+(k+1)} , W_{2(k+1)} , B_{1+2(k+1)}, \ldots$$

are balls of an $(\alpha(\beta\alpha)^k, \beta)$-play. If White can win the $(\alpha,\beta)$-game, he certainly can with the $(\alpha(\beta\alpha)^k, \beta)$-game.

Exercise. Suppose $0 < \alpha' \leqq \alpha < 1$ , $0 < \beta \leqq \beta' < 1$ and $\alpha\beta = \alpha'\beta'$. Prove that *every* $(\alpha,\beta)$-*winning set is* $(\alpha',\beta')$-winning.

The following is an open

Question. Suppose that $0 < \alpha' \leqq \alpha < 1$ and $0 < \beta \leqq \beta' < 1$ . Is it true that every $(\alpha,\beta)$-winning set is $(\alpha',\beta')$-winning? The author's guess is that in general it is not true.

Definition. Given $\alpha$ in $0 < \alpha < 1$ , a set $S$ is called $\alpha$-*winning* if it is $(\alpha,\beta)$-winning for every $\beta$ in $0 < \beta < 1$ .

It follows from Theorem 1A that an $\alpha$-winning set has the power of the continuum. In fact it was shown by Schmidt (1966) that such a set has Hausdorff dimension $n$ .

LEMMA 1D. *The intersection of countably many* $\alpha$-*winning sets is* $\alpha$-*winning*.

Proof. Suppose that $\alpha$-winning sets $S_1, S_2, S_3$ are given. Also suppose that $\beta$ is given, $0 < \beta < 1$. White plays as follows:

For his 1st, 3rd, 5th,... move, he uses an $(\alpha,\beta\alpha\beta)$-winning strategy for $S_1$ . Hence

$$\bigcap_{k=1}^{\infty} B_k \in S_1 \quad .$$

For his 2nd, 6th, 10th,... move, White uses an $(\alpha,\beta(\alpha\beta)^3)$-winning strategy for $S_2$. Hence

$$\bigcap_{k=1}^{\infty} B_k \in S_2 .$$

Similarly, for his 4th, 12th, 20th,... move, he uses an $(\alpha,\beta(\alpha\beta)^7)$-winning strategy for $S_3$. Hence

$$\bigcap_{k=1}^{\infty} B_k \in S_3 .$$

Continuing in this manner, White can play so that

$$\bigcap_{k=1}^{\infty} B_k \in \bigcap_{j=1}^{\infty} S_j ,$$

and the lemma is proved.

## §2. Badly Approximable n-tuples and $(\alpha,\beta)$-Games.

We first give some literature.

(i) Cassels (1956) proved:

*Suppose that* $\lambda_1,\ldots,\lambda_k$ *are* $k$ *real numbers. Then there exist continuum many* $\alpha$ *such that each of*

$$\alpha + \lambda_1, \alpha + \lambda_2, \ldots, \alpha + \lambda_k$$

*is badly approximable.*

(ii) Davenport (1964) extended Cassel's method to show:

*Suppose that* $I$ *is a nondegenerate real interval and that* $f_1(x),\ldots,f_k(x)$ *are* $k$ *real-valued functions, each of which has a nonvanishing derivative on* $I$. *Then there exist continuum many* $\alpha$ *such that each of*

$$f_1(\alpha), f_2(\alpha), \ldots, f_k(\alpha)$$

*is badly approximable.*

(iii) Schmidt (1965a) proved:

*There exist continuum many* $\alpha$ *such that every irrational which is algebraically dependent on* $\alpha$ *is badly approximable.*

(iv) Davenport (1954).

In this paper, Davenport showed the existence of continuum many badly approximable pairs $(\alpha_1,\alpha_2)$.

(v) Cassels (1955b).

Here Cassels proved the existence of continuum many badly approximable n-tuples.

(vi) Davenport (1962).

Davenport gave a simple construction of continuum many badly approximable n-tuples by extending Cassels' method (i).

(vii) Schmidt (1966).

$(\alpha,\beta)$-games are introduced and applied to badly approximable n-tuples.

(viii) Schmidt (1969).

In this paper, Schmidt established the existence of continuum many badly approximable systems of linear forms for any positive integers m,n.

Here we prove

THEOREM 2A. *Suppose that* $0 < \alpha < 1$ , $0 < \beta < 1$ , $2\alpha < 1 + \alpha\beta$. *Then the set* S *of badly approximable* n-*tuples is* $(\alpha,\beta)$-*winning. In fact the image of* S *under a non-singular linear transformation is* $(\alpha,\beta)$-*winning.*

In view of Theorem 1A and Lemma 1D, we get

COROLLARY 2B. *Let* $T_1, T_2, \ldots$ *be non-singular linear transformations, and let* $\underline{a}_1, \underline{a}_2, \ldots$ *be n-tuples. The set of n-tuples* $\underline{x}$ *such that*

$$T_i(\underline{x}) + \underline{a}_i$$

*is badly approximable for* $i = 1,2,\ldots$ *is* $\alpha$*-winning for* $0 < \alpha < \frac{1}{2}$ *, and hence has the power of the continuum.*

In particular this includes the results of (i), (iv), (v), (vi). The general result (viii) on badly approximable systems of linear forms is more difficult.

*Proof of the Theorem.* For notational convenience, we take $n = 2$. We are to show that, for any nonsingular linear transformation $T$ , the set of $\underline{x}$ such that $T\underline{x}$ is badly approximable is $(\alpha,\beta)$-winning. We put

$$\|T\| = \sup_{\underline{x} \neq \underline{0}} \frac{|T\underline{x}|}{|\underline{x}|} .$$

Let $\gamma = 1 + \alpha\beta - 2\alpha > 0$ , and let $t$ be an integer, necessarily positive, with $(\alpha\beta)^t < \frac{\gamma}{2}$ . Define $R$ by

$$R = (\alpha\beta)^{-\frac{2}{3}t} > 1 .$$

We assume, without loss of generality, that Black starts the game with a ball $B_1$ of radius $\rho_1$ so small that

$$6\rho_1^2 R^3 (1 + \|T\|)^2 < 1.$$

Finally, we put

$$\delta = \min\left(\rho_1 , \frac{\rho_1\gamma}{4\|T^{-1}\|}\right) .$$

To prove the theorem, it suffices to show that White can play such that, for each nonnegative integer $n$ ,

$$(2.1) \qquad \max\left(|y_1 - \frac{p_1}{q}| , |y_2 - \frac{p_2}{q}|\right) > \frac{\delta}{q^{3/2}}$$

for every $(y_1,y_2) = T(x_1,x_2)$ with $(x_1,x_2) \in B_{nt+1}$ and with every ordered pair of rationals $\left(\frac{p_1}{q} , \frac{p_2}{q}\right)$ having $1 \le q < R^n$ . For then if $(x_1,x_2) \in \bigcap_{k=1}^{\infty} B_k$ and if $(y_1,y_2) = T(x_1,x_2)$ , it follows that (2.1) holds for <u>every</u> positive integer $q$ , whence $(y_1,y_2)$ is badly approximable.

We proceed now by induction on $n$ . If $n = 0$ , the inequality $1 \le q < R^n$ has no solutions $q$ , so that the statement above is true. Suppose, then, that $B_1,B_{1+t},\ldots,B_{1+(k-1)t}$ already have the desired properties. We shall show that, in his next $t$ moves, White can play so that $B_{1+kt}$ also will have the desired properties.

Inequality (2.1) is certainly satisfied for all $\underline{x} \in B_{1+(k-1)t}$ and all rational pairs $\left(\frac{p_1}{q} , \frac{p_2}{q}\right)$ with $1 \le q < R^{k-1}$ . Hence, White has to "worry" only about pairs with $R^{k-1} \le q < R^k$ . Let us call a rational pair $\left(\frac{p_1}{q} , \frac{p_2}{q}\right)$ with $R^{k-1} \le q < R^k$ <u>dangerous</u> if (2.1) is violated for some $\underline{y} = T\underline{x}$ with $\underline{x} \in B_{1+(k-1)t}$ . If there are no dangerous points, the induction is complete and we are done. If dangerous points exist, <u>we claim that they lie on a line</u>.

To prove this claim, suppose that $\left(\frac{p_1}{q} , \frac{p_2}{q}\right) , \left(\frac{p_1'}{q'} , \frac{p_2'}{q'}\right) , \left(\frac{p_1''}{q} , \frac{p_2''}{q}\right)$ are dangerous points. Since $\left(\frac{p_1}{q} , \frac{p_2}{q}\right)$ is dangerous,

$$\max\left(\left|y_1 - \frac{p_1}{q}\right| , \left|y_2 - \frac{p_2}{q}\right|\right) \leq \frac{\delta}{q^{3/2}} \tag{2.2}$$

for some $\underline{y} = T\underline{x}$ with $\underline{x} \in B_{1+(k-1)t}$. Let $\underline{c} = (c_1,c_2)$ denote the center of $B_{1+(k-1)t}$, and put $\underline{d} = (d_1,d_2) = T\underline{c}$.

$B_{1+(k-1)t}$ has radius $\rho_1(\alpha\beta)^{t(k-1)} = \rho_1 R^{-\frac{3}{2}(k-1)}$, so $|\underline{c} - \underline{x}| \leq \rho_1 R^{-\frac{3}{2}(k-1)}$, and therefore

$$\begin{aligned} |\underline{d} - \underline{y}| = |T(\underline{c} - \underline{x})| &< \|T\|\cdot|\underline{c} - \underline{x}| \\ &\leq \|T\|\rho_1 R^{-\frac{3}{2}(k-1)} \quad . \end{aligned} \tag{2.3}$$

It follows that

$$\begin{aligned} |qd_1 - p_1| &\leq |qy_1 - p_1| + q|d_1 - y_1| \\ &\leq \frac{\delta}{q^{1/2}} + q\|T\|\rho_1 R^{-\frac{3}{2}(k-1)} && \text{(by (2.2),(2.3))} \\ &< \frac{\delta R^{1/2}}{R^{k/2}} + \frac{\|T\|\rho_1 R^{3/2}}{R^{k/2}} && \text{(using } R^{k-1} \leq q < R^k\text{)} \\ &< \frac{\rho_1 R^{3/2}(1 + \|T\|)}{R^{k/2}} \quad ; \end{aligned}$$

in the same way, we obtain

$$|qd_2 - p_2| < \frac{\rho_1 R^{3/2}(1 + \|T\|)}{R^{k/2}} \quad .$$

Similar estimates hold for $q'$, $p_1'$, $p_2'$ and for $q''$, $p_1''$, $p_2''$.

Consider the matrix

$$M = \begin{pmatrix} 1 & \frac{p_1}{q} & \frac{p_2}{q} \\ 1 & \frac{p_1'}{q'} & \frac{p_2'}{q'} \\ 1 & \frac{p_1''}{q''} & \frac{p_2''}{q''} \end{pmatrix}.$$

To prove the claim, it is enough to prove that the determinant of $M$ is $0$. This will happen if the matrix

$$M_1 = \begin{pmatrix} q & p_1 & p_2 \\ q' & p_1' & p_2' \\ q'' & p_1'' & p_2'' \end{pmatrix}$$

has determinant $0$. Now $\det M_1 = \det M_2$, where

$$M_2 = \begin{pmatrix} q & qd_1-p_1 & qd_2-p_2 \\ q' & q'd_1-p_1' & q'd_2-p_2' \\ q'' & q''d_1-p_1'' & q''d_2-p_2'' \end{pmatrix} .$$

Using our estimates, we see that

$$|\det M_2| < 6R^k \left( \frac{\rho_1 R^{3/2}(1 + \|T\|)}{R^{k/2}} \right)^2 = 6\rho_1^2 R^3 (1 + \|T\|)^2 < 1 ,$$

by our assumption that $\rho_1$ be small. Since each entry in $M_1$ is an integer and since $|\det M_1| < 1$, $\det M_1 = 0$. Hence $\det M = 0$, and the claim is established.

We have shown that all dangerous points lie on a line $L$. Choose a vector $\underline{\underline{u}}$ of length $1$, orthogonal to the line $T^{-1}L$, which when applied to the center of $B_{1+(k-1)t}$, points away from $T^{-1}L$:

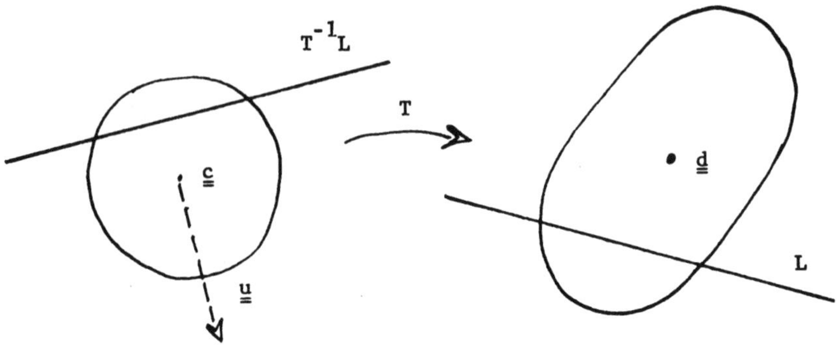


$B_{1+(k-1)t}$ $\qquad\qquad$ $T(B_{1+(k-1)t})$

Applying Lemma 1B, but with the roles of White and Black interchanged, we see that White can play so that every point $\underline{\underline{x}} \in B_{1+kt}$ has a distance from $T^{-1}L$ which is (with $\rho_\ell$ denoting the radius of $B_\ell$ )

$$\geq \frac{\gamma}{2}\,\rho_{1+(k-1)t} = \frac{\gamma}{2}(\alpha\beta)^{(k-1)t}\rho_1 = \frac{\gamma}{2}\rho_1 R^{-\frac{3}{2}(k-1)} .$$

Suppose now that $\underline{\underline{p}} = \left(\frac{p_1}{q}, \frac{p_2}{q}\right)$ is dangerous and that $\underline{\underline{y}} = T\underline{\underline{x}}$ , where $\underline{\underline{x}} \in B_{1+kt}$. Then $T^{-1}\underline{\underline{p}} \in T^{-1}L$ , so

$$|\underline{\underline{x}} - T^{-1}\underline{\underline{p}}| \geq \frac{\gamma}{2}\,\rho_1 R^{-\frac{3}{2}(k-1)} .$$

It follows that

$$\max\left(\left|y_1 - \frac{p_1}{q}\right|, \left|y_2 - \frac{p_2}{q}\right|\right) \geq \frac{1}{\sqrt{2}}|\underline{\underline{y}} - \underline{\underline{p}}| = \frac{1}{\sqrt{2}}|T(\underline{\underline{x}} - T^{-1}\underline{\underline{p}})| \geq \frac{1}{\sqrt{2}} \cdot \frac{1}{\|T^{-1}\|}|\underline{\underline{x}} - T^{-1}\underline{\underline{p}}|$$

$$> \frac{1}{2\|T^{-1}\|} \cdot \frac{\gamma}{2}\,\rho_1 R^{-\frac{3}{2}(k-1)} \geq \frac{\gamma\rho_1}{4\|T^{-1}\|} \cdot \frac{1}{q^{3/2}}$$

$$\geq \frac{\delta}{q^{3/2}} .$$

Thus (2.1) is satisfied for dangerous pairs $\left(\frac{p_1}{q}, \frac{p_2}{q}\right)$ as well, and $B_{1+kt}$ has the desired properties. This completes the induction, and the theorem is proved.

Second Part: Measures

References: Khintchine (1924, 1926a)
Cassels (1950a, 1950b)
Schmidt (1960)
Sprindžuk (1977).

## §3. Statement of Results

THEOREM 3A. (Khintchine (1926a)). *Let* $\psi_1(q),\dots,\psi_n(q)$ *be functions of the positive integer* $q$ *with* $0 < \psi_i(q) \leq 1 \quad (i = 1,\dots,n)$ , *and suppose that*

$$\psi(q) = \prod_{i=1}^{n} \psi_i(q)$$

*is non-increasing. If*

$$\sum_{q=1}^{\infty} \psi(q) \tag{3.1}$$

*is convergent, then for almost all*[†)] n-*tuples* $(\alpha_1,\dots,\alpha_n)$ , *there are only finitely many* $q$ *with*

$$\{q\alpha_i\} < \psi_i(q) \qquad (i = 1,\dots,n). \tag{3.2}$$

*But if the sum* (3.1) *is divergent, then for almost all* $(\alpha_1,\dots,\alpha_n)$ *there are infinitely many* $q$ *with* (3.2).

The sum (3.1) is divergent for $\psi_i(q) = \varepsilon q^{-1/n} \quad (i = 1,\dots,n)$. So for any $\varepsilon > 0$ ,

$$\{q\alpha_i\} < \varepsilon q^{-1/n} \qquad (i = 1,\dots,n) \tag{3.3}$$

---

[†)] in the sense of Lebesgue measure.

has infinitely many solutions for almost all $(\alpha_1,\dots,\alpha_n)$ , or put differently, <u>almost no n-tuple is badly approximable</u>. In fact even

$$\{q\alpha_i\} < q^{-1/n}(\log q)^{-1/n} \qquad (i = 1,\dots,n) \tag{3.4}$$

has infinitely many solutions almost always. On the other hand,

$$\{q\alpha_i\} < q^{-1/n - \varepsilon} \qquad (i = 1,\dots,n)$$

has almost always†) only finitely many solutions, and since $\|x\| = \min(\{x\}, \{-x\})$, it follows on using all the n-tuples $(\pm\alpha_1,\dots,\pm\alpha_n)$ that

$$\|q\alpha_i\| < q^{-1/n - \varepsilon} \qquad (i = 1,\dots,n) \tag{3.5}$$

almost always has only finitely many solutions. If $\alpha_1,\dots,\alpha_n$ is called <u>very well approximable</u> if for some $\varepsilon > 0$ the system (3.5) has infinitely many solutions, then it follows that <u>almost no n-tuple is very well approximable</u>.

The second (and more difficult) assertion of Theorem 3A follows from

<u>THEOREM 3B</u>. (Schmidt 1960) <u>Make the same assumptions as in Theorem 3A, and suppose that</u> (3.1) <u>is divergent</u>. <u>Write</u> $N(h,\alpha_1,\dots,\alpha_n)$ <u>for the number of solutions of</u> (3.2) <u>with</u> $1 \leq q \leq h$ , <u>and put</u>

$$\Psi(h) = \sum_{q=1}^{h} \psi(q) ,$$

$$\Omega(h) = \sum_{q=1}^{h} \psi(q)q^{-1} .$$

<u>Then for</u> $\varepsilon > 0$ ,

†) For more subtle results involving the Hausdorff dimension see Jarník (1929) and Besicovitch (1934) .

throws of a die, the number of "two"s after $h$ throws will be asymptotically equal to $\frac{1}{6}h$ with probability 1. Essential here is that the events "two is thrown at the $q$-th move", and "two is thrown at the $r$-th move" where $r \neq q$, are independent. More generally one shows that if $E_1, E_2, \ldots$ is a sequence of events, where $E_q$ has probability $\mu(E_q)$ and where (4.2) is divergent, and where $E_q, E_r$ with $q \neq r$ are "independent" in the sense that $\mu(E_q \cap E_r) = \mu(E_q)\mu(E_r)$, then it is almost certain that of the first $h$ events

$$\sim \mu(E_1) + \ldots + \mu(E_h)$$

actually take place.

A difficulty in our present situation is that $E_q$, $E_r$ with $q \neq r$ are not independent. However, it will be shown that if $q$ and $r$ are not too close in a certain sense, then $E_q, E_r$ are nearby independent.

Another point to be observed in the proof will be the following. If $h$ is large, it will be relatively easy to show that the exceptional set $G_h$ of $\underline{\underline{\alpha}}$ for which $N(h,\underline{\underline{\alpha}}) = N(h,\alpha_1,\ldots,\alpha_n)$ if far from $\Psi(h)$ has a small measure $\mu(G_h)$. But this is not enough, we need to show that $\mu(\bigcup_{h \geq h_0} G_h)$ is small. To overcome this difficulty one sets

$$N(u,v;\underline{\underline{\alpha}}) \tag{5.1}$$

for the number of $q$ in $u < q \leq v$ with (3.2), and

$$\Psi(u,v) = \sum_{q=u+1}^{v} \psi(q) , \tag{5.2}$$

and one shows that for large $u,v$, one has $N(u,v;\underline{\underline{\alpha}}) \sim \Psi(u,v)$ with great probability. If this is true for $(u,v]$ †) and for $(v,w]$, then this is also true for $(u,w]$. The most economical way to obtain every

†) This denotes the half open interval.

interval $(u,v]$ will be to build them up of intervals $(a2^j,(a+1)2^j]$.

We will begin with the case $n = 1$ of Theorem 3B and will indicate the necessary changes for $n > 1$ in §10. We shall follow Schmidt (1960) rather closely.

§6. On certain intervals. Let $\omega(h)$ , $h \geqq 1$ , be a non-negative integral-valued function which increases to infinity. We write $\omega(0) = 0$ and define $S'$ to be the set consisting of 0 and of all integers $h > 0$ such that $\omega(h-1) < \omega(h)$. We define $S''$ to be the set of integers $h \geqq 0$ having $\omega(h) < \omega(h+1)$. Finally, $S$ is the set of values of $\omega(h)$ , $h \geqq 0$ .

Next, we define for fixed $t > 0$ intervals of order $t$ to be the half-open intervals

$$(u2^t + v_1 \ , \ (u+1)2^t + v_2] \ ,$$

where $u$ , $v_1$ , $v_2$ are non-negative integers such that $v_1 < 2^t$ and $v_1$ , $v_2$ are the smallest non-negative integers satisfying $u2^t + v_1 \in S$ , $(u + 1)2^t + v_2 \in S$. (It is possible, of course, that for given $u,t$ there exists no such $v_1$.) The intervals of order $t$ cover the positive axis exactly once.

LEMMA 6A. *Every interval $(0,x], x \in S$ , can be expressed as disjoint union of intervals $\cup I_i$ of the type described above, where no two of the intervals $I_i$ are of the same order.*

*Proof.* Write $x$ in the binary scale,

$$x = \sum_{t=0}^{w} b_t 2^t \ ,$$

where $b_t$ equals 0 or 1, but $b_w = 1$. There exists an interval $(0,j_1]$ of order $w$ with $j_1 \leqq x$. If $j_1 = x$, then we are through. If not, and if

$$j_1 = \sum_{t=0}^{w} b_t^{(2)} 2^t ,$$

then $b_w^{(2)} = b_w = 1$ and there exists a largest integer $w_2$ having

$$b_{w_2}^{(2)} < b_{w_2} .$$

Hence there exists an interval $(j_1,j_2]$ of order $w_2$, $j_2 \leqq x$. If $j_2 = x$, then $(0,x] = (0,j_1] \cup (j_1,j_2]$. Otherwise, if

$$j_2 = \sum_{t=0}^{w} b_t^{(3)} 2^t , \; b_w^{(3)} = b_w, \ldots, b_{w_2}^{(3)} = b_{w_2} ,$$

then there exists a largest $w_3 < w_2$, having

$$b_{w_3}^{(3)} < b_{w_3} .$$

We proceed as before. Since $j_1 < j_2 < \ldots$, we finally arrive at $j_f = x$ and $(0,x] = (0,j_1] \cup \ldots \cup (j_{f-1},j_f]$. The orders of the intervals are $w > w_2 > \ldots > w_f > 0$.

§7. Sums involving a function $\varphi(k,q)$. Let $k,q$ be positive and write $\varphi(k,q)$ for the number of integers $x$, $0 \leqq x < q$, so that g.c.d.$(x,q) \leqq k$.

LEMMA 7A.

$$\sum_{q=1}^{v} \varphi(k,q)q^{-1} = v + O(vk^{-1} + \log v \log k).$$

In this section, the inequality indicated by the O-symbol holds for all values of all variables involved. Further log 1 is redefined to be equal to 1 .

Proof. Clearly,

$$\varphi(k,q) = \sum_{\substack{w \mid q \\ w \leq k}} \varphi\left(\frac{q}{w}\right),$$

where $\varphi(x)$ is the Euler $\varphi$ - function. Using the well-known relation

$$\varphi(x) = x \sum_{y \mid x} \mu(y) y^{-1} ,$$

involving the Moebius function we obtain

$$\sum_{q=1}^{v} \varphi(k,q) q^{-1} = \sum_{q=1}^{v} q^{-1} \sum_{\substack{w \mid q \\ w \leq k}} qw^{-1} \sum_{y \mid qw^{-1}} \mu(y) y^{-1}$$

$$= \sum_{w=1}^{\min(k,v)} w^{-1} \sum_{y=1}^{[(v/w)]} \mu(y) y^{-1} \sum_{q=1}^{[(v/yw)]} 1 ,$$

where $[\alpha]$ is the integer part of $\alpha$ . Thus

$$\sum_{q=1}^{v} \varphi(k,q) q^{-1} = v \sum_{w=1}^{\min(k,v)} w^{-2} \sum_{y=1}^{[(v/w)]} \mu(y) y^{-2} + O(\log v \log k)$$

$$= v \sum_{w=1}^{\min(k,v)} w^{-2} \zeta(2)^{-1} + O\left(\sum_{w=1}^{\min(k,v)} w^{-1}\right) + O(\log v \log k)$$

$$= v + O(vk^{-1} + \log v \log k).$$

LEMMA 7B.

$$\sum_{q=1}^{v} \psi(q)\varphi(k,q)q^{-1} = \Psi(v) + O\bigl(\Psi(v)k^{-1} + \Omega(v)\log k\bigr).$$

Proof. Put $\Pi(k,0) = 0$ and

$$\Pi(k,r) = \sum_{q=1}^{r} \varphi(k,q)q^{-1}$$

for $r \geq 1$ . Lemma 7A yields

$$\Pi(k,r) = r + O(rk^{-1} + \log r \log k). \tag{7.1}$$

Using partial summation we obtain

$$\begin{aligned} &\sum_{q=1}^{v} \psi(q)\varphi(k,q)q^{-1} \\ &= \sum_{q=1}^{v} \psi(q)(\Pi(k,q) - \Pi(k,q-1)) \\ &= \sum_{q=1}^{v-1} \Pi(k,q)(\psi(q) - \psi(q+1)) + \Pi(k,v)\psi(v) \\ &= \sum_{q=1}^{v-1} q(\psi(q) - \psi(q+1)) + v\psi(v) + R(k,v) \\ &= \Psi(v) + R(k,v) , \end{aligned} \tag{7.2}$$

where, according to (7.1),

$$\begin{aligned} &R(k,v) \\ &= O\left(\sum_{q=1}^{v-1} (qk^{-1} + \log q \log k)(\psi(q) - \psi(q+1)\right) + O(vk^{-1} + \log v \log k)\psi(v) \\ &= O\left(\Psi(v)k^{-1} + \log k \sum_{q=2}^{v} \psi(q)(\log q - \log(q-1)) + (\log k)\psi(1)\right) . \end{aligned} \tag{7.3}$$

Now

$$\sum_{q=2}^{v} \psi(q)(\log q - \log(q-1))$$

(7.4) $$= O\left(\sum_{q=2}^{v} \psi(q)\log\left(1 + \frac{1}{q-1}\right)\right)$$

$$= O(\Omega(v)) .$$

Lemma 7B is a consequence of (7.2), (7.3), and (7.4).

§8. Bounds for certain integrals. We introduce the following functions and integrals.

$$\beta(q,\alpha) = \begin{cases} 1 , \text{ if } \ 0 \leqq \alpha < \psi(q) , \\ 0 \ \text{ otherwise,} \end{cases}$$

$$\gamma(q,\alpha) = \sum_{p} \beta(q,q\alpha - p) ,$$

$$\gamma(k,q,\alpha) = \sum_{\substack{p \\ \text{g.c.d.}(p,q) \leqq k}} \beta(q,q\alpha - p) ,$$

$$I(q) = \int_0^1 \gamma(q,\alpha)d\alpha ,$$

$$I(k;q) = \int_0^1 \gamma(k,q,\alpha)d\alpha ,$$

$$I(k;q,r) = \int_0^1 \gamma(k,q,\alpha)\gamma(k,r,\alpha)d\alpha .$$

We observe that

$$N(v,\alpha) = \sum_{q=1}^{v} \gamma(q,\alpha) ,$$

and for $0 \leqq u < v$ we put

$$N(k;u,v;\alpha) = \sum_{q=u+1}^{v} \gamma(k,q,\alpha) .$$

LEMMA 8A.

(8.1) $$I(q) = \psi(q) , \quad I(k;q) = \psi(q)\varphi(k,q)q^{-1} ,$$

(8.2) $$I(k;q,r) \leqq \psi(q)\psi(r) + A(k;q,r)\psi(q)q^{-1} ,$$

*where* $A(k;q,r)$ *is the number of solutions* $p,s$ *of*

(8.3) $$qs - rp = 0 \quad \textit{with} \quad 0 \leqq p < q$$

*having*

$$\text{g.c.d. } (p,q) \leqq k , \quad \text{g.c.d. } (s,r) \leqq k .$$

We remark that in the notation of §4 and §5,

$$\gamma(q,\alpha) = \begin{cases} 1 & \text{if } \alpha \in E_q , \\ 0 & \text{otherwise.} \end{cases}$$

Hence $\mu(E_q) = I(q) = \psi(q)$ , and the independence of $E_q$ , $E_r$ for $q \neq r$ would mean that

(8.4) $$\int_0^1 \gamma(q,\alpha)\gamma(r,\alpha)d\alpha = \psi(q)\psi(r) .$$

However, this is not true, and as a technical device we have to introduce $I(k;q,r)$ to replace the integral on the left of (8.4), and instead of (8.4) we have (8.2)

*Proof*. $I(q) = \psi(q)$ is rather trivial, while the second half of (8.1) follows from

$$I(k,q) = \sum_{\substack{p \\ \text{g.c.d.}(p,q) \le k}} \int_0^1 \beta(q,\alpha q - p)d\alpha$$

$$= \varphi(k,q)q^{-1} \int_{-\infty}^{\infty} \beta(q,\alpha)d\alpha \ .$$

As for $I(k;q,r)$, we have

$$I(k;q,r) = \sum_{\substack{p,\ \text{g.c.d.}(p,q) \le k \\ s,\ \text{g.c.d.}(s,r) \le k}} \int_0^1 \beta(q,\alpha q - p)\beta(r,\alpha r - s)d\alpha \ .$$

We split this sum into two parts,

$$I(k;q,r) = I_0(k;q,r) + I_1(k;q,r) \ ,$$

where $I_0$ consists of the terms with $qs - rp \neq 0$ . Then

$$I_0(k;q,r) \le \sum_{\substack{p,s \\ qs-rp \neq 0}} \int_0^1 \beta(q,\alpha q - p)\beta(r,\alpha r - s)d\alpha$$

(8.5) $$= \sum_{\substack{p,s \\ qs-rp \neq 0}} \int_{-(p/q)}^{1-(p/q)} \beta(q,q\alpha')\beta\left(r,r\alpha' - \frac{qs - rp}{q}\right) d\alpha' \ .$$

To find an estimate for this sum, write $q = q'd, r = r'd$ , where $d = \text{g.c.d.}(q,r)$. We have $qs - rp = dh$ where $h = q's - r'p$ . For given $h$ , $p$ is determined modulo $q'$ , and has therefore $d$ possibilities modulo $q$. Thus for given $h$ , the intervals $[-(p/q) , 1 - (p/q))$ will cover the real axis $d$ times. Hence we get

$I_0(k;q,r)$

$$\leq d \sum_{h \neq 0} \int_{-\infty}^{\infty} \beta(q,q\alpha')\beta\left(r, r\alpha' - \frac{hd}{q}\right)d\alpha'$$

$$\leq d \int_{-\infty}^{\infty}\int_{-\infty}^{\infty} \beta(q,q\alpha')\beta(r,r\alpha' - \lambda dq^{-1})d\alpha'\, d\lambda$$

$$= \psi(q)\psi(r) .$$

In changing from the summation over $h$ to the continuous parameter $\lambda$ we used the fact that the function

$$\int_{-\infty}^{\infty} \beta(q,q\alpha')\beta(r,r\alpha' - \lambda dq^{-1})d\alpha'$$

is monotonically decreasing in $\lambda$ when $\lambda \geq 0$ , and monotonically increasing when $\lambda \leq 0$ .

To prove Lemma 8A it remains to give an upper bound for $I_1(k;q,r)$. In analogy to (8.5) we find that

$$I_1(k;q,r) = \sum_{\substack{p,\ \text{g.c.d.}(p,q) \leq k \\ s,\ \text{g.c.d.}(s,r) \leq k \\ qs - rp = 0}} \int_{-(p/q)}^{1-(p/q)} \beta(q,q\alpha')\beta(r,r\alpha')d\alpha'$$

$$\leq A(k;q,r)\psi(q)q^{-1} .$$

<u>LEMMA 8B</u>.

$$\int_0^1 N(v,\alpha)d\alpha = \Psi(v) ,$$

$$\int_0^1 N(k;u,v;\alpha)d\alpha = \sum_{q=u+1}^{v} \psi(q)\varphi(k,q)q^{-1} ,$$

$$\int_0^1 N^2(k;u,v;\alpha)d\alpha \leq \Psi^2(u,v) + 2 \sum_{q=u+1}^{v} \psi(q)d_k(q) ,$$

*where* $\Psi(u,v)$ *is defined in* (5.2) *and where* $d_k(q)$ *is the number of divisors of* $q$ *not exceeding* $k$ .

*Proof*. The first two assertions follow from (8.1). As an immediate consequence of (8.2) we have

$$\int_0^1 N^2(k;u,v;\alpha)d\alpha \leq \Psi^2(u,v) + 2 \sum_{u<r\leq q\leq r} A(k;q,r)\psi(q)q^{-1} .$$

Now

$$\sum_{r=1}^{q} A(k;q,r)$$

is equal to the number of solutions $r$ , $p$ , $s$ of

$$qs - rp = 0 \quad , \quad 0 \leq p < q \ , \quad 1 \leq r \leq q \ ,$$
$$\text{g.c.d.}(p,q) \leq k \ , \ \text{g.c.d.}(s,r) \leq k \ .$$

Define $a,b$ by

$$\frac{a}{b} = \frac{p}{q} = \frac{s}{r} \quad , \quad \text{g.c.d. } (a,b) = 1 \ .$$

Then $b$ divides $q$ and $qb^{-1}$ divides both $p$ and $q$ , so that g.c.d.$(p,q) \leq k$ implies $qb^{-1} \leq k$ . Thus the number of possible choices for $b$ is $d_k(q)$. Furthermore, there are $\varphi(b) \leq b$ possibilities for $a$ and $qb^{-1}$ possibilities for $r$ , once $b$ is given. Hence

$$\sum_{r=1}^{q} A(k;q,r) \leq qd_k(q)$$

and

$$\sum_{u<r\leq q\leq v} A(k;q,r)\psi(q)q^{-1} \leq \sum_{q=u+1}^{v} \psi(q)d_k(q) \quad .$$

§9. Proof of Theorem 3B. $(n = 1)$. Write $\omega(h) = [\Psi(h)\Omega(h)]$ and define $S$, $S'$, $S''$ as in §6. For $s \geq 1$ let $L_s$ be the set of pairs $(u,v), u \in S'$, $v \in S''$, so that $(\omega(u),\omega(v)]$ is an interval of any order $t$ with respect to $\omega$ (see §6), and $\omega(v) \leq 2^s$. From now on, the numbers $k,s$ are always connected by the relation

$$k = 2^s \quad . \tag{9.1}$$

From here on, we make heavy use of the methods developed in Cassels (1950b). Write $h^* = h^*(s)$ for the largest integer $h^*$ having $\omega(h^*) \leq 2^s$.

LEMMA 9A.

$$0 \leq \int_0^1 (N(h^*,\alpha) - N(k;0,h^*;\alpha))d\alpha = O(s\ 2^{s/2}) \tag{9.2}$$

$$\sum_{(u,v)\in L_s} \int_0^1 (N(k;u,v;\alpha) - \Psi(u,v))^2 d\alpha = O(s^2 2^s) \quad . \tag{9.3}$$

Proof. The first two equations of Lemma 8B give

$$\begin{aligned} &\int_0^1 (N(h^*,\alpha) - N(k;0,h^*,\alpha))d\alpha \\ &\qquad = \Psi(h^*) - \sum_{q=1}^{h^*} \psi(q)\varphi(k,q)q^{-1} \\ &\qquad = O(\Psi(h^*)k^{-1} + \Omega(h^*)\log k)\ , \end{aligned}$$

according to Lemma 7B. Since

$$\Omega(h^*) = O(2^{s/2}) \quad ,$$

(9.2) follows.

Using Lemma 8B again we see that a single integral in (9.3) does not exceed

$$2\sum_{q=u+1}^{v} \psi(q)d_k(q) + 2\Psi(u,v)\Big(\Psi(u,v) - \sum_{q=u+1}^{v} \psi(q)\varphi(k,q)q^{-1}\Big) .$$

We first take the sum over those pairs $(u,v) \in L_s$ where $(\omega(u),\omega(v)]$ is an interval of fixed order $t$. Since intervals of order $t$ cover the positive axis exactly once, we obtain the upper bound

$$2\sum_{q=1}^{h^*} \psi(q)d_k(q) + 2\Psi(h^*)\Big(\Psi(h^*) - \sum_{q=1}^{h^*} \psi(q)\varphi(k,q)q^{-1}\Big) .$$

We observe that since $\psi(q)$ is non-increasing

$$\sum_{q=1}^{h^*} \psi(q)d_k(q) = \sum_{t=1}^{k} \sum_{q_1=1}^{[h^*/t]} \psi(tq_1) \leq \sum_{t=1}^{k} t^{-1}\sum_{q=1}^{h^*} \psi(q) = O(2^s \log k),$$

and using Lemma 7B we find the upper bound

$$O(2^s \log k) + O\big(\Psi^2(h^*)k^{-1} + \Psi(h^*)\Omega(h^*)\log k\big) = O(s2^s) .$$

Summing over $t$ and observing $t \leq s$ we obtain (9.3).

<u>LEMMA 9B</u>. <u>There is a sequence of subsets</u> $\sigma_1,\sigma_2,\ldots$ <u>of the unit-interval with measures</u>

$$\mu_s = \int_{\sigma_s} d\alpha = O(s^{-1-\varepsilon})$$

<u>such that</u>

$$N(h,\alpha) = \Psi(h) + O(2^{s/2}\, s^{2+\varepsilon})$$

<u>for any</u> $h$ <u>with</u> $\omega(h) \leq 2^s$ , $h \in S'$ , <u>and any</u> $\alpha$ <u>in</u> $0 \leq \alpha < 1$ <u>which is not in</u> $\sigma_s$ .

Proof. We define $\sigma_s$ to be the set of all $\alpha$ in $0 \leq \alpha < 1$, for which not both of the following two inequalities hold:

$$0 \leq N(h^*, \alpha) - N(k;0,h^*,\alpha) \leq s^{2+\varepsilon} 2^{s/2} \tag{9.4}$$

$$\sum_{(u,v)\in L_s} (N(k;\,u,v;\alpha) - \Psi(u,v))^2 \leq s^{3+\varepsilon} 2^s \ . \tag{9.5}$$

As a consequence of Lemma 9A,

$$\mu_s = O(s^{-1-\varepsilon}) \ .$$

If $h \leq h^*$, $h \in S'$, then the interval $(0,\omega(h)]$ is the union of at most $s$ intervals $(\omega(u),\omega(v)]$, where $(u,v) \in L_s$. So

$$N(k;0,h;\alpha) - \Psi(h) = \sum \big(N(k;u,v;\alpha) - \Psi(u,v)\big),$$

where the sum is over at most $s$ pairs $(u,v) \in L_s$. This fact, together with (9.5) and Cauchy's inequality yields for $0 \leq \alpha < 1$, $\alpha \notin \sigma_s$,

$$(N(k;0,h;\alpha) - \Psi(h))^2 \leq s^{4+\varepsilon} 2^s \ .$$

This last equation together with (9.4)(which by monotonicity must be true with $h^*$ replaced by $h \leq h^*$), gives Lemma 9B.

Proof of Theorem 3B. $(n = 1)$. Since $\sum s^{-1-\varepsilon}$ is convergent, we may appeal to Lemma 4A to see that almost every $\alpha$ lies outside $\sigma_s$ for $s \geq s_0 = s_0(\alpha)$. Let $\alpha$ have this property, and assume $h$ to be so large that $\omega(h) \geq 2^{s_0}$. Choose $s$ so that $2^{s-1} \leq \omega(h) < 2^s$.

Suppose $h \in S'$. Then we have with Lemma 9B,

$$N(h,\alpha) = \Psi(h) + O(2^{\frac{1}{2}s} s^{2+\varepsilon})$$

$$= \Psi(h) + O(\Psi^{\frac{1}{2}}(h)\Omega^{\frac{1}{2}}(h)\log^{2+\varepsilon}\Psi(h)).$$

Hence Theorem 3B holds for $h \in S'$ . By the same argument we can prove the Theorem for $h \in S''$.

To any $h$ there exist $h'$ , $h''$ with $h' \in S'$ , $h'' \in S''$ having $h' \leqq h \leqq h''$ and

$$\omega(h') = \omega(h) = \omega(h'').$$

Then

$$|\Psi(h)\Omega(h) - \Psi(h')\Omega(h')| \leqq 1 ,$$

and

$$|\Psi(h) - \Psi(h')| \leqq \Omega(h)^{-1} \leqq \Omega(1)^{-1} = \psi(1)^{-1} ,$$

and similarly for $\Psi(h'')$. Since

$$N(h',\alpha) \leqq N(h,\alpha) \leqq N(h'',\alpha) ,$$

the case $n = 1$ of Theorem 3B follows.

§10. The case $n \geqq 2$. Using

$$v - \sum_{q=1}^{v} \varphi^n(k,q)q^{-n}$$

$$= \sum_{q=1}^{v} (q^n - \varphi^n(k,q))q^{-n}$$

$$\leqq n \sum_{q=1}^{v} (q - \varphi(k,q))q^{n-1}q^{-n}$$

$$= n(v - \sum_{q=1}^{v} \varphi(k,q)q^{-1}),$$

we easily generalize Lemmas 7A, 7B to

$$\sum_{q=1}^{v} \varphi^n(k,q)q^{-n} = v + O(vk^{-1} + \log k \log v),$$

$$\sum_{q=1}^{v} \psi(q)\varphi^n(k,q)q^{-n} = \Psi(v) + O(\Psi(v)k^{-1} + \Omega(v)\log k).$$

In analogy to $\beta(q,\alpha)$ of §8 we define $\beta(q,\alpha_1,\ldots,\alpha_n)$ to be the characteristic function of the box

$$0 \leq \alpha_i < \psi_i(q) \qquad (i = 1,\ldots,n),$$

and put

$$\gamma(q,\alpha_1,\ldots,\alpha_n) = \sum_{p_1,\ldots p_n} \beta(q,q\alpha_1 - p_1,\ldots,q\alpha_n - p_n),$$

$$\gamma(k;q,\alpha_1,\ldots,\alpha_n) = \sum_{\substack{p_i, \text{g.c.d.}(p_i,q) \leq k \\ (i=1,\ldots,n)}} \beta(q,q\alpha_1 - p_1,\ldots,q\alpha_n - p_n).$$

$I(q)$, $I(k,q)$, $I(k;q,r)$ are now n-dimensional integrals. To find an upper bound for

$$I(k;q,r) = \sum_{\substack{p_i, \text{ g.c.d.}(p_i,q) \leq k \\ s_i, \text{ g.c.d.}(s_i,r) \leq k \\ (i=1,\ldots,n)}} \int_0^1 \cdots \int_0^1 \beta(q,q\alpha_1 - p_1,\ldots) \beta(r,r\alpha_1 - s_1,\ldots)d\alpha_1 \ldots d\alpha_n,$$

we split this sum into $n + 1$ parts,

$$I(k;q,r) = I_0 + \ldots + I_n,$$

where $I_j$ consists of the terms with exactly $j$ indices $i_1, \ldots, i_j$ having $qs_i - rp_i = 0$. We find that

$$I_0(k;q,r) \leqq \psi(q)\psi(r)$$

and

$$I_j(k;q,r) \leqq c^{(j)} A^j(k;q,r)\psi(q)q^{-j}$$
$$\leqq c^{(j)} A(k;q,r)\psi(q)q^{-1} .$$

There are no other modifications of any depth.

## IV. Integer Points in Parallelepipeds.

(A chapter in the Geometry of Numbers).

References: Minkowski (1896 & 1910, 1907)

Cassels (1959),

Lekkerkerker (1969),

Mahler (1939, 1955).

### §1. Minkowski's Theorem on Successive Minima.

Let $K$ be a compact convex set in $E^n$, symmetric about $\underline{\underline{0}}$, and suppose that $\underline{\underline{0}}$ lies in the interior of $K$. We call such a set a convex body. Let $V(K)$ denote the volume of $K$. (By the volume of $K$ we mean the Riemann integral of the characteristic function of $K$. It can be proved that every convex body has a volume in this sense. Alternatively, the existence of the volume of $K$ may be added as a hypothesis.)

Let $\lambda_1 = \lambda_1(K)$ be the infimum of those numbers $\lambda \geq 0$ such that $\lambda K$ contains an integer point different from $\underline{\underline{0}}$. By our rather narrow definition of convex bodies, this infimum is in fact a minimum. It is easily seen that $0 < \lambda_1 < \infty$. Put $\tilde{\lambda} = 2V(K)^{-1/n}$. Then $\tilde{\lambda}K$ is a convex body with volume $2^n$. By Minkowski's Convex Body Theorem (Theorem 2B of Chapter II), $\tilde{\lambda}K$ contains an integer point $\neq \underline{\underline{0}}$. Therefore, $\lambda_1 \leq 2V(K)^{-1/n}$, or

$$\lambda_1^n V(K) \leq 2^n \quad . \tag{1.1}$$

For each integer $j$ with $1 \leq j \leq n$, let $\lambda_j = \lambda_j(K)$ be the infimum of all $\lambda \geq 0$ such that $\lambda K$ contains $j$ linearly independent integer points. Clearly each $\lambda_j$ is actually a minimum, and

$$0 < \lambda_1 \leq \lambda_2 \leq \cdots \leq \lambda_n < \infty \quad .$$

We call $\lambda_1, \lambda_2, \cdots, \lambda_n$ the <u>successive minima</u> of K . $\lambda_1$ is the <u>first minimum of</u> K , $\lambda_2$ is the <u>second minimum of</u> K , etc.

<u>Examples</u>. Let K denote the closed rectangle with center (0,0) , length 4 and width 1 , having sides parallel to the coordinate axes. Then $\lambda_1(K) = \frac{1}{2}$ and $\lambda_2(K) = 2$ .

The closed disc L with radius $\frac{1}{2}$ and center (0,0) satisfies $\lambda_1(L) = \lambda_2(L) = 2$ .

<u>THEOREM 1A</u>*. (Minkowski's Second Convex Body Theorem. Minkowski (1907)). <u>Suppose that</u> K <u>is a convex body in</u> $E^n$ . <u>Then</u>

(1.2) $$\frac{2^n}{n!} \le \lambda_1 \lambda_2 \cdots \lambda_n V(K) \le 2^n .$$

Note that the right hand inequality here sharpens (1.1) . We shall not prove this theorem here but shall prove relatively weak special cases in Lemmas 1C, 1D, 1E and in Corollary 2B .

The cube $|x_i| \leqq 1$ $(i = 1, \ldots, n)$ has $\lambda_1 = \ldots = \lambda_n = 1$ and $V(K) = 2^n$ , so that the right inequality in (1.2) is best possible. The "octahedron" 0 consisting of points with $|x_1| + \cdots + |x_n| \leqq 1$ has $\lambda_1 = \cdots = \lambda_n = 1$ and $V(0) = 2^n/n!$ , so that the left inequality in (1.2) is best possible.

<u>Remark</u>. By definition, and since K is compact, $\lambda_1 K$ contains an integer point $\underline{\underline{g}}_1 \neq \underline{\underline{0}}$ . Also $\lambda_2 K$ contains an integer point $\underline{\underline{g}}_2$ such that $\underline{\underline{g}}_1, \underline{\underline{g}}_2$ are linearly independent, $\lambda_3 K$ contains an integer point $\underline{\underline{g}}_3$ such that $\underline{\underline{g}}_1, \underline{\underline{g}}_2, \underline{\underline{g}}_3$ are linearly independent, etc. In this way, we obtain linearly independent integer points $\underline{\underline{g}}_1, \underline{\underline{g}}_2, \cdots, \underline{\underline{g}}_n$ such that

$$\underline{\underline{g}}_i \in \lambda_i K \qquad (1 \le i \le n) .$$

Now $\underline{\underline{g}}_1$ may be replaced by $-\underline{\underline{g}}_1$, hence it is not unique. Occasionally, more precisely when $\lambda_1 = \lambda_2$, there are more than these two possibilities for $\underline{\underline{g}}_1$. Generally, there is only a finite number of choices for $\underline{\underline{g}}_1,\ldots,\underline{\underline{g}}_n$.

LEMMA 1B. *Let* $K$ *be a convex body in* $E^n$, *and let* $\underline{\underline{g}}_1,\cdots,\underline{\underline{g}}_n$ *be linearly independent integer points with* $\underline{\underline{g}}_i \in \lambda_i K$ $(1 \le i \le n)$. *Suppose that* $J$ *is an integer with* $2 \le J \le n$ *and that* $\underline{\underline{g}}$ *is an integer point in* $\lambda K$, *where* $\lambda < \lambda_J$. *Then*

$$\underline{\underline{g}} = u_1\underline{\underline{g}}_1 + \cdots + u_{J-1}\underline{\underline{g}}_{J-1} \tag{1.3}$$

*with real coefficients* $u_1,\cdots,u_{J-1}$.

*Proof*. If $\lambda < \lambda_1$, then $\underline{\underline{g}} = \underline{\underline{0}}$ and (1.3) is satisfied with $u_1 = \cdots = u_{J-1} = 0$. Otherwise, there is an integer $j$ with $2 \le j \le J$ such that $\lambda_{j-1} \le \lambda < \lambda_j$. Now the points $\underline{\underline{g}}_1,\cdots,\underline{\underline{g}}_{j-1},\underline{\underline{g}}$ all lie in $\lambda K$. Since $\lambda < \lambda_j$, these points must be linearly dependent. But $\underline{\underline{g}}_1,\cdots,\underline{\underline{g}}_{j-1}$ are linearly independent, so $\underline{\underline{g}}$ is necessarily a linear combination of $\underline{\underline{g}}_1,\cdots,\underline{\underline{g}}_{j-1}$, and therefore a linear combination of $\underline{\underline{g}}_1,\cdots,\underline{\underline{g}}_{j-1},\cdots,\underline{\underline{g}}_{J-1}$.

LEMMA 1C. *Let* $K$ *be a convex body in* $E^n$. *Then the left inequality* (1.2) *holds, or*

$$\frac{2^n}{n!} \le \lambda_1\lambda_2\cdots\lambda_n V(K) .$$

*Proof*. Let $\underline{\underline{g}}_1,\cdots,\underline{\underline{g}}_n$ be linearly independent integer points with $\underline{\underline{g}}_i \in \lambda_i K$ $(1 \le i \le n)$. Then

$$\underline{\underline{h}}_i = \lambda_i^{-1}\underline{\underline{g}}_i \in K \qquad (1 \le i \le n).$$

The points $\pm\underline{\underline{h}}_1,\cdots,\pm\underline{\underline{h}}_n$ lie in $K$, hence their convex hull $H$ lies in

$K$ . It is easily seen that $H$ consists of all points $t_1\underline{h}_1+\cdots+t_n\underline{h}_n$ with $|t_1|+\cdots+|t_n| \le 1$ .

If $\underline{h}_1,\cdots,\underline{h}_n$ were the basis vectors $\underline{e}_1 = (1,0,\ldots,0),\ldots,\underline{e}_n = (0,\ldots,0,1)$, then $H$ would be the octahedron $0$ , consisting of the points $(t_1,\ldots,t_n) \in E^n$ with $|t_1|+\cdots+|t_n| \le 1$ , and having volume

$$V(0) = \frac{2^n}{n!} \ .$$

In general, $H$ is obtained from the octahedron $0$ by a linear transformation of determinant $\det(\underline{h}_1,\ldots,\underline{h}_n)$ . Therefore

$$V(H) = V(0)|\det(\underline{h}_1,\ldots,\underline{h}_n)| = \frac{2^n}{n!}\frac{|\det(\underline{g}_1,\ldots,\underline{g}_n)|}{\lambda_1\lambda_2\cdots\lambda_n} \ge \frac{2^n}{n!\lambda_1\lambda_2\cdots\lambda_n} \quad ;$$

here $|\det(\underline{g}_1,\cdots,\underline{g}_n)| \ge 1$ since $\underline{g}_1,\cdots,\underline{g}_n$ are linearly independent integer points. Thus

$$V(K) \ge V(H) \ge \frac{2^n}{n!\lambda_1\lambda_2\cdots\lambda_n} \quad ,$$

and the lemma follows.

Definitions. The unit ball (in $E^n$) is the set of all points $\underline{x} = (x_1,x_2,\cdots,x_n)$ with $x_1^2+\cdots+x_n^2 \le 1$ . An ellipsoid is the image of the unit ball under a nonsingular linear transformation.

LEMMA 1D. (Minkowski (1896)). *If* $\mathcal{E}$ *is an ellipsoid, then*

$$\lambda_1\cdots\lambda_n V(\mathcal{E}) \le 2^n \ .$$

Proof. Let $\underline{e}_1 = (1,0,\ldots,0),\ldots,\underline{e}_n = (0,\ldots,0,1)$ be the standard basis vectors in $E^n$ . The unit ball $B$ consists of all points $t_1\underline{e}_1+\cdots+t_n\underline{e}_n$ with $t_1^2+\cdots+t_n^2 \le 1$ . Further $\mathcal{E} = T(B)$ , where $T$ is some nonsingular linear transformation. Put $\underline{f}_i = T\underline{e}_i$ $(1 \le i \le n)$; then

$\mathbb{E}$ consists of all points $t_1\underline{\underline{f}}_1+\cdots+t_n\underline{\underline{f}}_n$ with $t_1^2+\cdots+t_n^2 \le 1$ .

We define a new inner product on $E^n$ as follows: given arbitrary points $\underline{\underline{x}} = x_1\underline{\underline{f}}_1+\cdots+x_n\underline{\underline{f}}_n$ and $\underline{\underline{x}}' = x_1'\underline{\underline{f}}_1+\cdots+x_n'\underline{\underline{f}}_n$ in $E^n$ (the $\underline{\underline{f}}_i$-s form a basis since $T$ is nonsingular), we put

$$\underline{\underline{x}} \otimes \underline{\underline{x}}' = x_1x_1'+\cdots+x_nx_n' \quad .$$

With this inner product, we obtain a Euclidean space. For notational convenience, we write

$$|\underline{\underline{x}}|^{\otimes} = \sqrt{\underline{\underline{x}} \otimes \underline{\underline{x}}} \quad .$$

Then $\mathbb{E}$ consists of all points $\underline{\underline{x}} \in E^n$ with $|\underline{\underline{x}}|^{\otimes} \le 1$ .

Let $\underline{\underline{g}}_1,\ldots,\underline{\underline{g}}_n$ be linearly independent integer points with $\underline{\underline{g}}_i \in \lambda_i\mathbb{E}$ $(1 \le i \le n)$. Applying the Gram-Schmidt orthonormalization process to $\underline{\underline{g}}_1,\ldots,\underline{\underline{g}}_n$ , we obtain vectors $\underline{\underline{c}}_1,\ldots,\underline{\underline{c}}_n$ in $E^n$ and real numbers $t_{ij}$ $(1 \le j \le i \le n)$ with $\underline{\underline{c}}_i \otimes \underline{\underline{c}}_j = \delta_{ij}$ $(1 \le i,j \le n)$ and

$$\underline{\underline{g}}_i = t_{i1}\underline{\underline{c}}_1+\cdots+t_{ii}\underline{\underline{c}}_i \qquad (1 \le i \le n).$$

Let $T^*$ be the linear transformation with $T^*\underline{\underline{c}}_i = \lambda_i\underline{\underline{c}}_i$ $(1 \le i \le n)$ , and put $\mathbb{E}^* = T^*(\mathbb{E})$ . To complete the proof of the lemma, it suffices to show that

$$(1.4) \qquad \lambda_1(\mathbb{E}^*) \ge 1 \ ;$$

for then

$$\begin{aligned}\lambda_1\lambda_2\cdots\lambda_n V(\mathbb{E}) &= |\det T^*|\, V(\mathbb{E}) = V(\mathbb{E}^*)\\ &\le (\lambda_1(\mathbb{E}^*))^n V(E^*)\\ &\le 2^n \qquad \text{(by (1.1))}.\end{aligned}$$

In order to establish (1.4) , suppose that

$$\underline{\underline{x}} = u_1\underline{\underline{g}}_1 + \cdots + u_n\underline{\underline{g}}_n$$

is an arbitrary point in $\mathcal{K}^*$ . Then

$$(1.5)\qquad \begin{aligned} \underline{\underline{x}} &= \sum_{i=1}^{n}\left(\sum_{j=i}^{n} t_{ji}u_j\right)\underline{\underline{c}}_i = \sum_{i=1}^{n}\left(\lambda_i^{-1}\sum_{j=i}^{n} t_{ji}u_j\right)T^*\underline{\underline{c}}_i \\ &= T^*\left(\sum_{i=1}^{n} \lambda_i^{-1}\left(\sum_{j=i}^{n} t_{ji}u_j\right)\underline{\underline{c}}_i\right) , \end{aligned}$$

and $\underline{\underline{y}} = \sum_{i=1}^{n} \lambda_i^{-1}\left(\sum_{j=i}^{n} t_{ji}u_j\right)\underline{\underline{c}}_i$ lies in $\mathcal{K}$ . Necessarily, $|\underline{\underline{y}}|^{\otimes} \le 1$ , so that

$$(1.6)\qquad \sum_{i=1}^{n} \lambda_i^{-2}\left(\sum_{j=i}^{n} t_{ji}u_j\right)^2 \le 1 .$$

We now further suppose that $\underline{\underline{x}} \ne \underline{\underline{0}}$ is an integer point in $\mathcal{K}^*$ . There is an integer $J$ , $1 \le J \le N$ , such that

$$\underline{\underline{x}} = u_1\underline{\underline{g}}_1 + \cdots + u_J\underline{\underline{g}}_J$$

with $u_J \ne 0$ . By Lemma 1B , $\underline{\underline{x}} \notin \lambda\mathcal{K}$ whenever $\lambda < \lambda_J$ . Therefore for such an $\underline{\underline{x}}$ ,

$$(1.7)\qquad |\underline{\underline{x}}|^{\otimes} \ge \lambda_J .$$

We have

$$\begin{aligned} \sum_{i=1}^{n}\lambda_i^{-2}\left(\sum_{j=i}^{n}t_{ji}u_j\right)^2 &= \sum_{i=1}^{J}\lambda_i^{-2}\left(\sum_{j=i}^{n}t_{ji}u_j\right)^2 \ge \lambda_J^{-2}\sum_{i=1}^{J}\left(\sum_{j=i}^{n}t_{ji}u_j\right)^2 \\ &= \lambda_J^{-2}\sum_{i=1}^{n}\left(\sum_{j=i}^{n}t_{ji}u_j\right)^2 = \lambda_J^{-2}(|\underline{\underline{x}}|^{\otimes})^2 \qquad \text{(by (1.5))} \\ &\ge \lambda_J^{-2}\cdot\lambda_J^{2} = 1 \qquad \text{(by (1.7))} . \end{aligned}$$

This last result, together with (1.6), shows that no integer point $\underline{\underline{x}} \neq \underline{\underline{0}}$ belongs to the interior of $\not{E}^*$. Therefore $\lambda_1(\not{E}^*) \geq 1$, and the lemma follows.

LEMMA 1E. Let $\Pi$ be a closed parallelepiped in $E^n$, symmetric about $\underline{\underline{0}}$. Then

$$\frac{2^n}{n!} \leq \lambda_1\lambda_2\cdots\lambda_n V(\Pi) \leq 2^n\, n!$$

Proof. In view of Lemma 1C, only the right-hand inequality requires proof. If C denotes the cube

$$|x_i| \leq 1 \,, \qquad (1 \leq i \leq n)$$

then $\Pi = T(C)$ for some nonsingular linear transformation $T$. Let $\not{E} = T(B)$, where $B$ is the unit ball in $E^n$. Now $B \subseteq C$, so $\not{E} \subseteq \Pi$, whence

$$\lambda_i(\Pi) \leq \lambda_i(\not{E}) \qquad (1 \leq i \leq n).$$

Then

$$\lambda_1(\Pi)\cdots\lambda_n(\Pi)V(\Pi) \leq \lambda_1(\not{E})\cdots\lambda_n(\not{E})V(\not{E})\,\frac{V(\Pi)}{V(\not{E})}$$

$$\leq 2^n\,\frac{V(\Pi)}{V(\not{E})} \qquad \text{(By Lemma 1D)}$$

$$= 2^n\,\frac{V(C)}{V(B)} = \frac{4^n}{V(B)} \,.$$

$B$ contains the octahedron $|x_1|+\cdots+|x_n| \leq 1$ of volume $\frac{2^n}{n!}$, so $V(B) \geq 2^n/n!$ and

$$\lambda_1(\Pi)\cdots\lambda_n(\Pi)V(\Pi) \leq 2^n\, n!$$

Remarks. In our applications, $\Pi$ will often be defined by inequalities

$$|L_i(\underline{x})| \leq 1 \quad , \qquad (1 \leq i \leq n)$$

where $L_i(\underline{x}) = \alpha_{i1}x_1+\cdots+\alpha_{in}x_n$ $(1 \leq i \leq n)$ are $n$ linear forms of determinant $\pm 1$. In this case $V(\Pi) = 2^n$, so Lemma 1E becomes

$$\frac{1}{n!} \leq \lambda_1\lambda_2\cdots\lambda_n \leq n! \quad .$$

Even more briefly, we have

$$(1.8) \qquad 1 \ll \lambda_1\lambda_2\cdots\lambda_n \ll 1 \quad .$$

Here and throughout this chapter, the constants in $\ll$ depend only on the dimension, which will usually be $n$, but will be $\ell$ in §5.

The following remark will be useful. Any $n$ points $\underline{x}_1,\ldots,\underline{x}_n$ in a parallelepiped $\Pi$ span the octahedron of points $t_1\underline{x}_1+\cdots+t_n\underline{x}_n$ with $|t_1|+\cdots+|t_n| \leq 1$ of volume $(2^n/n!)|\det(\underline{x}_1,\ldots,\underline{x}_n)|$. Since this volume cannot exceed $V(\Pi)$, we get $|\det(\underline{x}_1,\ldots,\underline{x}_n)| \leq n!2^{-n}V(\Pi)$. The $n$ integer points $\underline{g}_1,\ldots,\underline{g}_n$ of Lemma 1B are of the type $\underline{g}_i = \lambda_i\underline{x}_i$ with $\underline{x}_i$ in $\Pi$ $(1 \leq i \leq n)$. Hence

$$(1.9) \qquad |\det(\underline{g}_1,\ldots,\underline{g}_n)| \leq \lambda_1\cdots\lambda_n\, n!\, 2^{-n}\, V(\Pi) \leq (n!)^2 \ll 1 \quad .$$

In most of our applications we will need only parallelepipeds. In the next section, which may at first be omitted, we will show that an argument as was given in Lemma 1E for parallelepipeds may in fact be given for any convex body $K$.

## §2. Jordan's Theorem.

THEOREM 2A. Let $K$ be a symmetric convex body. Then there is an ellipsoid $\mathcal{E}$ with

$$(2.1) \qquad \mathcal{E} \subseteq K \subseteq \sqrt{n}\,\mathcal{E} \quad .$$

COROLLARY 2B. The successive minima of K satisfy

$$\lambda_1 \cdots \lambda_n V(K) \leq 2^n n^{n/2} .$$

Proof of Corollary. $\lambda_i = \lambda_i(K) \leq \lambda_i(\mathcal{E})$ . By Lemma 1D ,

$$\lambda_1(K) \ldots \lambda_n(K) V(K) \leq \lambda_1(\mathcal{E}) \ldots \lambda_n(\mathcal{E}) V(\mathcal{E}) (V(K)/V(\mathcal{E}))$$

$$\leq 2^n V(K)/V(\mathcal{E})$$

$$\leq 2^n n^{n/2} .$$

Proof of Theorem 2A. Compactness arguments show that there is an ellipsoid $\mathcal{E}_0 \subseteq K$ of maximum volume. This ellipsoid will be shown to satisfy (2.1) . Without loss of generality we may suppose that $\mathcal{E}_0 = B$ , the unit ball. It will suffice to show that a point $\underline{\underline{x}}$ with $|\underline{\underline{x}}| > \sqrt{n}$ cannot lie in K , so without loss of generality it will suffice to show that the point $\underline{\underline{p}} = (p,0,\ldots,0)$ with $p > \sqrt{n}$ cannot lie in K . If we suppose to the contrary that $\underline{\underline{p}}$ lies in K , then the convex hull H of $B, \underline{\underline{p}}, -\underline{\underline{p}}$ is contained in K .

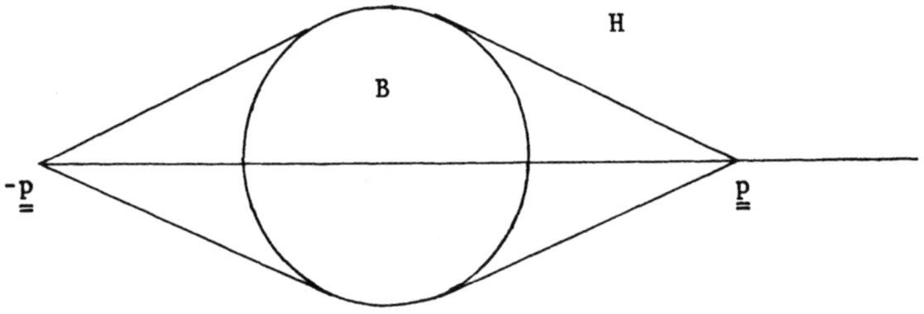


A short computation shows that the ellipsoid

$$\mathcal{E}_1 : \left(\frac{x_1}{a}\right)^2 + \left(\frac{x_2}{b}\right)^2 + \cdots + \left(\frac{x_n}{b}\right)^2 \leq 1$$

will be contained in H and therefore in K if

$$(2.2) \qquad \frac{a^2}{p^2} + b^2\left(1 - \frac{1}{p^2}\right) \leq 1 \quad \text{and if} \quad a \geq b \quad .$$

If we set

$$a = p/\sqrt{n} \quad \text{and} \quad b = \sqrt{\frac{p^2 - a^2}{p^2 - 1}} \quad ,$$

the first condition (2.2) holds with equality, and the second is true since $p > \sqrt{n}$ .

The function $f(x) = x^n (x-1)^{1-n}$ is increasing for $x \geq n$ , and therefore $f(p^2) > f(n)$ . Now

$$\begin{aligned}(V(\mathcal{E}_1)/V(B))^2 = a^2 b^{2(n-1)} &= \frac{p^2}{n}\left(\frac{p^2(1-(1/n))}{p^2-1}\right)^{n-1} \\ &= \frac{p^{2n}}{(p^2-1)^{n-1}} \Bigg/ \frac{n^n}{(n-1)^{n-1}} = f(p^2)/f(n) > 1 \ ,\end{aligned}$$

contradicting the maximality of $B$ .

For the subject of this section see also John (1948).

## §3. Davenport's Lemma.

THEOREM 3A. (Essentially Davenport (1937)). *Let $L_1(\underline{x}),\ldots,L_n(\underline{x})$ be $n$ linear forms of determinant 1 , and let $\lambda_1,\ldots,\lambda_n$ be the successive minima of the parallelepiped $\Pi$ defined by*

$$|L_i(\underline{x})| \leq 1 \qquad (i = 1,\ldots,n) \ .$$

*Suppose $\rho_1,\ldots,\rho_n$ are real numbers with*

$$(3.1) \qquad \rho_1 \geq \rho_2 \geq \cdots \geq \rho_n > 0 \ ,$$

$$(3.2) \qquad \rho_1\lambda_1 \leq \rho_2\lambda_2 \leq \cdots \leq \rho_n\lambda_n \ ,$$

$$(3.3) \qquad \rho_1\rho_2 \cdots \rho_n = 1 \ .$$

*Then there is a permutation of the linear forms $L_1,\ldots,L_n$ such that* _after this permutation the new parallelepiped_ $\Pi'$ _defined by_

$$|\rho_i L_i(\underline{x})| \leq 1 \qquad (i = 1,\ldots,n)$$

_has successive minima_ $\lambda_1',\ldots,\lambda_n'$ _with_

(3.4) $$2^{-n}\lambda_i\rho_i \leq \lambda_i' \leq 2^{n^2}(n!)^2\lambda_i\rho_i \qquad (i = 1,\ldots,n) .$$

_Moreover, let_ $\underline{g}_1,\ldots,\underline{g}_n$ _be linearly independent integer points with_ $\underline{g}_i \in \lambda_i\Pi$ , _i.e. with_ $|L_j(\underline{g}_i)| \leq \lambda_i$ $(1 \leq i,j \leq n)$ . _Then every integer point_ $\underline{x}$ _which is not in the subspace_ $S_{i-1}$ _spanned by_ $\underline{g}_1,\ldots,\underline{g}_{i-1}$ _has_

(3.5) $$\max(|\rho_1 L_1(\underline{x})|,\ldots,|\rho_n L_n(\underline{x})|) \geq 2^{-n}\rho_i\lambda_i .$$

_Proof_. Put $N(\underline{x}) = \max(|L_1(\underline{x})|,\ldots,|L_n(\underline{x})|)$ . The $n$ linearly independent integer points $\underline{g}_1,\ldots,\underline{g}_n$ with $\underline{g}_j \in \lambda_j\Pi$ satisfy

$$N(\underline{g}_j) = \lambda_j \qquad (1 \leq j \leq n) .$$

Put $S_0 = \{\underline{0}\}$ , and let $S_i$ be the subspace spanned by $\underline{g}_1,\ldots,\underline{g}_i$ $(1 \leq i \leq n)$ . In $S_i$ , the linear forms $L_1,\ldots,L_n$ satisfy $n-i$ independent linear conditions. In particular, there is a nontrivial relation:

$$\alpha_1 L_1 +\cdots+ \alpha_n L_n \equiv 0 \quad \text{in} \quad S_{n-1} .$$

Order the forms in such a way that the coefficient $\alpha_n$ is largest in absolute value. In $S_{n-2}$ , the forms $L_1,\ldots,L_n$ satisfy a further nontrivial linear condition, independent of the condition above. In fact, there is such a condition of the type

$$\beta_1 L_1 +\cdots+ \beta_{n-1}L_{n-1} \equiv 0 \quad \text{in} \quad S_{n-2} .$$

Reorder the forms $L_1,\ldots,L_n$ such that $\beta_{n-1}$ has the largest absolute value among $\beta_1,\dots,\beta_{n-1}$ . Etc.

If $\underline{\underline{x}} \in S_{n-1}$ , then

$$\alpha_n L_n(\underline{\underline{x}}) = -\alpha_1 L_1(\underline{\underline{x}}) - \cdots - \alpha_{n-1} L_{n-1}(\underline{\underline{x}}) \quad ,$$

whence

$$|L_n(\underline{\underline{x}})| \le |L_1(\underline{\underline{x}})| + \cdots + |L_{n-1}(\underline{\underline{x}})| \quad ,$$

since

$$\left|\frac{\alpha_i}{\alpha_n}\right| \le 1 \qquad (1 \le i \le n) \ .$$

Thus

$$|L_1(\underline{\underline{x}})| + \cdots + |L_{n-1}(\underline{\underline{x}})| \ge \tfrac{1}{2}(|L_1(\underline{\underline{x}})| + \cdots + |L_n(\underline{\underline{x}})|)$$

for $\underline{\underline{x}} \in S_{n-1}$ . Similarly,

$$|L_1(\underline{\underline{x}})| + \cdots + |L_{n-2}(\underline{\underline{x}})| \ge \tfrac{1}{2}(|L_1(\underline{\underline{x}})| + \cdots + |L_{n-1}(\underline{\underline{x}})|)$$

$$\ge \frac{1}{4}\ (|L_1(\underline{\underline{x}})| + \cdots + |L_n(\underline{\underline{x}})|)$$

for $\underline{\underline{x}} \in S_{n-2}$ . (The last inequality is true since $S_{n-2} \subseteq S_{n-1}$ .) Continuing in this manner, we see that

$$(3.6) \qquad |L_1(\underline{\underline{x}})| + \cdots + |L_j(\underline{\underline{x}})| \ge 2^{j-n}(|L_1(\underline{\underline{x}})| + \cdots + |L_n(\underline{\underline{x}})|)$$

whenever $\underline{\underline{x}} \in S_j$ $(1 \le j \le n)$ .

Now suppose that $\underline{\underline{x}}$ is an integer point and that $\underline{\underline{x}} \notin S_{i-1}$ . Then there is an integer $j$ with $i \le j \le n$ such that $\underline{\underline{x}} \in S_j$ and $\underline{\underline{x}} \notin S_{j-1}$ . Since $\underline{\underline{x}} \notin S_{j-1}$ , $N(\underline{\underline{x}}) \ge \lambda_j$ by Lemma 1B . Because $\underline{\underline{x}} \in S_j$ , we have

$$\begin{aligned}
\max(\rho_1|L_1(\underline{\underline{x}})|,\ldots,\rho_n|L_n(\underline{\underline{x}})|) &\ge \max(\rho_1|L_1(\underline{\underline{x}})|,\ldots,\rho_j|L_j(\underline{\underline{x}})|) \\
&\ge \rho_j \max(|L_1(\underline{\underline{x}})|,\ldots,|L_j(\underline{\underline{x}})|) \qquad \text{(by (3.1))} \\
&\ge \frac{\rho_j}{j}(|L_1(\underline{\underline{x}})|+\cdots+|L_j(\underline{\underline{x}})|) \\
&\ge \frac{2^{j-n}}{j}\rho_j(|L_1(\underline{\underline{x}})|+\cdots+|L_n(\underline{\underline{x}})|) \qquad \text{(by (3.6))} \\
&\ge 2^{-n}\rho_j N(\underline{\underline{x}}) \ge 2^{-n}\rho_j\lambda_j \\
&\ge 2^{-n}\rho_i\lambda_i \qquad \text{(by (3.2))} .
\end{aligned}$$

Thus (3.5) is established. It follows that

$$\lambda_i' \ge 2^{-n}\rho_i\lambda_i \qquad (1 \le i \le n) , \tag{3.7}$$

which is the first half of (3.4) .

In view of (3.3) , $V(\Pi') = V(\Pi) = 2^n$ , whence

$$\frac{1}{n!} \leqq \lambda_1\lambda_2\cdots\lambda_n \leqq n!$$

and

$$\frac{1}{n!} \leqq \lambda_1'\lambda_2'\cdots\lambda_n' \leqq n!$$

by Lemma 1E . Therefore

$$\begin{aligned}
\lambda_i' \leqq \frac{n!}{\lambda_1'\cdots\lambda_{i-1}'\lambda_{i+1}'\cdots\lambda_n'} &\leqq \frac{2^{n(n-1)}\, n!}{\rho_1\lambda_1\cdots\rho_{i-1}\lambda_{i-1}\rho_{i+1}\lambda_{i+1}\cdots\rho_n\lambda_n} \qquad \text{(by (3.7))} \\
&< 2^{n^2}(n!)^2\rho_i\lambda_i \qquad ,
\end{aligned}$$

and the second inequality (3.4) holds.

## §4. Reciprocal Parallelepipeds.

Let $\underline{\underline{a}}_1,\ldots,\underline{\underline{a}}_n$ be a basis of $E^n$ of determinant $D$ . There are unique vectors $\underline{\underline{a}}_1^*,\ldots,\underline{\underline{a}}_n^*$ such that the inner product

$$\underline{\underline{a}}_i\underline{\underline{a}}_j^* = \delta_{ij} \qquad (1 \leq i,j \leq n) ,$$

the Kronecker symbol. Now $\underline{\underline{a}}_1^*,\ldots,\underline{\underline{a}}_n^*$ is called the reciprocal basis to the given basis; its determinant is $D^{-1}$ .

A parallelepiped $\Pi$ is determined by inequalities

$$(4.1) \qquad |\underline{\underline{a}}_i\underline{\underline{x}}| \leq 1 \qquad (i = 1,\ldots,n) ,$$

and conversely $\underline{\underline{a}}_1,\ldots,\underline{\underline{a}}_n$ , except for their ordering and replacement of $\underline{\underline{a}}_i$ by $-\underline{\underline{a}}_i$ , are determined by $\Pi$ . (The proof of the second statement may be done as an exercise). The reciprocal†) parallelepiped $\Pi^*$ is defined by

$$(4.2) \qquad |\underline{\underline{a}}_i^*\underline{\underline{x}}| \leq 1 \qquad (i = 1,\ldots,n) .$$

THEOREM 4A. (Mahler 1939). *Let $\Pi,\Pi^*$ be reciprocal parallelepipeds, with successive minima $\lambda_1,\ldots,\lambda_n$ and $\lambda_1^*,\ldots,\lambda_n^*$ , respectively. Then*

$$(4.3) \qquad \lambda_i^* \ll \lambda_{n+1-i}^{-1} \ll \lambda_i^* ,$$

*with the constants in $\ll$ depending only on $n$ . Moreover, if $\Pi,\Pi^*$ are respectively defined by (4.1), (4.2), if $\underline{\underline{g}}_1,\ldots,\underline{\underline{g}}_n$ are linearly independent integer points with $\underline{\underline{g}}_j \in \lambda_j\Pi$ $(j = 1,\ldots,n)$ and if $\underline{\underline{g}}_1^*,\ldots,\underline{\underline{g}}_n^*$ are reciprocal to $\underline{\underline{g}}_1,\ldots,\underline{\underline{g}}_n$ , then*

$$(4.4) \qquad |\underline{\underline{a}}_i^*\underline{\underline{g}}_j^*| \ll \lambda_j^{-1} \qquad (1 \leq i,j \leq n) .$$

*Proof*. The hypothesis $\underline{\underline{g}}_j \in \lambda_j\Pi$ means that

$$(4.5) \qquad |\underline{\underline{a}}_i\underline{\underline{g}}_j| \leq \lambda_j \qquad (1 \leq i,j \leq n) .$$

---

†) Also called *polar* parallelepiped.

Setting $E = |\det(\underline{g}_1,\ldots,\underline{g}_n)|$ we have

(4.6) $$1 \leq E \ll 1$$

by (1.9) .

The identity

$$\sum_{i=1}^{n} (\underline{a}_i\underline{x})(\underline{a}_i^*\underline{y}) = \underline{x}\underline{y}$$

is easily proved by the linearity of both sides in $\underline{x}$ and in $\underline{y}$ and by substituting the special values $\underline{x} = \underline{a}_1^*,\ldots,\underline{a}_n^*$ and $\underline{y} = \underline{a}_1,\ldots,\underline{a}_n$ . From the identity we obtain

$$\sum_{i=1}^{n} (\underline{a}_i\underline{g}_k)(\underline{a}_i^*\underline{g}_j^*) = \underline{g}_k\underline{g}_j^* = \delta_{kj} \qquad (1 \leq k,j \leq n) .$$

It follows that

$$\underline{a}_i^*\underline{g}_j^* = A_{ij}/\det A \qquad (1 \leq i,j \leq n) ,$$

where $A$ is the matrix with entries $\underline{a}_\ell\underline{g}_m$ $(1 \leq \ell,m \leq n)$ , and where $A_{ij}$ is the cofactor of $\underline{a}_i\underline{g}_j$ in this matrix. It is clear that $|\det A| = DE$ , and (4.5) yields $|A_{ij}| \ll \lambda_1 \cdots \lambda_{j-1}\lambda_{j+1} \cdots \lambda_n$ . Thus

(4.7) $$|\underline{a}_i^*\underline{g}_j^*| \ll \lambda_1\lambda_2\cdots\lambda_n D^{-1}E^{-1}\lambda_j^{-1} \ll \lambda_j^{-1} ,$$

by (4.6) and since $\lambda_1 \cdots \lambda_n D^{-1} = 2^{-n}\lambda_1 \cdots \lambda_n V(\Pi) \ll 1$ by Lemma 1E . Thus (4.4) is true.

The points $\underline{g}_1^*,\ldots,\underline{g}_n^*$ are not necessarily integer points, but it is easily seen that $E\underline{g}_1^*,\ldots,E\underline{g}_n^*$ are integer points, and these points have

$$|\underline{a}_i^*(E\underline{g}_j^*)| \ll \lambda_j^{-1} \qquad (1 \leq i,j \leq n)$$

by (4.6), (4.7) . Considering the integer point $E\underline{\underline{g}}_n^*$ we see that $\lambda_1^* \ll \lambda_n^{-1}$. Considering $E\underline{\underline{g}}_n^*$, $E\underline{\underline{g}}_{n-1}^*$, we find that $\lambda_2^* \ll \lambda_{n-1}^{-1}$, etc. So generally $\lambda_i^* \ll \lambda_{n+1-i}^{-1}$ $(i = 1,\dots,n)$, or $\lambda_i^*\lambda_{n+1-i} \ll 1$. Since

$$\lambda_1 \cdots \lambda_n \lambda_1^* \cdots \lambda_n^* = 4^{-n}\lambda_1 \cdots \lambda_n V(\Pi)\lambda_1^* \cdots \lambda_n^* V(\Pi^*) \gg 1 ,$$

the assertion (4.3) follows.

COROLLARY 4B. *Suppose that* $V(\Pi) = 2^n$ (*i.e. that the determinant* $D = 1$ ). *Then*

$$\lambda_1^* \ll \lambda_1^{1/(n-1)} .$$

*Proof*. We have $1 \ll \lambda_1 \cdots \lambda_n \leqq \lambda_1 \lambda_n^{n-1}$ and $\lambda_1^* \ll \lambda_n^{-1} \ll \lambda_1^{1/(n-1)}$ .

## §5. Khintchine's Transference Principle.

Let $\alpha_{ij}$ $(1 \le i \le n, 1 \le j \le m)$ be nm real numbers. Throughout this section, we consider linear forms

$$L_i(\underline{\underline{x}}) = \alpha_{i1}x_1 + \cdots + \alpha_{im}x_m , \qquad (1 \le i \le n)$$

and the *dual* forms

$$M_j(\underline{\underline{u}}) = \alpha_{1j}u_1 + \cdots + \alpha_{nj}u_n , \qquad (1 \le j \le m)$$

where $\underline{\underline{x}} = (x_1,\dots,x_m)$ and $\underline{\underline{u}} = (u_1,\dots,u_n)$ . Put

$$\mathcal{L}(\underline{\underline{x}}) = (L_1(\underline{\underline{x}}),\dots,L_n(\underline{\underline{x}})) ,$$

$$\mathcal{M}(\underline{\underline{u}}) = (M_1(\underline{\underline{u}}),\dots,M_m(\underline{\underline{u}})) .$$

Also, put $\ell = m+n$ .

THEOREM 5A. *Suppose that there is an integer point* $(\underline{\underline{x}},\underline{\underline{y}}) = (x_1,\dots,x_m,y_1,\dots,y_n) \ne (\underline{0},\underline{0})$ *in* $E^\ell$ *with*

$$\overline{|\underline{x}|} \le X \quad \underline{\text{and}} \quad \overline{|\mathcal{L}(\underline{x})-\underline{y}|} \le Y \ .$$

Then there is an integer point $(\underline{u},\underline{v}) = (u_1,\ldots,u_n,v_1,\ldots,v_m) \neq (\underline{0},\underline{0})$ in $E^{\ell}$ such that

$$\overline{|\underline{u}|} \ll U \quad \underline{\text{and}} \quad \overline{|\mathcal{M}(\underline{u}) - \underline{v}|} \ll V \ ,$$

where

$$U = X^{m/(\ell-1)} Y^{(1-m)/(\ell-1)} \quad , \quad V = X^{(1-n)/(\ell-1)} Y^{n/(\ell-1)} \ .$$

(Here and throughout this section, the constants in $\ll$ depend only on $\ell$ .)

Proof. Consider the following vectors in $E^{\ell}$ :

$$\underline{a}_i = (\delta_{i1},\ldots,\delta_{im},0,\ldots,0) \qquad (1 \le i \le m) \ ,$$

$$\underline{a}_{m+i} = (\alpha_{i1},\ldots,\alpha_{im},\delta_{i1},\ldots,\delta_{in}) \qquad (1 \le i \le n) \ ,$$

$$\underline{a}^*_j = (\delta_{j1},\ldots,\delta_{jm},-\alpha_{1j},\ldots,-\alpha_{nj}) \qquad (1 \le j \le m) \ ,$$

$$\underline{a}^*_{m+j} = (0,\ldots,0,\delta_{j1},\ldots,\delta_{jn}) \qquad (1 \le j \le n) \ ,$$

where $\delta_{ij}$ is the Kronecker symbol. It is easily shown that

$$\underline{a}_i \underline{a}^*_j = \delta_{ij} \qquad (1 \le i,j \le \ell) \ .$$

Now we write

$$\underline{b}_i = X^{-1}(X^m Y^n)^{1/\ell} \underline{a}_i \qquad (1 \le i \le m) \ ,$$

$$\underline{b}_i = Y^{-1}(X^m Y^n)^{1/\ell} \underline{a}_i \qquad (m < i \le \ell) \ ,$$

$$\underline{b}^*_j = X(X^m Y^n)^{-1/\ell} \underline{a}^*_j \qquad (1 \le j \le m) \ ,$$

$$\underline{b}^*_j = Y(X^m Y^n)^{-1/\ell} \underline{a}_j \qquad (m < j \le \ell) \ .$$

Then again

$$\underline{\underline{b}}_i \underline{\underline{b}}_j^* = \delta_{ij} \qquad (1 \le i,j \le \ell) .$$

Thus the bases $\underline{\underline{b}}_1,\dots,\underline{\underline{b}}_\ell$ and $\underline{\underline{b}}_1^*,\dots,\underline{\underline{b}}_\ell^*$ are reciprocal. We observe that

$$\det(\underline{\underline{b}}_1,\dots,\underline{\underline{b}}_\ell) = \det(\underline{\underline{b}}_1^*,\dots,\underline{\underline{b}}_\ell^*) = 1 .$$

If $\Pi$ denotes the parallelepiped in $E^\ell$ defined by

$$|\underline{\underline{b}}_i \underline{\underline{z}}| \le 1 \qquad (1 \le i \le \ell) ,$$

then the reciprocal parallelepiped $\Pi^*$ is defined by

$$|\underline{\underline{b}}_j^* \underline{\underline{z}}| \le 1 \qquad (1 \le j \le \ell) .$$

Put $\underline{\underline{z}} = (\underline{\underline{x}},-\underline{\underline{y}})$ , where $(\underline{\underline{x}},\underline{\underline{y}})$ is the integer point of the hypothesis. It is easily checked that

$$|\underline{\underline{b}}_i \underline{\underline{z}}| \le (X^m Y^n)^{1/\ell} \qquad (1 \le i \le \ell) .$$

Thus if $\lambda_1$ and $\lambda_1^*$ is the first minimum of $\Pi$ and of $\Pi^*$ , then

$$\lambda_1 \le (X^m Y^n)^{1/\ell} ,$$

whence

$$\lambda_1^* \ll (X^m Y^n)^{\frac{1}{\ell(\ell-1)}}$$

by Corollary 4B . It follows that there is an integer point $\underline{\underline{w}} = (\underline{\underline{v}},\underline{\underline{u}}) = (v_1,\dots,v_m,u_1,\dots,u_n) \ne (\underline{\underline{0}},\underline{\underline{0}})$ in $E^\ell$ with

$$|\underline{\underline{b}}_j^* \underline{\underline{w}}| \ll (X^m Y^n)^{\frac{1}{\ell(\ell-1)}} \qquad (1 \le j \le \ell) ,$$

and $\underline{\underline{w}}$ consequently satisfies

$$|\underline{\underline{a}}_j^* \underline{\underline{w}}| \ll X^{-1}(X^m Y^n)^{\frac{1}{\ell}+\frac{1}{\ell(\ell-1)}} = V \qquad (1 \le j \le m)$$

and

$$|\underline{\underline{a}}_j^* \underline{\underline{w}}| \ll Y^{-1}(X^m Y^n)^{\frac{1}{\ell}+\frac{1}{\ell(\ell-1)}} = U \qquad (m < j \le \ell) \; .$$

Since

$$\overline{|\mathfrak{M}(\underline{\underline{u}})-\underline{\underline{v}}|} = \max_{1 \le j \le m} |\underline{\underline{a}}_j^* \underline{\underline{w}}|$$

and

$$\overline{|\underline{\underline{u}}|} = \max_{m < j \le \ell} |\underline{\underline{a}}_j^* \underline{\underline{w}}| \; ,$$

the proof of the theorem is complete.

<u>Remark</u>. Suppose that the integer point $\underline{\underline{x}}$ of the theorem is different from $\underline{\underline{0}}$, and suppose that $Y$ is small in terms of $m$ and $n$. Then $X \geqq 1$, and $V \leqq Y^{n/(\ell-1)}$ is small, and

$$\overline{|\mathfrak{M}(\underline{\underline{u}}) - \underline{\underline{v}}|} < 1 \; .$$

If we had $\underline{\underline{u}} = \underline{\underline{0}}$, then also $\underline{\underline{v}} = \underline{\underline{0}}$, a contradiction. Therefore, <u>if $\underline{\underline{x}} \neq \underline{\underline{0}}$ and $Y$ is small, then $\underline{\underline{u}} \neq \underline{\underline{0}}$</u>.

<u>THEOREM 5B</u>. $L_1,\ldots,L_n$ <u>is a badly approximable system of linear forms if and only if $M_1,\ldots,M_m$ is a badly approximable system</u>.

<u>Remark</u>. In the special case $m = 1$, sufficiency was proved by Perron (1921b), while necessity was established by Khintchine (1926b). In the case $m = 1$, or $n = 1$, the theorem asserts that an $n$-tuple is badly approximable if and only if the corresponding linear form $\alpha_1 x_1 + \cdots + \alpha_n x_n$

is badly approximable. See Ch. II, §4.

Proof of the Theorem. By symmetry, it is enough to prove the sufficiency. Suppose, then, that $M_1, \ldots, M_m$ is a badly approximable system, and that $\underline{x} \neq \underline{0}$, $\underline{y}$ are integer points in $E^m, E^n$ respectively. Put $Y = |\mathfrak{L}(\underline{x}) - \underline{y}|$ and $X = |\underline{x}|$. It is required to show that

$$X^m Y^n \gg 1 \quad . \tag{5.1}$$

It is no restriction to assume that $Y$ is small. In view of Theorem 5A and the remark following the proof of that theorem, there are integer points $\underline{u} \neq \underline{0}$, $\underline{v}$ in $E^n, E^m$ respectively such that the conclusions of Theorem 5A hold. Since $M_1, \ldots, M_m$ is a badly approximable system,

$$U^n V^m \gg 1 \quad ,$$

and hence

$$U^{n(\ell-1)} V^{m(\ell-1)} \gg 1 \quad .$$

But this is equivalent to (5.1), and the theorem is proved.

THEOREM 5C. ("Khintchine's Transference Principle") Let $\omega$ be the supremum of the numbers $\eta \geq 0$ such that there are infinitely many integer points $\underline{x} \neq \underline{0}$ with

$$|\mathfrak{L}(\underline{x}) - \underline{y}| < \frac{1}{|\underline{x}|^{\frac{m}{n}(1+\eta)}} \tag{5.2}$$

for some integer point $\underline{y}$. Similarly, let $\omega^*$ be the supremum of the numbers $\eta^* \geq 0$ such that there are infinitely many integer points $\underline{u} \neq \underline{0}$ with

$$(5.3) \qquad \overline{|\mathfrak{M}(\mathbf{u})-\mathbf{v}|} < \frac{1}{\overline{|\mathbf{u}|}^{\frac{n}{m}(1+\eta^*)}}$$

*for some integer point* $\mathbf{v}$ . *Then*

$$(5.4) \qquad \omega^* \geqq \frac{\omega}{(m-1)\omega + m+n-1} \quad ,$$

*and*

$$(5.5) \qquad \omega \geqq \frac{\omega^*}{(n-1)\omega^* + m+n-1} \quad .$$

*In particular,* $\omega = 0$ *precisely if* $\omega^* = 0$ .

The special case $m = 1$ of the theorem was proved by Khintchine, who established inequality (5.5) in (1925) and inequality (5.4) in (1926b). The general case is due to Dyson (1947b). Jarnik (1959) showed (5.4), (5.5) to be best possible.

*Proof of the Theorem.* We prove only (5.4) . We may assume that $\omega > 0$ , and in the case $m > 1$ that $\omega^* < \frac{1}{m-1}$ , since otherwise the result is obvious. Suppose, then, that $0 < \eta < \omega$ . Suppose also that $\eta^* > \omega^*$ , and that $\eta^* < \frac{1}{m-1}$ if $m > 1$ . Since $\eta < \omega$ , there are infinitely many integer points $\mathbf{x} \neq \mathbf{0}$ satisfying (5.2) .

If we put $X = \overline{|\mathbf{x}|}$ and $Y = \min_{\mathbf{y}} \overline{|\mathfrak{L}(\mathbf{x})-\mathbf{y}|}$ , where the minimum is taken over all integer points $\mathbf{y} \in E^n$ , then $\mathbf{x} \neq \mathbf{0}$ satisfies (5.2) for some integer point $\mathbf{y}$ if and only if

$$(5.6) \qquad YX^{\frac{m}{n}(1+\eta)} < 1 \quad .$$

Here $Y$ will be small if we take $X$ sufficiently large. (This is possible, since (5.2) is satisfied for infinitely many integer points $\mathbf{x}$ .) By Theorem 5B and the remark following its proof, there exist integer points

$\underline{\underline{u}} \neq \underline{\underline{0}}$ and $\underline{\underline{v}}$ with $\overline{|\underline{\underline{u}}|} \ll U$ and $\overline{|\mathfrak{m}(\underline{\underline{u}})-\underline{\underline{v}}|} \ll V$ . Since $\eta^* > \omega^*$ , (5.3) is satisfied for only finitely many $\underline{\underline{u}}$ , and therefore

$$VU^{\frac{n}{m}(1+\eta^*)} \gg 1 \tag{5.7}$$

if X , hence U , is large.

From (5.7) , we obtain

$$V^{m(\ell-1)}U^{n(\ell-1)(1+\eta^*)} \gg 1 \quad ,$$

which yields

$$X^{m+mn\eta^*}Y^{n-n(m-1)\eta^*} \gg 1 \quad .$$

The exponent of Y here is positive since $\eta^* < \frac{1}{m-1}$ , and therefore

$$X^{\frac{m}{n}\cdot\frac{1+n\eta^*}{1-(m-1)\eta^*}}Y \gg 1 \quad .$$

Comparing this with (5.6) and taking X sufficiently large, we conclude that

$$1 + \eta \leq \frac{1+n\eta^*}{1-(m-1)\eta^*} \quad .$$

It follows that

$$\eta^* \geqq \frac{\eta}{(m-1)\eta + m+n-1} \quad .$$

Since $\eta$ and $\eta^*$ may be chosen arbitrarily close to $\omega$ and $\omega^*$ , respectively, inequality (5.4) is established.

A system of linear forms is called <u>very well approximable</u> if $\omega > 0$ . It follows from Theorem 5C that a system of linear forms is very well approximable if and only if the dual system is. In particular, in the case $m = 1$ , an n-tuple $(\alpha_1,\ldots,\alpha_n)$ is called <u>very well approximable</u>

if for some $\varepsilon > 0$ there are infinitely many $q > 0$ having

$$\max(\|q\alpha_1\|,\ldots,\|q\alpha_n\|) < q^{-(1/n)-\varepsilon},$$

and a form $\alpha_1 q_1 + \cdots + \alpha_n q_n$ is very well approximable if for some $\omega$ there are infinitely many $n$-tuples $(q_1,\ldots,q_n)$ with

$$|\alpha_1 q_1 + \cdots + \alpha_n q_n| < q^{-n-\omega} ,$$

where $q = \max(|q_1|,\ldots,|q_n|) \neq 0$. It now follows from the theorem that <u>an</u> $n$-tuple $(\alpha_1,\ldots,\alpha_n)$ <u>is very well approximable precisely if the corresponding linear form</u> $\alpha_1 q_1 + \cdots + \alpha_n q_n$ is <u>very well approximable</u>. It was shown in §3,4 of Ch. III that almost no $n$-tuple is very well approximable, and it now follows that almost no linear form is very well approximable. For further transference theorems see Wang, Yu and Zhu (1979) and Schmidt and Wang (1979).

## §6. The Grassmann Algebra.

For a more thorough discussion see Greub (1967).

Let $\underline{\underline{e}}_1 = (1,0,\ldots,0),\ldots, \underline{\underline{e}}_n = (0,0,\ldots,1)$ be the standard basis vectors of $\mathbb{R}^n$.

Let $\mathbb{R}^n_0$ be the vector space of real numbers. It is a one dimensional Euclidean space, and it has $1$ as a basis vector. For $1 \leqq p \leqq n$, consider all formal expressions

(6.1) $$\underline{\underline{e}}_{i_1} \wedge \underline{\underline{e}}_{i_2} \wedge \cdots \wedge \underline{\underline{e}}_{i_p}$$

with integers $1 \leq i_1 < i_2 < \cdots < i_p \leq n$. There are

(6.2) $$\ell = \binom{n}{p}$$

such expressions. Let $\mathbb{R}^n_p$ be the $\ell$-dimensional vector space generated by the symbols (6.1). The elements $\underline{\underline{X}}^{(p)}$ of $\mathbb{R}^n_p$ will be called <u>p-vectors</u>. We shall denote p-vectors by $\underline{\underline{X}}^{(p)}$, $\underline{\underline{Y}}^{(p)}$ etc. or by $\underline{\underline{X}}_\sigma$, $\underline{\underline{Y}}_\sigma$ etc. Note that

$\mathbb{R}^n_1 = \mathbb{R}^n$ , the ordinary n-space, and denote vectors of this space by $\underline{\underline{x}}$, $\underline{\underline{y}}$, etc., as usual.

Write $C(n,p)$ for the set of p-tuples of integers $i_1,\ldots,i_p$ with $1 \le i_1 < \cdots < i_p \le n$ . For $\sigma = \{i_1 < \cdots < i_p\}$ in $C(n,p)$ let $\underline{\underline{E}}_\sigma$ be the vector (6.1) . The $\ell$ vectors $\underline{\underline{E}}_\sigma$ form a basis of $\mathbb{R}^n_p$ . Put

$$\underline{\underline{E}}_\sigma \underline{\underline{E}}_\tau = \delta_{\sigma\tau} = \begin{cases} 1 & \text{if } \sigma = \tau , \\ 0 & \text{otherwise} \end{cases}$$

for any elements $\sigma,\tau$ of $C(n,p)$ , and use linearity to extend this to an inner produce on $\mathbb{R}^n_p$ . With this inner product, $\mathbb{R}^n_p$ becomes a Euclidean space.

For convenience of notation we shall allow more general expressions $\underline{\underline{e}}_{j_1} \wedge \underline{\underline{e}}_{j_2} \wedge \cdots \wedge \underline{\underline{e}}_{j_p}$ , where $1 \le j_k \le n$ $(k = 1,\ldots,p)$ with arbitrary $p \ge 1$ , with the convention that such an expression is $\underline{\underline{0}}$ if $j_h = j_k$ for some $h \ne k$ (so, in particular, it will be $\underline{\underline{0}}$ if $p > n$) , and that it is plus or minus (6.1) if $j_1,\ldots,j_p$ is obtained from a p-tuple $i_1 < i_2 < \cdots < i_p$ by an even or by an odd permutation.

The direct sum $G_n = \mathbb{R}^n_0 \oplus \mathbb{R}^n_1 \oplus \cdots \oplus \mathbb{R}^n_n$ is a vector space of dimension $2^n$ . We define a product $\wedge$ in $G_n$ by the formulas $1 \wedge 1 = 1$ ,

$$(6.3) \quad 1 \wedge (\underline{\underline{e}}_{i_1} \wedge \cdots \wedge \underline{\underline{e}}_{i_p}) = (\underline{\underline{e}}_{i_1} \wedge \cdots \wedge \underline{\underline{e}}_{i_p}) \wedge 1 = \underline{\underline{e}}_{i_1} \wedge \cdots \wedge \underline{\underline{e}}_{i_p}$$

and

$$(6.4) \quad (\underline{\underline{e}}_{i_1} \wedge \cdots \wedge \underline{\underline{e}}_{i_p}) \wedge (\underline{\underline{e}}_{j_1} \wedge \cdots \wedge \underline{\underline{e}}_{j_q}) = \underline{\underline{e}}_{i_1} \wedge \cdots \wedge \underline{\underline{e}}_{i_p} \wedge \underline{\underline{e}}_{j_1} \wedge \cdots \wedge \underline{\underline{e}}_{j_q} ,$$

and by extending the definition using linearity to any two vectors of $G_n$ . With this product $G_n$ becomes an associative algebra, the <u>Grassmann</u> or <u>exterior</u> <u>algebra</u> of $\mathbb{R}^n$ .

For any vectors $\underline{\underline{x}}_1,\ldots,\underline{\underline{x}}_p$ in $\mathbb{R}^n$ with $1 \le p \le n$ , the vector

$$(6.5) \qquad \underline{\underline{X}}^{(p)} = \underline{\underline{x}}_1 \wedge \cdots \wedge \underline{\underline{x}}_p$$

is a p-vector. A p-vector of this type is called decomposable.

LEMMA 6A. Suppose that $\underline{x}_i = (\xi_{i1},\ldots,\xi_{in}) = \sum_{j=1}^{n}\xi_{ij}\underline{e}_j \quad (1 \leq i \leq p)$. Then the vector $\underline{X}^{(p)}$ given by (6.5) is

$$\underline{X}^{(p)} = \sum \xi_\sigma \underline{E}_\sigma \ , \tag{6.6}$$

where the summation is extended over $\sigma$ in $C(n,p)$, and where the coefficient $\xi_\sigma$ is the $(p \times p)$-determinant $|\xi_{ij}|$ with $1 \leq i \leq p$ and $j \in \sigma$.

For example, if $n = 4$ and $p = 2$, we have

$$\underline{x}_1 \wedge \underline{x}_2 = \begin{vmatrix} \xi_{11} & \xi_{12} \\ \xi_{21} & \xi_{22} \end{vmatrix} \underline{E}_{12} + \begin{vmatrix} \xi_{11} & \xi_{13} \\ \xi_{21} & \xi_{23} \end{vmatrix} \underline{E}_{13} + \begin{vmatrix} \xi_{11} & \xi_{14} \\ \xi_{21} & \xi_{24} \end{vmatrix} \underline{E}_{14}$$
$$+ \begin{vmatrix} \xi_{12} & \xi_{13} \\ \xi_{22} & \xi_{23} \end{vmatrix} \underline{E}_{23} + \begin{vmatrix} \xi_{12} & \xi_{14} \\ \xi_{22} & \xi_{24} \end{vmatrix} \underline{E}_{24} + \begin{vmatrix} \xi_{13} & \xi_{14} \\ \xi_{23} & \xi_{24} \end{vmatrix} \underline{E}_{34} \ .$$

Proof. Both sides of (6.6) are linear in each $\underline{x}_i$, and hence we may suppose that $\underline{x}_1,\ldots,\underline{x}_p$ are among the basis vectors $\underline{e}_1,\ldots,\underline{e}_n$. If two among $\underline{x}_1,\ldots,\underline{x}_p$ are the same $\underline{e}_i$, then both sides of (6.6) vanish. So we may suppose that $\underline{x}_1,\ldots,\underline{x}_p$ are distinct vectors among $\underline{e}_1,\ldots,\underline{e}_n$. Both sides of (6.6) now change into + or − themselves if we permute $\underline{x}_1,\ldots,\underline{x}_p$ by an even or by an odd permutation. Hence it suffices to consider the special case when $\underline{x}_1 = \underline{e}_{i_1},\ldots,\underline{x}_p = \underline{e}_{i_p}$ where $i_1 < i_2 < \cdots < i_p$. Then $\underline{X}^{(p)} = \underline{E}_\tau$ with $\tau = \{i_1 < \cdots < i_p\}$; on the other hand $\xi_\sigma = 1$ if $\sigma = \tau$ and $\xi_\sigma = 0$ if $\sigma \neq \tau$.

LEMMA 6B. The vector (6.5) is $\underline{0}$ if and only if $\underline{x}_1,\ldots,\underline{x}_p$ are linearly dependent.

Proof. This is an obvious consequence of Lemma 6A.

LEMMA 6C. *Suppose* $\underline{\underline{x}}_1,\ldots,\underline{\underline{x}}_p$ *and* $\underline{\underline{y}}_1,\ldots,\underline{\underline{y}}_p$ *are sets of* $p$ *independent points in* $\mathbb{R}^n$ . *The points* $\underline{\underline{X}}^{(p)} = \underline{\underline{x}}_1 \wedge \cdots \wedge \underline{\underline{x}}_p$ *and* $\underline{\underline{Y}}^{(p)} = \underline{\underline{y}}_1 \wedge \cdots \wedge \underline{\underline{y}}_p$ *of* $R^n_p$ *are proportional, i.e.*,

$$\underline{\underline{Y}}^{(p)} = \lambda \underline{\underline{X}}^{(p)} \quad , \tag{6.7}$$

*if and only if* $\underline{\underline{x}}_1,\ldots,\underline{\underline{x}}_p$ *and* $\underline{\underline{y}}_1,\ldots,\underline{\underline{y}}_p$ *span the same subspace in* $\mathbb{R}^n$ .

*Proof*. Let $S^p$ be the subspace spanned by $\underline{\underline{x}}_1,\ldots,\underline{\underline{x}}_p$ . If $\underline{\underline{y}}_1,\ldots,\underline{\underline{y}}_p$ lie in $S^p$ , then clearly (6.7) holds. Conversely, suppose we have (6.7) . Since $\underline{\underline{y}}_1,\ldots,\underline{\underline{y}}_p$ are independent, we have $\lambda \neq 0$ . It is clear that

$$\underline{\underline{x}} \wedge (\underline{\underline{x}}_1 \wedge \cdots \wedge \underline{\underline{x}}_p) = \underline{\underline{x}} \wedge \underline{\underline{X}}^{(p)} = \underline{\underline{0}}$$

precisely if $\underline{\underline{x}}$ lies in $S^p$ . Since $\underline{\underline{y}}_i \wedge \underline{\underline{X}}^{(p)} = \lambda^{-1}(\underline{\underline{y}}_i \wedge \underline{\underline{Y}}^{(p)}) = 0$ , we see that $\underline{\underline{y}}_i$ lies in $S^p$ $(1 \le i \le p)$ .

LEMMA 6D. (Laplace identity). *For* $\underline{\underline{x}}_1,\ldots,\underline{\underline{x}}_p,\underline{\underline{y}}_1,\ldots,\underline{\underline{y}}_p$ *in* $\mathbb{R}^n$ ,

$$(\underline{\underline{x}}_1 \wedge \cdots \wedge \underline{\underline{x}}_p) \cdot (\underline{\underline{y}}_1 \wedge \cdots \wedge \underline{\underline{y}}_p) = \begin{vmatrix} \underline{\underline{x}}_1\underline{\underline{y}}_1 & \underline{\underline{x}}_1\underline{\underline{y}}_2 & \cdots & \underline{\underline{x}}_1\underline{\underline{y}}_p \\ \underline{\underline{x}}_2\underline{\underline{y}}_1 & \underline{\underline{x}}_2\underline{\underline{y}}_2 & \cdots & \underline{\underline{x}}_2\underline{\underline{y}}_p \\ \vdots & \vdots & & \\ \underline{\underline{x}}_p\underline{\underline{y}}_1 & \underline{\underline{x}}_p\underline{\underline{y}}_2 & \cdots & \underline{\underline{x}}_p\underline{\underline{y}}_p \end{vmatrix} , \tag{6.8}$$

*where the raised dot on the L.H.S. denotes the inner product in* $\mathbb{R}^n_p$ , *while juxtaposition in the determinant on the right denotes the inner product in* $R^n = R^n_1$ .

*Proof*. Exercise. Use linearity in $\underline{\underline{x}}_1,\ldots,\underline{\underline{y}}_p$ .

LEMMA 6E. *Let* $\underline{\underline{x}}_1,\ldots,\underline{\underline{x}}_n$ *be in* $\mathbb{R}^n$ . *For each* $\tau = \{j_1 < \cdots < j_p\} \in C(n,p)$ *put*

$$\underline{\underline{X}}_\tau = \underline{\underline{x}}_{j_1} \wedge \underline{\underline{x}}_{j_2} \wedge \cdots \wedge \underline{\underline{x}}_{j_p} \quad .$$

Then

(6.9) $$\det_{\tau \in C(n,p)} (\underline{\underline{X}}_\tau) = (\det(\underline{\underline{x}}_1, \ldots, \underline{\underline{x}}_n))^{\ell p/n} \quad .$$

(It is understood that the $\tau$'s in (6.9) are ordered lexicographically. Observe that $\ell p/n = \binom{n-1}{p-1}$ ).

Remark. The entries in the matrix $(\underline{\underline{X}}_\tau)$ are of the form

$$X_{\tau\sigma} = \det(x_{ij}) \quad , \qquad (i \in \tau \, , \, j \in \sigma)$$

where $\sigma \in C(n,p)$ . The matrix $(\underline{\underline{X}}_\tau) = (X_{\tau\sigma})$ is called the p-th compound of the matrix $(\underline{\underline{x}}_1, \ldots, \underline{\underline{x}}_n)$ . For example, if $n = 4$ and $p = 2$ , the $(6 \times 6)$-matrix

$$\begin{pmatrix} \begin{vmatrix} x_{11} & x_{12} \\ x_{21} & x_{22} \end{vmatrix} & \begin{vmatrix} x_{11} & x_{13} \\ x_{21} & x_{23} \end{vmatrix} & \cdots & \begin{vmatrix} x_{13} & x_{14} \\ x_{23} & x_{24} \end{vmatrix} \\ \begin{vmatrix} x_{11} & x_{12} \\ x_{31} & x_{32} \end{vmatrix} & \begin{vmatrix} x_{11} & x_{13} \\ x_{31} & x_{33} \end{vmatrix} & \cdots & \begin{vmatrix} x_{13} & x_{14} \\ x_{33} & x_{34} \end{vmatrix} \\ \vdots & \vdots & & \vdots \\ \begin{vmatrix} x_{31} & x_{32} \\ x_{41} & x_{42} \end{vmatrix} & \begin{vmatrix} x_{31} & x_{33} \\ x_{41} & x_{43} \end{vmatrix} & \cdots & \begin{vmatrix} x_{33} & x_{34} \\ x_{43} & x_{44} \end{vmatrix} \end{pmatrix}$$

is the second compound of the matrix

$$\begin{pmatrix} x_{11} & x_{12} & x_{13} & x_{14} \\ x_{21} & x_{22} & x_{23} & x_{24} \\ x_{31} & x_{32} & x_{33} & x_{34} \\ x_{41} & x_{42} & x_{43} & x_{44} \end{pmatrix}$$

<u>Proof</u>. Consider the following two types of operations on n-tuples of vectors in $\mathbb{R}^n$ :

$$\text{(I)} \quad T_i(\lambda): \ (\underline{\underline{a}}_1,\dots,\underline{\underline{a}}_i,\dots,\underline{\underline{a}}_n) \mapsto (\underline{\underline{a}}_1,\dots,\lambda\underline{\underline{a}}_i,\dots,\underline{\underline{a}}_n),$$

$$\text{(II)} \quad U_{ij}(\lambda): \ (\underline{\underline{a}}_1,\dots,\underline{\underline{a}}_i,\dots,\underline{\underline{a}}_j,\dots,\underline{\underline{a}}_n) \mapsto (\underline{\underline{a}}_1,\dots,\underline{\underline{a}}_i,\dots,\underline{\underline{a}}_j+\lambda\underline{\underline{a}}_i,\dots,\underline{\underline{a}}_n) \ .$$

Let $\underline{\underline{x}}_1,\dots,\underline{\underline{x}}_n$ be arbitrary vectors in $\mathbb{R}^n$ . From elementary linear algebra, there are finitely many operations of types (I) and (II) , say $0_1,\dots,0_R$ , such that

$$0_R 0_{R-1}\cdots 0_1 : (\underline{\underline{e}}_1,\dots,\underline{\underline{e}}_n) \mapsto (\underline{\underline{x}}_1,\dots,\underline{\underline{x}}_n) \ .$$

If $(\underline{\underline{x}}_1,\dots,\underline{\underline{x}}_n) = (\underline{\underline{e}}_1,\dots,\underline{\underline{e}}_n)$ , then both sides of (6.9) are 1 . To prove the lemma, it therefore suffices to show that both sides of (6.9) change in the same way if we apply operations of the type (I) or (II) to $(\underline{\underline{x}}_1,\dots,\underline{\underline{x}}_n)$ .

If we apply $T_i(\lambda)$ to $(\underline{\underline{x}}_1,\dots,\underline{\underline{x}}_n)$ , the R.H.S. of (6.9) is multiplied by $\lambda^{\ell p/n}$ . A factor of $\lambda$ is introduced by this operation in each entry of the $\tau$-th row of the matrix $(X_{\tau\sigma})$ precisely when $i \in \tau$ . Since there are $\ell = \binom{n}{p}$ elements of $C(n,p)$ , there are $\binom{n-1}{p-1} = \frac{\ell p}{n}$ elements of $C(n,p)$ containing i . Thus the L.H.S. of (6.9) is multiplied by $\lambda^{\ell p/n}$ if we apply $T_i(\lambda)$ .

In applying operations of type (II) , we see that the R.H.S. of (6.9) remains unchanged. Let us examine the effect of the operation $U_{ij}(\lambda)$ on the $\tau$-th row

$$\underline{\underline{X}}_\tau = \underline{\underline{x}}_{i_1} \wedge \underline{\underline{x}}_{i_2} \wedge \cdots \wedge \underline{\underline{x}}_{i_p} \qquad (\tau = \{i_1 < i_2 < \cdots < i_p\})$$

of the matrix in the L.H.S. of (6.9) . If $j \notin \tau$ , or if $j \in \tau$ and $i \in \tau$ , then it is clear that $\underline{\underline{X}}_\tau$ remains unchanged. Suppose, then, that

$j \in \tau$ and $i \notin \tau$. After applying $U_{ij}(\lambda)$, $\mathbf{X}_\tau$ becomes $\mathbf{X}_\tau \pm \lambda \mathbf{X}_{\bar{\tau}}$ where $\bar{\tau}$ is the element of $C(n,p)$ obtained from $\tau$ by replacing $j$ with $i$. Note that $\mathbf{X}_{\bar{\tau}}$ is unaffected by the operation (II). It follows that the L.H.S. of (6.9) remains unchanged under operations of the type (II), and the lemma is proved.

LEMMA 6F. *Let $\mathbf{a}_1,\ldots,\mathbf{a}_n$ be a basis of $\mathbb{R}^n$. Suppose $1 \le p \le n$, and for $\sigma = \{i_1 < \cdots < i_p\}$ in $C(n,p)$ put*

(6.10) $$\mathbf{A}_\sigma = \mathbf{a}_{i_1} \wedge \cdots \wedge \mathbf{a}_{i_p} .$$

*Then the $\ell = \binom{n}{p}$ vectors $\mathbf{A}_\sigma$ with $\sigma \in C(n,p)$ are a basis of $\mathbb{R}^n_p$. If $\mathbf{a}_1,\ldots,\mathbf{a}_n$ have determinant 1, then so do the vectors $\mathbf{A}_\sigma$.*

*Moreover, let $\mathbf{a}_1^*,\ldots,\mathbf{a}_n^*$ be the basis of $\mathbb{R}^n$ which is reciprocal to $\mathbf{a}_1,\ldots,\mathbf{a}_n$, and for $\sigma = \{i_1 < \cdots < i_p\}$ in $C(n,p)$ put*

(6.11) $$(\mathbf{A}^*)_\sigma = \mathbf{a}_{i_1}^* \wedge \cdots \wedge \mathbf{a}_{i_p}^* .$$

*Then*

(6.12) $$\mathbf{A}_\sigma (\mathbf{A}^*)_\tau = \delta_{\sigma\tau} ,$$

*i.e. the basis $(\mathbf{A}^*)_\sigma$ with $\sigma \in C(n,p)$ is reciprocal to the basis $\mathbf{A}_\sigma$ with $\sigma \in C(n,p)$.*

*Proof*. Only (6.12) requires a proof. But (6.12) is a consequence of the Laplace identity.

## §7. Mahler's Theory of Compound Sets.

Let $\mathbf{a}_1,\ldots,\mathbf{a}_n$ be points of $\mathbb{R}^n$ with determinant 1. The inequalities

$$|\mathbf{a}_i \mathbf{x}| \le 1 \qquad (1 \le i \le n)$$

define a parallelepiped $\Pi$ of volume $2^n$ . Suppose $1 \leq p \leq n$ . For $\sigma = \{i_1 < \cdots < i_p\}$ in $C(n,p)$ define $\underset{=}{A}_\sigma$ by (6.10) . By Lemma 6F the vectors $\underset{=}{A}_\sigma$ have determinant 1 , and the inequalities

$$|\underset{=}{A}_\sigma \underset{=}{X}^{(p)}| \leq 1 \qquad (\sigma \in C(n,p))$$

define a parallelepiped $\Pi^{(p)}$ in $\mathbb{R}_p^n$ of volume $2^\ell$ where $\ell = \binom{n}{p}$ . This parallelepiped is closely related to Mahler's p-th compound body of $\Pi$ and will be called the p-th pseudocompound of $\Pi$ . (Mahler (1955) had more generally defined compounds of symmetric convex sets).

Let $\lambda_1,\ldots,\lambda_n$ be the successive minima of $\Pi$ , and for $\tau$ in $C(n,p)$ put

$$\lambda_\tau = \prod_{i \in \tau} \lambda_i \quad . \tag{7.1}$$

There is an ordering $\tau_1,\tau_2,\ldots,\tau_\ell$ of the elements of $C(n,p)$ such that

$$\lambda_{\tau_1} \leq \lambda_{\tau_2} \leq \cdots \leq \lambda_{\tau_\ell} \quad . \tag{7.2}$$

Now let $\underset{=}{g}_1,\ldots,\underset{=}{g}_n$ be independent integer points such that $\underset{=}{g}_j$ lies in $\lambda_j\Pi$ , i.e., that

$$|\underset{=}{a}_i \underset{=}{g}_j| \leq \lambda_j \qquad (1 \leq i,j \leq n) \quad . \tag{7.3}$$

For $\tau = \{j_1 < \cdots < j_p\}$ in $C(n,p)$ , put $\underset{=}{G}_\tau = \underset{=}{g}_{j_1} \wedge \cdots \wedge \underset{=}{g}_{j_p}$ . By Laplace's identity (6.8) , $\underset{=}{A}_\sigma \underset{=}{G}_\tau$ is equal to the $p \times p$ - determinant $|\underset{=}{a}_i \underset{=}{g}_j|$ with $i \in \sigma$ , $j \in \tau$ . Hence in view of (7.3) we obtain

$$|\underset{=}{A}_\sigma \underset{=}{G}_\tau| \leq p!\lambda_\tau \qquad (\sigma,\tau \in C(n,p)) \quad . \tag{7.4}$$

Thus we obtain the following result due to Mahler (1955) .

<u>THEOREM 7A</u>. <u>Suppose the parallelepiped</u> $\Pi$ <u>with minima</u> $\lambda_1,\ldots,\lambda_n$

*and the p-th pseudocompound* $\Pi^{(p)}$ *with minima* $\nu_1,\ldots,\nu_\ell$ *are defined as above. Also let* $\underline{\underline{g}}_1,\ldots,\underline{\underline{g}}_n$ *and* $\underline{\underline{G}}_\tau$ *be points as above. Then* (7.4) *holds and we have*

$$(7.5) \qquad \lambda_{\tau_i} \ll \nu_i \ll \lambda_{\tau_i} \qquad (1 \le i \le \ell) \quad .$$

*Here the constants in* $\ll$ *depend only on* n.

*Proof.* The points $\underline{\underline{G}}_\tau$ with $\tau \in C(n,p)$ are $\ell$ independent integer points, and hence we have $\nu_i \le p!\lambda_{\tau_i}$ by (7.2) and (7.4) . The lower bound for $\nu_i$ follows from $1 \ll \nu_1 \cdots \nu_\ell \ll 1$ and from $1 \ll \Pi_{i=1}^{\ell}\lambda_{\tau_i} = (\lambda_1 \cdots \lambda_n)^t \ll 1$ , where $t = \binom{n-1}{p-1}$ .

*Exercises.* (Will not be needed in sequel). Let $\underline{\underline{a}}_1,\ldots,\underline{\underline{a}}_n$ of $\mathbb{R}^n$ have determinant 1 . (1) Given $\sigma = \{i_1 < \cdots < i_p\}$ in $C(n,p)$ , let $\sigma'$ be the element of $C(n,n-p)$ which is the complement of $\sigma$ in $\{1,\ldots,n\}$ . Define $\underline{\underline{A}}_\sigma$ by (6.10) and write $\underline{\underline{A}}_\sigma = \{A_{\sigma\tau}\}$ with $\tau \in C(n,p)$ , so that each $A_{\sigma\tau}$ is a certain $(p \times p)$-determinant. Let $\underline{\underline{\hat{A}}}_\sigma$ be the vector $\underline{\underline{\hat{A}}}_\sigma = \{\hat{A}_{\sigma\tau}\}$ with

$$\hat{A}_{\sigma\tau} = (-1)^\sigma(-1)^\tau A_{\sigma'\tau'} \quad ,$$

where $(-1)^\sigma = (-1)^{i_1+\cdots+i_p}$ if $\sigma = \{i_1 < \cdots < i_p\}$ . Expand the determinant $\det(\underline{\underline{a}}_1,\ldots,\underline{\underline{a}}_n)$ according to Laplace†) to show that

$$\underline{\underline{A}}_{\sigma_1}\underline{\underline{\hat{A}}}_{\sigma_2} = \begin{cases} 1 & \text{if } \sigma_1 = \sigma_2 \ , \\ 0 & \text{otherwise} \ . \end{cases}$$

(2) Show that the map which sends $\underline{\underline{X}} = \{X_\sigma\}$ in $\mathbb{R}^n_p$ into $\underline{\underline{Y}} = \{Y_{\sigma'}\}$ in $\mathbb{R}^n_{n-p}$ with $Y_{\sigma'} = (-1)^\sigma X_\sigma$ , sends the parallelepiped $(\Pi^{(p)})^*$ (i.e. the reciprocal of $\Pi^{(p)}$) onto $\Pi^{(n-p)}$ .

†) Not the Laplace identity of Lemma 6D .

(3) Take $p = 1$ and deduce (4.3) in Theorem 4A from (7.5) in Theorem 7A.

## §8. Point Lattices.

Given $n$ linearly independent points $\underline{b}_1, \ldots, \underline{b}_n$ in $\mathbb{R}^n$, the set of points

$$c_1\underline{b}_1 + \cdots + c_n\underline{b}_n$$

with integer coefficients $c_1, \ldots, c_n$ is called a lattice, and $\underline{b}_1, \ldots, \underline{b}_n$ is called a basis of this lattice. The basis of a lattice is not unique. If $\underline{b}'_1, \ldots, \underline{b}'_n$ is another basis, then $\underline{b}'_i = \sum_{j=1}^{n} c_{ij}\underline{b}_j$ $(1 \leq i \leq n)$ and $\underline{b}_i = \sum_{j=1}^{n} c'_{ij}\underline{b}'_j$ $(1 \leq i \leq n)$ with integers $c_{ij}$ and $c'_{ij}$. Since the matrices $(c_{ij})$ and $(c'_{ij})$ are inverses of each other, they are unimodular, i.e. they have determinant $1$ or $-1$.

Therefore if our space is equipped with a Euclidean metric, if it is a Euclidean space $E^n$, then the volume of the parallelepiped of points $x_1\underline{b}_1 + \cdots + x_n\underline{b}_n$ with $0 \leq x_i < 1$ is independent of the basis and depends only on the lattice. Given a lattice $\Lambda$ this volume is denoted by $d(\Lambda)$ and is called the determinant of the lattice. It equals $|\det(b_{ij})|$, if $\underline{b}_i = (b_{i1}, \ldots, b_{in})$ in a Cartesian coordinate system.

Given a Cartesian coordinate system, the set of integer points forms a lattice $\Lambda_0$, the fundamental lattice. Every lattice $\Lambda$ is of the form $\Lambda = T\Lambda_0$ where $T$ is a non-singular linear transformation, and $d(\Lambda)$ is equal to the determinant of $T$.

If $K$ is a symmetric convex body, we define the successive minima $\lambda_1, \ldots, \lambda_n$ of $K$ with respect to a lattice $\Lambda$ by the property that $\lambda_i$ is the smallest positive number such that $\lambda_i K$ contains $i$ linearly independent points of $\Lambda$ . The results of this chapter so far were about integer points, i.e. about $\Lambda_0$ , but they can all be generalized to arbitrary lattices. In the proofs we simply have to transform $\Lambda$ back to $\Lambda_0$ ; then a convex body of volume $V(K)$ will be mapped into a convex body of volume $V(K)d(\Lambda)^{-1}$ . Theorem 1A* now becomes

$$(8.1) \qquad \frac{2^n}{n!} d(\Lambda) \leq \lambda_1 \cdots \lambda_n V(K) \leq 2^n d(\Lambda) \quad .$$

We have not proved Theorem 1A* , but we did prove Lemma 1C and Corollary 2B , so that in the general case we have proved

$$(8.2) \qquad \frac{2^n}{n!} d(\Lambda) \leq \lambda_1 \cdots \lambda_n V(K) \leq 2^n n^{n/2} d(\Lambda) \quad .$$

THEOREM 8A. *Necessary and sufficient for a subset* $\Lambda$ *of* $\mathbb{R}^n$ *to be a lattice is that the following three conditions hold.*

(i) $\Lambda$ *forms a group under vector addition,*

(ii) $\Lambda$ *contains* $n$ *linearly independent points,*

(iii) $\Lambda$ *is discrete, i.e., it contains no limit point.*

*Proof.* The necessity of the conditions follows from the definition. The sufficiency is easy when $n = 1$ : By (ii), (iii) there is a point $\underline{b}_1 \in \Lambda$ , $\underline{b}_1 \neq \underline{0}$ , which is closest to $\underline{0}$ . Then it is easily seen that every $\underline{b} \in \Lambda$ is of the form $\underline{b} = c_1\underline{b}_1$ with integral $c_1$ , and hence $\Lambda$ is a lattice with basis $\underline{b}_1$ . When $n > 1$ we may choose our coordinate system so that $\Lambda$ contains $n-1$ linearly independent points on the subspace $\mathbb{R}^{n-1}$ of $\mathbb{R}^n$ where $x_n = 0$ . Then $\Lambda' = \Lambda \cap \mathbb{R}^{n-1}$ is a lattice by induction. Let $\underline{b}_1,\ldots,\underline{b}_{n-1}$ be a basis of $\Lambda'$ . There is a point $\underline{b}_n = (b_{n1},\ldots,b_{nn})$ in $\Lambda$ with $b_{nn} > 0$ , which is clear, and in fact there is such a point for which $b_{nn} > 0$ is minimal, which is less clear. But it does become clear when we observe that $\Lambda$

is discrete and that $b_{nn}$ remains unchanged if we add to $\underline{\underline{b}}_n$ an element of $\Lambda'$, so that we may restrict ourselves to $\underline{\underline{b}}_n$ for which $(b_{n1},\ldots,b_{n,n-1})$ lies in the parallelepiped of points $\lambda_1\underline{\underline{b}}_1+\cdots+\lambda_{n-1}\underline{\underline{b}}_{n-1}$ with $0 \leqq \lambda_i < 1$ $(i = 1,\ldots,n-1)$. Now if $\underline{\underline{g}} = (g_1,\ldots,g_n) \in \Lambda$, then also $\underline{\underline{g}}' = \underline{\underline{g}} - [g_n/b_{nn}]\underline{\underline{b}}_n$ lies in $\Lambda$. The n-th coordinate of $\underline{\underline{g}}'$ is $\geqq 0$ and less than $b_{nn}$, hence must be zero. Thus $\underline{\underline{g}}' \in \Lambda'$, and $\underline{\underline{g}}$ is a linear combination of $\underline{\underline{b}}_1,\ldots,\underline{\underline{b}}_{n-1},\underline{\underline{b}}_n$. Hence $\Lambda$ is a lattice with basis $\underline{\underline{b}}_1,\ldots,\underline{\underline{b}}_n$.

Suppose $S^d$ is a d-dimensional rational subspace of $E^n$, i.e. a subspace spanned by vectors with rational coordinates. Then $\Lambda_0 \cap S^d$ satisfies the conditions of Theorem 8A (with d in place of n) and we have

COROLLARY 8B. *The integer points on a rational subspace $S^d$ form a lattice on $S^d$.*

## V. Roth's Theorem

References: Thue (1908, 1909)

Roth (1955a)

Schmidt (1971d).

### §1. Liouville's Theorem

THEOREM 1A (Liouville (1844)). *Suppose* $\alpha$ *is a real algebraic number of degree* $d$ . *Then there is a constant* $c(\alpha) > 0$ *such that*

$$\left|\alpha - \frac{p}{q}\right| > \frac{c(\alpha)}{q^d}$$

*for every rational* $\frac{p}{q}$ *distinct from* $\alpha$ .

(In considering inequalities of this type here and elsewhere, we tacitly assume that $q > 0$).

This theorem was used by Liouville to construct transcendental numbers. For example, put $\alpha = \sum_{\nu=1}^{\infty} 2^{-\nu!}$ , $q(k) = 2^{k!}$ , $p(k) = 2^{k!} \sum_{\nu=1}^{k} 2^{-\nu!}$. Then

$$\left|\alpha - \frac{p(k)}{q(k)}\right| = \sum_{\nu=k+1}^{\infty} 2^{-\nu!} < 2 \cdot 2^{-(k+1)!} = 2(q(k))^{-k-1} .$$

Hence for any $d$ and any constant $c > 0$ one has

$$\left|\alpha - \frac{p(k)}{q(k)}\right| < \frac{c}{(q(k))^d}$$

if $k$ is large. By Liouville's Theorem, $\alpha$ cannot be algebraic of any degree $d$ , and hence $\alpha$ is transcendental.

For the sake of later refinements we shall break the extremely simple proof of Liouville's Theorem into three parts (a), (b), and (c).

*Proof*. (a) Let $P(X)$ be the *defining polynomial* of $\alpha$ , i.e., the polynomial of degree $d$ with root $\alpha$ which has coprime integer coefficients and a positive leading coefficient.

(b) Taylor's formula yields

$$|P(\frac{p}{q})| = |\sum_{i=1}^{d} (\frac{p}{q} - \alpha)^i \frac{1}{i!} P^{(i)}(\alpha)| \leqq \frac{1}{c(\alpha)} |\frac{p}{q} - \alpha|$$

if $|\frac{p}{q} - \alpha| \leqq 1$.

(c) Unless $d = 1$ and $\frac{p}{q} = \alpha$, we have $P(\frac{p}{q}) \neq 0$, whence $|P(\frac{p}{q})| \geqq \frac{1}{q^d}$, and combining this with (b) we obtain Liouville's Theorem if $|\frac{p}{q} - \alpha| \leqq 1$. The theorem is obvious if $|\frac{p}{q} - \alpha| > 1$.

## §2. Roth's Theorem and its History.

Suppose $\alpha$ is a real algebraic number of degree $d \geqq 2$. Liouville's Theorem implies that the inequality

$$|\alpha - \frac{p}{q}| < \frac{1}{q^{\mu}} \tag{2.1}$$

has only finitely many rational solutions $\frac{p}{q}$ if $\mu > d$. The great Norwegian mathematician Thue (1909)[†)] showed that (2.1) has only finitely many solutions if $\mu > \frac{1}{2}d + 1$. Then Siegel (1921) in his thesis showed that this is already true if $\mu > 2\sqrt{d}$. (Siegel's result was a little better, with a more complicated function in place of $2\sqrt{d}$). A slight improvement to $\mu > \sqrt{2d}$ was made by Dyson (1947a). See also Gelfond (1952). Finally Roth (1955a) proved that (2.1) has only finitely many solutions if $\mu > 2$. For this result Roth was awarded the Field Prize

---

[†)] No. 12 in Thue's selected papers (1977).

in 1958. Roth's Theorem is as follows.

THEOREM 2A. *Suppose* $\alpha$ *is real and algebraic of degree* $d \geqq 2$. *Then for each* $\delta > 0$ , *the inequality*

$$|\alpha - \frac{p}{q}| < \frac{1}{q^{2+\delta}} \tag{2.2}$$

*has only finitely many solutions in rationals* $\frac{p}{q}$ .

Remarks. (i) The conclusion is trivially true for a complex $\alpha$ which is not real.

(ii) By Dirichlet's Theorem the exponent 2 in (2.2) is best possible. If $\alpha$ is of degree 2, then

$$|\alpha - \frac{p}{q}| > \frac{c(\alpha)}{q^2} \tag{2.3}$$

by Liouville's Theorem. In this case Liouville's Theorem is stronger than Roth's. Of course (2.3) follows already from Lemma 2E of Chapter I.

(iii) For no single $\alpha$ of degree $\geqq 3$ do we know whether (2.3) holds. It is very likely (see Ch. IV, §3) that in fact (2.3) is false for every such $\alpha$ , i.e., that no such $\alpha$ is badly approximable, or, put differently, that such $\alpha$ has unbounded partial quotients in its continued fraction.

(iv) An apparently very difficult conjecture (Lang (1965)) is that for $\alpha$ of degree $\geqq 3$,

$$|\alpha - \frac{p}{q}| < \frac{1}{q^2(\log q)^{\kappa}}$$

has only finitely many solutions if $\kappa > 1$ , or at least if $\kappa > \kappa_0(\alpha)$.

We now make a preliminary remark about the proof. Suppose we tried to modify the proof of Liouville's Theorem as follows. In step (a) we pick a polynomial $P(X)$ with rational integer coefficients which has a root at $\alpha$ of order $i$ and which has degree $r$. Next, in step (b) we suppose that (2.1) holds, and Taylor's expansion

$$P(\frac{p}{q}) = \sum_{j=i}^{r} (\frac{p}{q} - \alpha)^i \frac{1}{j!} P^{(j)}(\alpha)$$

yields $|P(\frac{p}{q})| \leqq cq^{-\mu i}$. Finally (c) we have $P(\frac{p}{q}) \neq 0$ whence $|P(\frac{p}{q})| \geqq q^{-r}$ for all but finitely many rationals $\frac{p}{q}$. Hence if (2.1) has infinitely many solutions, then $\mu i \leqq r$ or $\mu \leqq (\frac{i}{r})^{-1}$. Hence one should try to make $\frac{i}{r}$ as large as possible. But it is clear that always $\frac{i}{r} \leqq \frac{1}{d}$, and that $\frac{i}{r} = \frac{1}{d}$ if $P(X)$ is a power of the defining polynomial of $\alpha$. Hence this method only gives $\mu \leqq d$, i.e., nothing better than Liouville's result.

In order to improve on this estimate, Thue uses a polynomial $X_2Q(X_1) - P(X_1)$ in two variables, and Siegel uses a more general polynomial $P(X_1, X_2)$ in two variables, while Schneider (1936) and Roth use a polynomial $P(X_1,\ldots,X_m)$ in many variables.

Now if $\frac{p_1}{q_1},\ldots,\frac{p_m}{q_m}$ are very good rational approximations, one substitutes these into $P(X_1,\ldots,X_m)$. A major problem is that it is difficult to ascertain that $P(\frac{p_1}{q_1},\ldots,\frac{p_m}{q_m}) \neq 0$. This difficulty is overcome by "Roth's Lemma", which requires however that $q_1 < q_2 < \ldots < q_m$. A consequence is that we reach a contradiction only if we have at least m very good rational approximations. In the case of Thue's and Siegel's result one has $m = 2$ and hence one needs two very good rational approximations. Davenport (1968) used Thue's approach to show that for cubic $\alpha$ and for $\mu > 1 + \sqrt{3}$, solutions to (2.1) have $q \leqq c_1(\alpha,\mu)$ with an explicit $c_1(\alpha,\mu)$, with one possible exception $\frac{p}{q}$, and Schinzel (1967) used Siegel's approach to do the same for $\alpha$

of degree $d \geqq 3$ and for $\mu > 3\sqrt{d/2}$ .

But the method of Thue-Siegel-Roth does not give a bound which holds without exception, and hence it is non-effective. It gives no way to find all the solutions of (2.1) . Effective results, which however are weaker than the results of Thue (and hence a fortiori of Roth) were given by Feldman (1971) with Baker's method.

THEOREM 2B*. *Suppose that $\alpha$ is algebraic of degree $d \geqq 3$ . Then there are explicit constants $\mu_0(\alpha) < d$ and $c_2(\alpha)$ such that every solution of (2.1) with $\mu \geqq \mu_0$ has $q \leqq c_2(\alpha)$* .

This Theorem will not be proved in these Notes.

We remark that the method of Thue-Siegel-Roth does permit to estimate the *number* of solutions of (2.1) , and this is carried out by Davenport and Roth (1955).

Our proof of Roth's Theorem in §§ 4-11 will follow rather closely the exposition in Cassels (1957). Generalisations will be presented in chapters VI and VIII.

## §3. Thue's Equation.

THEOREM 3A. (Thue (1908)). *Suppose $F(X,Y)$ is a binary form with rational coefficients, and with at least 3 distinct linear factors (with algebraic coefficients). Then if $m$ is non-zero, the diophantine equation*

$$F(x,y) = m \tag{3.1}$$

*has only finitely many solutions in rational integers $x,y$* .

An equation of the type (3.1) will henceforth be called a *Thue equation*.

<u>Proof</u>. We may factor $F(X,Y)$ as

$$F(X,Y) = a(\gamma_1 X + \delta_1 Y)^{e_1} \cdots (\gamma_s X + \delta_s Y)^{e_s} \tag{3.2}$$

with $s \geqq 3$ and with real or complex algebraic $\gamma_1,\delta_1,\ldots,\gamma_s,\delta_s$ such that forms $\gamma_i X + \delta_i Y$ and $\gamma_j X + \delta_j Y$ with $i \neq j$ are linearly independent, with the further convention that each $\gamma_i$ is either 1 or 0, and that $\delta_i = 1$ if $\gamma_i = 0$. By rearranging the factors we may suppose that

$$0 < |\gamma_1 x + \delta_1 y| \leqq \cdots \leqq |\gamma_s x + \delta_s y| \quad .$$

Since $\gamma_1 X + \delta_1 Y$ and $\gamma_2 X + \delta_2 Y$ are linearly independent,

$$|\gamma_s x + \delta_s y| \geqq \cdots \geqq |\gamma_2 x + \delta_2 y| \geqq \frac{1}{2}(|\gamma_1 x + \delta_1 y| + |\gamma_2 x + \delta_2 y|)$$

$$\geqq c_1 \max(|x|,|y|) = c_1 \boxed{\underline{\underline{x}}} \quad ,$$

say, and

$$|F(x,y)| \geqq c_2 |\gamma_1 x + \delta_1 y|^{e_1} \boxed{\underline{\underline{x}}}^{d-e_1} \quad ,$$

where $d$ is the degree of $F$. If $\gamma_1 X + \delta_1 Y$ has rational coefficients, we have $|\gamma_1 x + \delta_1 y| \geqq c_3$, and since $d > e_1$ it follows that $|F(x,y)|$ tends to infinity with $\boxed{\underline{\underline{x}}}$. If $\gamma_1 = 1$ and $\delta_1$ is algebraic of degree $\ell \geqq 2$ it follows from Thue's result mentioned in §2 (hence a fortiori from Roth's Theorem) that for $\delta > 0$,

$$|\gamma_1 x + \delta_1 y| \geqq c_4(\delta_1,\delta) \boxed{\underline{\underline{x}}}^{-(\ell/2)-\delta} \quad ,$$

so that

$$|F(x,y)| \geqq c_5 \boxed{\underline{\underline{x}}}^{d-e_1((\ell/2)+1+\delta)} \quad .$$

Together with $\gamma_1 X + \delta_1 Y = X + \delta_1 Y$ , the form $F$ must have the conjugate factors, each with multiplicity $e_1$ , so that $d \geqq \ell e_1$ , and if $\ell = 2$ we even have $d \geqq 2e_1 + 1$ , since $F$ has at least 3 distinct factors. Since $\delta > 0$ may be picked arbitrarily small, we have $d > e_1((\ell/2)+1+\delta)$ , and again $|F(x,y)|$ tends to infinity with $\overline{|\underline{\underline{x}}|}$ .

By using the full strength of Roth's Theorem one obtains the following result, whose proof is left as an Exercise.

THEOREM 3B. *Suppose* $F(X,Y)$ *is a binary form of degree* $d \geqq 3$ *with rational coefficients and without multiple factors. Then for given* $\nu < d-2$ *there are only finitely many integer points* $\underline{\underline{x}} = (x,y)$ *with*

$$0 < |F(x,y)| < \overline{|\underline{\underline{x}}|}^{\nu} .$$

*In particular, if* $G(X,Y)$ *is a form of degree* $< d-2$ , *then the diophantine equation*

$$F(x,y) = G(x,y)$$

*has only finitely many solutions having*[†)] $F(x,y) \neq 0$ .

The method of proof of the above theorems by Thue's method is non-effective, i.e. it provides no bound for the size of the solutions. But Baker (1967/68) gave the bound

$$\overline{|\underline{\underline{x}}|} < \exp((dH)^{(10d)^5})$$

for the solutions $\underline{\underline{x}} = (x,y)$ of Thue's equation (3.1) , where $d$ is the degree of $F$ , where $F$ has distinct linear factors, and where the coefficients of $F$ as well as $m$ are rational integers of absolute value

---

†) There are infinitely many solutions of $F(x,y) = G(x,y) = 0$ precisely if $F,G$ have a common linear factor with rational coefficients.

at most H .

Thue's method <u>does</u> provide estimates for the <u>number</u> of solutions of his equations, and this was carried out with respect to Theorem 3A by Mahler (1933), and with respect to Theorem 3B by Davenport and Roth (1955). Siegel (1970) in the case $F(X,Y) = (\alpha X + \beta Y)^d + (\gamma X + \delta y)^d$ with $\alpha\delta - \beta\gamma \neq 0$ gives a bound which depends only on $d$ and on $m$. Note that for $d = 3$ every non-degenerate form is of this type.

## §4. Combinatorial Lemmas.

LEMMA 4A. *Suppose that* $r_1, \dots r_m$ *are positive integers and that* $0 < \varepsilon < 1$. *Then the number of* $m$-*tuples of integers* $i_1, \dots, i_m$ *with*

$$0 \leq i_h \leq r_h \qquad (1 \leq h \leq m)$$

*and with*

$$\left| \left( \sum_{h=1}^{m} \frac{i_h}{r_h} \right) - \frac{m}{2} \right| \geq \varepsilon m$$

*is at most*

$$(r_1 + 1) \cdots (r_m + 1) \cdot 2e^{-\varepsilon^2 m/4} .$$

Remarks. (a) The lemma may be given a probabilistic interpretation. Namely, regard the expressions $\frac{i_h}{r_h}$ as random variables with expectation $\frac{1}{2}$. By the Law of Large Numbers, $\sum \frac{i_h}{r_h} \approx \frac{m}{2}$ with high probability. (b) A slightly weaker version of this lemma has long been known: see, e.g., Schneider (1936). (c) Lemma 4A is an immediate consequence of Lemma 4C below. Our proof will be like that given in Mahler (1961),

Appendix A. Mahler attributes the proof given there to G. E. H. Reuter.

LEMMA 4B. *Suppose that* $n \geq 1$ , $r \geq 0$ *are integers. The number of* $n$-*tuples of nonnegative integers* $i_1,\ldots,i_n$ *with* $i_1+\cdots+i_n = r$ *is*

$$N(n,r) = \binom{r+n-1}{r} \quad . \tag{4.1}$$

*Proof*. The argument is by induction on $n$ and $r$ . We have

$$N(n,0) = 1 = \binom{n-1}{0}$$

for any $n \geq 1$ , so that (4.1) holds for $r = 0$ . Also,

$$N(1,r) = 1 = \binom{r}{r}$$

for any $r \geq 0$ , whence the lemma is true for $n = 1$ .

Now suppose that $r \geq 1$ , $n \geq 2$ are given, and assume that (4.1) holds for all pairs $(n',r')$ with $n' \leq n$ , $r' \leq r$ , and $(n',r') \neq (n,r)$ . We will prove (4.1) for the pair $(n,r)$ . Let $N^*(n,r)$ denote the number of $n$-tuples of nonnegative integers $i_1,\ldots,i_n$ with $i_1+\cdots+i_n \leq r$ . Then

$$N(n,r) = N^*(n-1,r) = N^*(n-1,r-1) + N(n-1,r) = N(n,r-1) + N(n-1,r)$$
$$= \binom{r+n-2}{r-1} + \binom{r+n-2}{r} = \binom{r+n-1}{r} \quad .$$

LEMMA 4C. *Suppose that* $r_1,\ldots,r_m$ *are positive integers and that* $0 < \varepsilon < 1$ . *Further suppose that* $n \geq 2$ *is an integer. Then the number of* $nm$-*tuples of nonnegative integers*

$$\begin{array}{c} i_{11},\ldots,i_{1n} \\ i_{21},\ldots,i_{2n} \\ \cdots \\ i_{m1},\ldots,i_{mn} \end{array}$$

<u>with</u>

$$\sum_{k=1}^{n} i_{hk} = r_h \qquad (1 \le h \le m) \tag{4.2}$$

<u>and with</u>

$$\left| \left( \sum_{h=1}^{m} \frac{i_{h1}}{r_h} \right) - \frac{m}{n} \right| \ge \varepsilon m$$

<u>is at most</u>

$$\binom{r_1 + n - 1}{r_1} \cdots \binom{r_m + n - 1}{r_m} \cdot 2e^{-\varepsilon^2 m/4} \quad .$$

<u>Remarks</u>. (a) Lemma 4C again may be given a probabilistic interpretation: the "random variables" $\frac{i_{hk}}{r_h}$ have expectation $\frac{1}{n}$ , so $\sum \frac{i_{h1}}{r_h} \approx \frac{m}{n}$ with high probability by the Law of Large Numbers. (b) Lemma 4A follows from Lemma 4C on taking $n = 2$ and considering 2m-tuples

$$\begin{array}{c} i_1, \ r_1 - i_1 \\ \cdots\cdots \\ i_m, \ r_m - i_m \quad . \end{array}$$

<u>Proof of Lemma 4C</u>. Let $M_+$ denote the number of nm-tuples with (4.2) and with

$$\left( \sum_{h=1}^{m} \frac{i_{h1}}{r_h} \right) - \frac{m}{n} \ge \varepsilon m \quad ;$$

and let $M_-$ denote the number of nm-tuples with (4.2) and with

$$\left( \sum_{h=1}^{m} \frac{i_{h1}}{r_h} \right) - \frac{m}{n} \le -\varepsilon m \quad .$$

To prove the lemma, it clearly suffices to show that

$$(4.3)\qquad M_{\pm} \le \binom{r_1 + n-1}{r_1} \cdots \binom{r_m + n-1}{r_m} e^{-\varepsilon^2 m/4} \quad .$$

Let integers $j$ and $c_j$ be given, with $1 \le j \le m$ and $0 \le c_j \le r_j$. Then write $f_j(c_j)$ for the number of $(n-1)$-tuples of nonnegative integers $i_{j2},\ldots,i_{jn}$ with $i_{j2}+\ldots+i_{jn} = r_j - c_j$. It is clear from the definitions of $M_+$ and $M_-$ that

$$(4.4)\qquad M_{\pm} = \sum_{\underline{\underline{c}}} f_1(c_1)\ldots f_m(c_m) \quad ,$$

where the sum is over all $\underline{\underline{c}} = (c_1,\ldots,c_m)$ with $0 \le c_j \le r_j \quad (1 \le j \le m)$ for which

$$\left(\sum_{h=1}^{m} \frac{c_h}{r_h}\right) - \frac{m}{n}$$

is $\ge \varepsilon m$ or $\le -\varepsilon m$, respectively. It follows that

$$(4.5)\qquad \begin{aligned} M_{\pm} e^{\varepsilon^2 m/2} &\le \sum_{c_1=0}^{r_1} \cdots \sum_{c_m=0}^{r_m} f_1(c_1)\ldots f_m(c_m) \exp\left(\pm \frac{\varepsilon}{2}\left(\left(\sum_{h=1}^{m}\frac{c_h}{r_h}\right) - \frac{m}{n}\right)\right) \\ &= \prod_{j=1}^{m}\left(\sum_{c_j=0}^{r_j} f_j(c_j)\exp\left(\pm\frac{\varepsilon}{2}\left(\frac{c_j}{r_j} - \frac{1}{n}\right)\right)\right) \quad . \end{aligned}$$

For the moment, we keep $j$ fixed.

It is clear from the definitions that

$$(4.6)\qquad \sum_{c=0}^{r_j} f_j(c) = \sum_{c+i_2+\cdots+i_n=r_j} 1 = N(n,r_j) = \binom{r_j + n - 1}{r_j} \quad .$$

Recall that $e^x \le 1 + x + x^2$ whenever $|x| \le 1$. We obtain

$$(4.7)\qquad \begin{aligned} \sum_{c=0}^{r_j} f_j(c)\exp\left(\pm\frac{\varepsilon}{2}\left(\frac{c}{r_j} - \frac{1}{n}\right)\right) &\le \sum_{c=0}^{r_j} f_j(c)\left(1 \pm \frac{\varepsilon}{2}\left(\frac{c}{r_j} - \frac{1}{n}\right) + \frac{\varepsilon^2}{4}\left(\frac{c}{r_j} - \frac{1}{n}\right)^2\right) \\ &\le \sum_{c=0}^{r_j} f_j(c)\left(1 + \frac{\varepsilon^2}{4}\right) \pm \frac{\varepsilon}{2r_j}\left(\sum_{c=0}^{r_j} c f_j(c) - \frac{r_j}{n}\sum_{c=0}^{r_j} f_j(c)\right) \\ &= \binom{r_j + n-1}{r_j}\left(1 + \frac{\varepsilon^2}{4}\right) \quad , \end{aligned}$$

the last equality following from (4.6) and from the fact that the coefficient of $\pm \frac{\varepsilon}{2r_j}$ in the next to the last line displayed above is 0: To show that the coefficient in question is indeed 0 , we observe that

$$f_j(c) = \sum_{\substack{i_2,\dots,i_n \\ i_2+\dots+i_n=r_j-c}} 1 = \sum_{\substack{i_1,\dots,i_n \\ i_1+\dots+i_n=r_j \\ i_1=c}} 1 \ ,$$

whence

$$cf_j(c) = \sum_{\substack{i_1,\dots,i_n \\ i_1+\dots+i_n=r_j \\ i_1=c}} i_1 \ .$$

Then

$$\sum_{c=0}^{r_j} cf_j(c) = \sum_{\substack{i_1,\dots,i_n \\ i_1+\dots+i_n=r_j}} i_1 = \frac{1}{n} \sum_{\substack{i_1,\dots,i_n \\ i_1+\dots+i_n=r_j}} (i_1+\dots+i_n)$$

$$= \frac{r_j}{n} \sum_{\substack{i_1,\dots,i_n \\ i_1+\dots+i_n=r_j}} 1 = \frac{r_j}{n} \sum_{c=0}^{r_j} f_j(c) \ ,$$

as claimed.

It follows from (4.5) and (4.7) that

$$M_{\pm} e^{\varepsilon^2 m/2} \le \binom{r_1+n-1}{r_1} \cdots \binom{r_m+n-1}{r_m} \left(1+\frac{\varepsilon^2}{4}\right)^m \le \binom{r_1+n-1}{r_1} \cdots \binom{r_m+n-1}{r_m} e^{\varepsilon^2 m/4} \ .$$

This inequality is equivalent to (4.3) , and the lemma is proved.

## §5. Further Auxiliary Lemmas.

Beginning with this section, we consider polynomials in $m$ variables with rational integer coefficients. We write

$$P(X_1,\dots,X_m) = \sum C(j_1,\dots,j_m)X_1^{j_1}\cdots X_m^{j_m} ,$$

where the sum is over all m-tuples of nonnegative integers $j_1,\dots,j_m$ ; all but finitely many of the coefficients $C(j_1,\dots,j_m)$ are 0 . We define the <u>height</u> of $P$ by

$$\lceil P \rceil = \max|C(j_1,\dots,j_m)| .$$

Finally, if $i_1,\dots,i_m$ are nonnegative integers, we write

$$P_{i_1\dots i_m} = \frac{1}{i_1!\dots i_m!}\frac{\partial^{i_1+\dots+i_m}}{\partial X_1^{i_1}\dots\partial X_m^{i_m}} P .$$

It will often be convenient to write $P_{\underline{\underline{i}}}$ instead of $P_{i_1\dots i_m}$ , where it is understood that $\underline{\underline{i}} = (i_1,\dots,i_m)$ .

<u>LEMMA 5A</u>. <u>If</u> $P$ <u>has rational integer coefficients</u>, <u>then so does</u> $P_{\underline{\underline{i}}}$ . <u>Furthermore</u>, <u>if</u> $P$ <u>has degree</u> $\le r_h$ <u>in the variable</u> $X_h$ $(1 \le h \le m)$ , <u>then</u>

$$\lceil P_{\underline{\underline{i}}} \rceil \le 2^{r_1+\dots+r_m}\lceil P \rceil .$$

<u>Proof</u>. We may write

$$P(X_1,\dots,X_m) = \sum_{j_1=0}^{r_1}\cdots\sum_{j_m=0}^{r_m} C(j_1,\dots,j_m)X_1^{j_1}\dots X_m^{j_m} ;$$

hence

$$(5.1) \qquad P_{\underline{\underline{i}}}(X_1,\dots,X_m) = \sum_{j_1=0}^{r_1}\cdots\sum_{j_m=0}^{r_m}\binom{j_1}{i_1}\cdots\binom{j_m}{i_m} C(j_1,\dots,j_m)X_1^{j_1-i_1}\dots X_m^{j_m-i_m} .$$

The new coefficients are again integers, since the binomial coefficients are integers. (We adopt the convention that $\binom{m}{n} = 0$ if $m < n$ .) Since

$$\binom{j_k}{i_k} \le 2^{j_k} \le 2^{r_k} \qquad (1 \le k \le m) ,$$

the second assertion of the lemma follows from (5.1) .

LEMMA 5B. (Siegel's Lemma). Let

$$L_j(\underline{z}) = \sum_{k=1}^{N} a_{jk} z_k \qquad (1 \le j \le M)$$

be $M$ linear forms with rational integer coefficients. Suppose that $N > M$ and that

$$|a_{jk}| \le A \quad , \qquad (1 \le j \le M, 1 \le k \le N)$$

where $A$ is a positive integer. Then there exists an integer point $\underline{z} = (z_1, \ldots, z_N) \ne \underline{0}$ with

(5.2) $$L_j(\underline{z}) = \underline{0} \qquad (1 \le j \le M)$$

and with

(5.3) $$\overline{|\underline{z}|} \le \left[ (NA)^{\frac{M}{N-M}} \right] = Z \quad ,$$

say.

Proof. Since $N > M$ , rational solutions of (5.2) with $\underline{z} \ne \underline{0}$ always exist. But if $\underline{z}$ is a solution of (5.2) , then so is $\lambda\underline{z}$ for any real $\lambda$ , and therefore integer points $\underline{z} \ne \underline{0}$ exist which satisfy (5.2).

It remains to show that (5.3) can be satisfied in addition to (5.2). Our proof of this is much like the proof of Dirichlet's Theorem. First, we have $Z + 1 > (NA)^{\frac{M}{N-M}}$ , whence $NA < (Z+1)^{\frac{N-M}{M}}$ , and therefore

$$NAZ + 1 \le NA(Z+1) < (Z+1)^{N/M} \quad .$$

For every integer point $\underline{z} = (z_1, \ldots, z_N)$ with

(5.4) $$0 \le z_i \le Z \quad , \qquad (1 \le i \le N)$$

we have

$$-B_j Z \le L_j(\underline{z}) \le C_j Z \quad , \qquad (1 \le j \le M)$$

where $-B_j$ and $C_j$ are the sums of the negative and the positive coefficients of $L_j(\underline{z})$ , respectively. Now $B_j + C_j \le NA$ , so that each $L_j(\underline{z})$ lies in an interval of length $\le NAZ$ . Therefore each $L_j(\underline{z})$ takes at most $NAZ + 1$ distinct values, and hence the M-tuple $L_1(\underline{z}),\ldots,L_M(\underline{z})$ takes at most

$$(NAZ + 1)^M < (Z + 1)^N$$

values.

On the other hand, the number of possibilities for $\underline{z}$ with (5.4) is $(Z + 1)^N$ . It follows that there are N-tuples $\underline{z}^{(1)} \neq \underline{z}^{(2)}$ with (5.4) and with

$$L_j(\underline{z}^{(1)}) = L_j(\underline{z}^{(2)}) \qquad (1 \le j \le M) \ .$$

The integer point $\underline{z} = \underline{z}^{(1)} - \underline{z}^{(2)}$ satisfies the conditions of the lemma.

As is well known, an algebraic <u>integer</u> $\alpha$ satisfies an equation $\alpha^d + a_1\alpha^{d-1} + \cdots + a_d = 0$ with rational integer coefficients $a_1,\ldots,a_d$ . If $\alpha$ is just algebraic of degree $d$ and satisfies $a_0\alpha^d + \cdots + a_d = 0$ , then $\beta = a_0\alpha$ is algebraic of degree $d$ and has $\beta^d + a_1\beta^{d-1} + a_2a_0\beta^{d-2} + \cdots + a_d a_0^{d-1} = 0$ , hence is an algebraic integer. Now if $|\alpha - (p/q)| < q^{-2-\delta}$ , then $|\beta - (a_0p/q)| < a_0q^{-2-\delta}$ . Hence if Roth's Theorem is true for $\beta$ , then it is true for $\alpha$ , and <u>it will suffice to prove the theorem for algebraic integers</u>. We also recall the well known fact (Hardy & Wright (1954), §14.2) that if $\alpha$ is an algebraic integer, then already the defining polynomial of $\alpha$ has its leading coefficient equal to 1 .

LEMMA 5C. Let $\alpha$ be an algebraic integer, with defining polynomial $Q(X) = X^d + a_1X^{d-1} + \cdots + a_{d-1}X + a_d$ . Then for each integer $\ell \geq 0$ , there are rational integers $a_1^{(\ell)},\ldots,a_d^{(\ell)}$ with

$$\alpha^\ell = a_1^{(\ell)}\alpha^{d-1}+\cdots+a_{d-1}^{(\ell)}\alpha + a_d^{(\ell)}$$

and with

$$|a_i^{(\ell)}| \leq (\lceil Q \rceil + 1)^\ell \qquad (1 \leq i \leq d) .$$

Proof. We proceed by induction on $\ell$ . The lemma is true trivially if $\ell < d$ . Suppose, then, that the lemma is true for $\ell - 1$ . We have

$$\begin{aligned}
\alpha^\ell = \alpha^{\ell-1}\cdot\alpha &= \left(a_1^{(\ell-1)}\alpha^{d-1}+\cdots+a_d^{(\ell-1)}\right)\alpha \\
&= a_1^{(\ell-1)}\alpha^d + a_2^{(\ell-1)}\alpha^{d-1}+\cdots+a_d^{(\ell-1)}\alpha \\
&= a_1^{(\ell-1)}\left(-a_1\alpha^{d-1}-\cdots-a_{d-1}\alpha - a_d\right) + a_2^{(\ell-1)}\alpha^{d-1}+\cdots+a_d^{(\ell-1)}\alpha \\
&= \left(a_2^{(\ell-1)}-a_1a_1^{(\ell-1)}\right)\alpha^{d-1} + \cdots + \left(a_d^{(\ell-1)}-a_{d-1}a_1^{(\ell-1)}\right)\alpha - a_da_1^{(\ell-1)} ,
\end{aligned}$$

so

$$\alpha^\ell = a_1^{(\ell)}\alpha^{d-1}+\cdots+a_{d-1}^{(\ell)}\alpha + a_d^{(\ell)}$$

in an obvious notation. For each $i$ , $1 \leq i \leq d$ , we have the estimate

$$|a_i^{(\ell)}| \leq (\lceil Q \rceil + 1)^{\ell-1} + \lceil Q \rceil (\lceil Q \rceil + 1)^{\ell-1} = (\lceil Q \rceil + 1)^\ell .$$

## §6. The Index of a Polynomial.

Let $P(X_1,\ldots,X_m)$ be a polynomial with rational integer coefficients, let $r_1,\ldots,r_m$ be positive integers, and let $(\alpha_1,\ldots,\alpha_m)$ be an arbitrary point of $\mathbb{R}^m$ .

Definition. Suppose at first that $P \not\equiv 0$ . The index of $P$ with respect to $(\alpha_1,\ldots,\alpha_m;r_1,\ldots,r_m)$ is the least value of

$$\frac{i_1}{r_1}+\cdots+\frac{i_m}{r_m}$$

for which $P_{i_1\ldots i_m}(\alpha_1,\ldots,\alpha_m)$ does not vanish. In particular, the index of $P$ is $0$ if $P(\alpha_1,\ldots,\alpha_m)\neq 0$ .

If $P\equiv 0$ , we define the index of $P$ to be $+\infty$ . In either case, we denote the index of $P$ by Ind $P$ .

LEMMA 6A. Let $(\alpha_1,\ldots,\alpha_m)$ and $r_1,\ldots,r_m$ be given as above. With respect to these parameters, we have:

(i) $\operatorname{Ind} P_{\underline{\underline{i}}} \geq \operatorname{Ind} P - \sum_{h=1}^{m}\frac{i_h}{r_h}$ ,

(ii) $\operatorname{Ind}(P^{(1)}+P^{(2)}) \geq \min(\operatorname{Ind} P^{(1)},\operatorname{Ind} P^{(2)})$ ,

(iii) $\operatorname{Ind}(P^{(1)}P^{(2)}) = \operatorname{Ind} P^{(1)} + \operatorname{Ind} P^{(2)}$ .

We remark that (ii) together with (iii) show that the index is a "valuation" of the ring of polynomials in $m$ variables.

Proof. (i). Let $T = P_{\underline{\underline{i}}}$ , and suppose that $T_{\underline{\underline{j}}}(\alpha_1,\ldots,\alpha_m)\neq 0$ . Then $P_{\underline{\underline{i}}+\underline{\underline{j}}}(\alpha_1,\ldots,\alpha_m)\neq 0$ , whence

$$\frac{i_1+j_1}{r_1}+\cdots+\frac{i_m+j_m}{r_m} \geq \operatorname{Ind} P \ .$$

Thus

$$\frac{j_1}{r_1}+\cdots+\frac{j_m}{r_m} \geq \operatorname{Ind} P - \sum_{h=1}^{m}\frac{i_h}{r_h} ,$$

and therefore

$$\operatorname{Ind} T \geq \operatorname{Ind} P - \sum_{h=1}^{m}\frac{i_h}{r_h} \ .$$

(ii). Suppose that $(P^{(1)} + P^{(2)})_{\underline{\underline{j}}}(\alpha_1,\ldots,\alpha_m) \neq 0$ . Then either $P^{(1)}_{\underline{\underline{j}}}(\alpha_1,\ldots,\alpha_m) \neq 0$ or $P^{(2)}_{\underline{\underline{j}}}(\alpha_1,\ldots,\alpha_m) \neq 0$ , hence

$$\text{either } \frac{j_1}{r_1}+\ldots+\frac{j_m}{r_m} \geq \operatorname{Ind} P^{(1)} \quad \text{or} \quad \frac{j_1}{r_1}+\ldots+\frac{j_m}{r_m} \geq \operatorname{Ind} P^{(2)} \quad .$$

Consequently,

$$\operatorname{Ind}(P^{(1)} + P^{(2)}) \geq \min(\operatorname{Ind} P^{(1)}, \operatorname{Ind} P^{(2)}) \quad .$$

(iii). In an obvious notation, we have

$$(6.1) \qquad (P^{(1)}P^{(2)})_{\underline{\underline{j}}} = \sum_{\underline{\underline{i}}+\underline{\underline{i}}'=\underline{\underline{j}}} C(\underline{\underline{i}},\underline{\underline{i}}')P^{(1)}_{\underline{\underline{i}}}P^{(2)}_{\underline{\underline{i}}'}$$

for any integer point $\underline{\underline{j}} = (j_1,\ldots,j_m)$ with $j_h \geq 0$ $(1 \leq h \leq m)$. (In fact it may be shown that $C(\underline{\underline{i}},\underline{\underline{i}}') = 1$ ; an obvious advantage of our $P_{\underline{\underline{i}}}$ over partial derivatives.)

Suppose now that $\underline{\underline{j}}$ is chosen so that $\sum_{h=1}^{m} \frac{j_h}{r_h} = \operatorname{Ind}(P^{(1)}P^{(2)})$ and $(P^{(1)}P^{(2)})_{\underline{\underline{j}}}(\alpha_1,\ldots,\alpha_m) \neq 0$ . By (6.1) , there exist $\underline{\underline{i}}$ and $\underline{\underline{i}}'$ with $\underline{\underline{i}} + \underline{\underline{i}}' = \underline{\underline{j}}$ such that $P^{(1)}_{\underline{\underline{i}}}(\alpha_1,\ldots,\alpha_m) \neq 0$ and $P^{(2)}_{\underline{\underline{i}}'}(\alpha_1,\ldots,\alpha_m) \neq 0$ . Then $\sum_{h=1}^{m} \frac{i_h}{r_h} \geq \operatorname{Ind} P^{(1)}$ and $\sum_{h=1}^{m} \frac{i'_h}{r_h} \geq \operatorname{Ind} P^{(2)}$ , whence

$$(6.2) \quad \operatorname{Ind}(P^{(1)}P^{(2)}) = \sum_{h=1}^{m} \frac{j_h}{r_h} = \sum_{h=1}^{m} \frac{i_h}{r_h} + \sum_{h=1}^{m} \frac{i'_h}{r_h} \geq \operatorname{Ind} P^{(1)} + \operatorname{Ind} P^{(2)} \quad .$$

Conversely, there exist m-tuples $\underline{\underline{i}}$ with $\sum_{h=1}^{m} \frac{i_h}{r_h} = \operatorname{Ind} P^{(1)}$ and $P^{(1)}_{\underline{\underline{i}}}(\alpha_1,\ldots,\alpha_m) \neq 0$ . Of these $\underline{\underline{i}}$ , assume that $\bar{\underline{\underline{i}}} = (\bar{i}_1,\ldots,\bar{i}_m)$ is the first lexicographically. Similarly, let $\bar{\underline{\underline{i}}}' = (\bar{i}'_1,\ldots,\bar{i}'_1)$ be the first m-tuple $\underline{\underline{i}}'$ lexicographically with $\sum_{h=1}^{m} \frac{i'_h}{r_h} = \operatorname{Ind} P^{(2)}$ and with $P^{(2)}_{\underline{\underline{i}}'}(\alpha_1,\ldots,\alpha_m) \neq 0$ . Putting $\underline{\underline{j}} = \bar{\underline{\underline{i}}} + \bar{\underline{\underline{i}}}'$ , we have

$$(P^{(1)}P^{(2)})_{\underline{\underline{j}}}(\alpha_1,\ldots,\alpha_m) = C(\bar{\underline{\underline{i}}},\bar{\underline{\underline{i}}}')P^{(1)}_{\bar{\underline{\underline{i}}}}(\alpha_1,\ldots,\alpha_m)P^{(2)}_{\bar{\underline{\underline{i}}}'}(\alpha_1,\ldots,\alpha_m) \neq 0$$

by (6.1) . This establishes the inequality reverse to (6.2) .

§7. The Index Theorem.

THEOREM 7A. Suppose that $\alpha$ is an algebraic integer of degree $d$ , $d \geq 2$ . Also suppose that $\varepsilon > 0$ , and that $m$ is an integer satisfying

$$m > 16\varepsilon^{-2}\log 4d \quad . \tag{7.1}$$

Let $r_1,\ldots,r_m$ be positive integers.

Then there is a polynomial $P(X_1,\ldots,X_m) \neq 0$ with rational integer coefficients such that

(i) $P$ has degree $\leq r_h$ in $X_h$ , $(1 \leq h \leq m)$

(ii) $P$ has index $\geq \frac{m}{2}(1-\varepsilon)$ with respect to $(\alpha,\alpha,\cdots,\alpha;r_1,\ldots,r_m)$ ,

and

(iii) $\overline{|P|} \leq B^{r_1+\cdots+r_m}$ , where $B = B(\alpha)$ .

Proof. We seek a polynomial

$$P(X_1,\ldots,X_m) = \sum_{j_1=0}^{r_1}\cdots\sum_{j_m=0}^{r_m} C(j_1,\ldots,j_m)X_1^{j_1}\cdots X_m^{j_m}$$

with rational integer coefficients $C(j_1,\ldots,j_m)$ such that (ii) and (iii) hold. The coefficients are

$$N = (r_1+1)\cdots(r_m+1)$$

integers to be determined. By (ii) , we need

$$P_{\underline{\underline{i}}}(\alpha,\alpha,\cdots,\alpha) = 0 \tag{7.2}$$

whenever

$$\left(\sum_{h=1}^{m}\frac{i_h}{r_h}\right) - \frac{m}{2} < -\frac{\varepsilon}{2}m \quad .$$

In view of Lemma 4A , the number of such m-tuples $\underline{\underline{i}}$ is at most

$(r_1+1)\cdots(r_m+1)\cdot 2e^{-\varepsilon^2 m/16}$ . It follows that the number of these conditions (7.2) is at most

$$N\cdot\frac{2}{4d}=\frac{N}{2d}$$

by (7.1) .

Each condition (7.2) is a linear equation in the coefficients $C(j_1,\ldots,j_m)$ . The coefficients of these equations will be rational integers times powers of $\alpha$ , hence will be algebraic. But each power of $\alpha$ is a linear combination of $1,\alpha,\ldots,\alpha^{d-1}$ with rational integer coefficients. Hence each condition (7.2) follows from d linear relations in $C(j_1,\ldots,j_m)$ with rational integer coefficients. Altogether, we obtain

$$M\leq d\cdot\frac{N}{2d}=\frac{N}{2}$$

linear equations for the $C(j_1,\ldots,j_m)$ with rational integer coefficients.

Let A be the maximum of the absolute values of these rational integer coefficients. For each $C(j_1,\ldots,j_m)$ in (7.2) , the coefficients in question have absolute value at most

$$\binom{j_1}{i_1}\cdots\binom{j_m}{i_m}(\lceil Q\rceil+1)^{\ell}\leq 2^{j_1+\cdots+j_m}(\lceil Q\rceil+1)^{\ell}$$

by Lemma 5C: Here $Q(X)$ is the defining polynomial for $\alpha$ , and $\ell=(j_1-i_1)+\cdots+(j_m-i_m)$ . Thus

$$A\leq(2(\lceil Q\rceil+1))^{r_1+\cdots+r_m} .$$

By Lemma 5B , our system of linear equations has a nontrivial integer solution with

$$|C(j_1,\dots,j_m)| \le Z \le (NA)^{\frac{M}{N-M}} \le NA$$

$$\le 2^{r_1+\cdots+r_m}(2(\lceil Q \rceil +1))^{r_1+\cdots+r_m} = B^{r_1+\cdots+r_m}$$

for each m-tuple $(j_1,\dots,j_m)$. The polynomial $P$ with these coefficients $C(j_1,\dots,j_m)$ satisfies

$$\lceil P \rceil \le B^{r_1+\cdots+r_m}$$

with $B = B(\alpha) = 4(\lceil Q \rceil +1)$ .

The construction of the polynomial $P$ in this section corresponds to part (a) in the proof of Liouville's Theorem. The next section will correspond to part (b) .

## §8. The index of $P(X_1,\dots,X_m)$ at rational points near $(\alpha,\alpha,\dots,\alpha)$ .

Suppose that $\alpha$ is an algebraic integer of degree $d$ , $d \ge 2$ . For any $\varepsilon > 0$ , let $m = m(\alpha,\varepsilon)$ be an integer satisfying

$$m > 16\varepsilon^{-2}\log 4d \quad .$$

Also, let $r_1,\dots,r_m$ be positive integers and let $P$ be a polynomial satisfying the conclusions of Theorem 7A .

THEOREM 8A. Suppose that $0 < \delta < 1$ , and that

(8.1) $$0 < \varepsilon < \frac{\delta}{36} \quad .$$

Let $\frac{p_1}{q_1},\dots,\frac{p_m}{q_m}$ be rational approximations to $\alpha$ , with

(8.2) $$|\alpha - \frac{p_h}{q_h}| < q_h^{-2-\delta} \qquad (1 \le h \le m)$$

and with

(8.3) $$q_h^{\delta} > D \quad , \qquad (1 \le h \le m)$$

<u>where</u> $D = D(\alpha) > 0$ . <u>Also,</u> <u>suppose</u> <u>that</u>

(8.4) $$r_1 \log q_1 \le r_h \log q_h \le (1+\varepsilon) r_1 \log q_1 \qquad (1 \le h \le m) .$$

<u>Then the index of</u> $P$ <u>with respect to</u> $\left(\frac{p_1}{q_1}, \ldots, \frac{p_m}{q_m}; r_1, \ldots, r_m\right)$ <u>is</u> $\ge \varepsilon m$ .

<u>Proof</u>. Suppose that $j_1, \ldots, j_m$ are nonnegative integers with

$$\sum_{h=1}^{m} \frac{j_h}{r_h} < \varepsilon m .$$

Put $T(X_1, \ldots, X_m) = P_{\underline{j}}(X_1, \ldots, X_m)$ , where $\underline{j} = (j_1, \ldots, j_m)$ . We have to show that $T\left(\frac{p_1}{q_1}, \ldots, \frac{p_m}{q_m}\right) = 0$ .

By the Index Theorem (Theorem 7A),

$$\overline{|P|} \le B^{r_1 + \cdots + r_m} ,$$

whence

$$\overline{|T|} \le (2B)^{r_1 + \cdots + r_m} ,$$

from Lemma 5A . Applying Lemma 5A again, we have

$$\overline{|T_{\underline{i}}|} \le (4B)^{r_1 + \cdots + r_m}$$

for any m-tuple $\underline{i} = (i_1, \ldots, i_m)$ of nonnegative integers. Hence in $T_{\underline{i}}(\alpha, \alpha, \cdots, \alpha)$ , each monomial has absolute value

$$\le (4B)^{r_1 + \cdots + r_m} (\max(1, |\alpha|))^{r_1 + \cdots + r_m} .$$

Since $T_{\underline{i}}$ is a sum of at most

$$(r_1 + 1) \ldots (r_m + 1) \le 2^{r_1 + \cdots + r_m}$$

monomials, we get

$$(8.5)\qquad |T_{\underline{\underline{i}}}(\alpha,\alpha,\cdots,\alpha)| \le (8B\max(1,|\alpha|))^{r_1+\cdots+r_m} = C^{r_1+\cdots+r_m},$$

where $C = C(\alpha)$.

By the Index Theorem, the index of $P$ with respect to $(\alpha,\alpha,\cdots,\alpha;r_1,\cdots,r_m)$ is $\ge \frac{m}{2}(1-\varepsilon)$. It follows from part (i) of Lemma 6A that the index of $T$ with respect to $(\alpha,\alpha,\cdots,\alpha;r_1,\cdots,r_m)$ is

$$\ge \frac{m}{2}(1-\varepsilon) - \sum_{h=1}^{m} \frac{j_h}{r_h} > \frac{m}{2}(1-3\varepsilon) .$$

By Taylor's formula,

$$T\left(\frac{p_1}{q_1},\cdots,\frac{p_m}{q_m}\right) = \sum_{i_1=0}^{r_1}\cdots\sum_{i_m=0}^{r_m} T_{i_1\cdots i_m}(\alpha,\alpha,\cdots,\alpha)\cdot\left(\frac{p_1}{q_1}-\alpha\right)^{i_1}\cdots\left(\frac{p_m}{q_m}-\alpha\right)^{i_m} .$$

From the paragraph above, the summands vanish unless $\sum_{h=1}^{m} \frac{i_h}{r_h} > \frac{m}{2}(1-3\varepsilon)$. In view of (8.2) and (8.5), we obtain

$$(8.6)\qquad \left|T\left(\frac{p_1}{q_1},\cdots,\frac{p_m}{q_m}\right)\right| \le \sum_{\underline{\underline{i}}}{}' C^{r_1+\cdots+r_m}\left(q_1^{i_1}q_2^{i_2}\cdots q_m^{i_m}\right)^{-2-\delta},$$

where $\Sigma'$ indicates that the sum is to be taken over all m-tuples of integers $i_h$ in $0 \leqq i_h \leqq r_h$ for which $\sum_{h=1}^{m}\frac{i_h}{r_h} > m(\frac{1}{2} - 2\varepsilon)$.

For such m-tuples, we have in view of (8.4), (8.1),

$$q_1^{i_1}q_2^{i_2}\cdots q_m^{i_m} = q_1^{r_1\frac{i_1}{r_1}} q_2^{r_2\frac{i_2}{r_2}}\cdots q_m^{r_m\frac{i_m}{r_m}} \ge q_1^{r_1\left(\frac{i_1}{r_1}+\cdots+\frac{i_m}{r_m}\right)}$$

$$> q_1^{r_1 m(\frac{1}{2}-2\varepsilon)} \ge \left(q_1^{r_1}q_2^{r_2}\cdots q_m^{r_m}\right)^{(\frac{1}{2}-2\varepsilon)/(1+\varepsilon)}$$

$$> \left(q_1^{r_1}q_2^{r_2}\cdots q_m^{r_m}\right)^{\frac{1}{2}(1-6\varepsilon)} .$$

The number of summands in (8.6) is $\le 2^{r_1+\cdots+r_m}$, so that

$$\left|T\left(\frac{p_1}{q_1},\cdots,\frac{p_m}{q_m}\right)\right| \le \prod_{h=1}^{m}\left(2C\, q_h^{-\frac{1}{2}(1-6\varepsilon)(2+\delta)}\right)^{r_h} .$$

Now

$$\tfrac{1}{2}(1-6\varepsilon)(2+\delta) > 1 + \tfrac{\delta}{2} - 9\varepsilon \qquad\qquad (\text{since } \delta < 1)$$

$$> 1 + \tfrac{\delta}{4}$$

by (8.1) ; therefore,

$$2Cq_h^{-\frac{1}{2}(1-6\varepsilon)(2+\delta)} < 2Cq_h^{-1-\frac{\delta}{4}} < q_h^{-1}$$

if $q_h^{\delta} > (2C)^4$ . This is true by (8.3) if we put $D = (2C)^4$ , say. It follows that

$$\left|T\left(\frac{p_1}{q_1},\ldots,\frac{p_m}{q_m}\right)\right| < \frac{1}{q_1^{r_1} q_2^{r_2} \cdots q_m^{r_m}} \quad .$$

Recall that P , hence T , has degree $\le r_h$ in $X_h$ $(1 \le h \le m)$ . Thus

$$T\left(\frac{p_1}{q_1},\ldots,\frac{p_m}{p_m}\right) = \frac{N}{q_1^{r_1} q_2^{r_2} \cdots q_m^{r_m}}$$

for some integer N . By the inequality above, N is necessarily 0 , whence

$$T\left(\frac{p_1}{q_1},\ldots,\frac{p_m}{q_m}\right) = 0 \quad .$$

Part (c) in the proof of Liouville's Theorem was trivial. But in the present context this part is most difficult, and will be resolved in Roth's Lemma in §10.

## §9. Generalized Wronskians.

Suppose that $\varphi_1,\ldots,\varphi_k$ are rational functions in m variables $X_1,\ldots,X_m$ . We consider differential operators

$$\Delta = \frac{\partial^{i_1+\cdots+i_m}}{\partial X_1^{i_1} \cdots \partial X_m^{i_m}} \quad .$$

The <u>order</u> of such a differential operator is $i_1+\cdots+i_m$ .

<u>Definition</u>. A <u>generalized Wronskian</u> of $\varphi_1,\cdots,\varphi_k$ is any determinant of the form

$$\det(\Delta_i\varphi_j) \qquad (1 \le i,j \le k) \tag{9.1}$$

where $\Delta_1,\ldots,\Delta_k$ are operators as above, with $\Delta_i$ of order $\le i-1$ $(1 \le i \le k)$ .

<u>Remark</u>. Take $m = 1$ . Then $\Delta_1$ is the identity operator, $\Delta_2$ is the identity operator or $\frac{\partial}{\partial X}$ , $\Delta_3$ is the identity operator or $\frac{\partial}{\partial X}$ or $\frac{\partial^2}{\partial X^2}$ , etc. A necessary condition for the generalized Wronskian (9.1) <u>not</u> to vanish identically is that $\Delta_1$ = identity operator, $\Delta_2 = \frac{\partial}{\partial X}$ , $\Delta_3 = \frac{\partial^2}{\partial X^2},\ldots$ . In this case, (9.1) becomes

$$\det\begin{pmatrix} \varphi_1 & \varphi_2 & \cdots & \varphi_k \\ \varphi_1' & \varphi_2' & \cdots & \varphi_k' \\ \vdots & \vdots & & \vdots \\ \varphi_1^{(k-1)} & \varphi_2^{(k-1)} & \cdots & \varphi_k^{(k-1)} \end{pmatrix},$$

which is the (ordinary) Wronskian of $\varphi_1,\ldots,\varphi_k$ .

<u>LEMMA 9A</u>. <u>Suppose that</u> $\varphi_1,\ldots,\varphi_k$ <u>are rational functions in</u> $X_1,\ldots,X_m$ <u>with real coefficients</u>, <u>and linearly independent over the reals</u>. <u>Then at least one generalized Wronskian of</u> $\varphi_1,\ldots,\varphi_k$ <u>is not identically zero.</u>

<u>Remark</u>. The converse of Lemma 9A is also true: namely, if $\varphi_1,\ldots,\varphi_k$ are linearly dependent over the reals, then all generalized Wronskians of

$\varphi_1,\ldots,\varphi_k$ vanish.

Proof. We use induction on $k$. If $k = 1$, then $\Delta_1$ is necessarily the identity operator, and the generalized Wronskian is $\varphi_1$. But $\varphi_1$ is linearly independent over the reals, so $\varphi_1 \neq 0$.

Suppose now that $\varphi_1,\ldots,\varphi_k$ are $k$ ($\geq 2$) rational functions satisfying the hypotheses of the lemma. Let $\Omega$ be any rational function in $X_1,\ldots,X_m$ with real coefficients, $\Omega \neq 0$. Consider the functions

$$\varphi_i^* = \Omega\varphi_i \qquad (1 \leq i \leq k).$$

Then $\varphi_1^*,\ldots,\varphi_k^*$ are again linearly independent over the reals. Any generalized Wronskian of $\varphi_1^*,\ldots,\varphi_k^*$ is a linear combination of general Wronskians of $\varphi_1,\ldots,\varphi_k$. (The coefficients in this linear combination are rational functions involving the partial derivatives of $\Omega$.) To prove the lemma, it therefore will suffice to show that some generalized Wronskian of $\varphi_1^*,\ldots,\varphi_k^*$ does not vanish trivially. If we put $\Omega = \varphi_1^{-1}$, then $\varphi_1^* = 1$, $\varphi_2^* = \dfrac{\varphi_2}{\varphi_1},\ldots,\varphi_k^* = \dfrac{\varphi_k}{\varphi_1}$. This argument shows that it is no restriction to assume that, in our list of given rational functions $\varphi_1,\ldots,\varphi_k$, the function $\varphi_1$ is identically $1$.

The set of all linear combinations

$$c_1\varphi_1 + \cdots + c_k\varphi_k$$

with real coefficients $c_1,\ldots,c_k$ forms a real vector space $V$ of dimension $k$. Since $k > 1$, and since $\varphi_1 = 1$ and $\varphi_2$ are linearly independent, $\varphi_2$ is not a constant. Thus $\dfrac{\partial\varphi_2}{\partial X_j} \neq 0$ for some $j$. Without loss of generality, we suppose that $\dfrac{\partial\varphi_2}{\partial X_1} \neq 0$. Let $W$ be the subspace of $V$ consisting of all elements $c_1\varphi_1 + \cdots + c_k\varphi_k$ with

$$\frac{\partial}{\partial X_1}(c_1\varphi_1+\cdots+c_k\varphi_k) = 0 \quad .$$

$W$ is not the zero subspace, since $\varphi_1 \in W$. Also, $W \neq V$, since $\varphi_2 \notin W$. Accordingly, if we put $t = \dim W$, then $1 \le t \le k-1$.

We choose rational functions $\psi_1,\ldots,\psi_k$ such that $\psi_1,\ldots,\psi_t$ is a basis of $W$ and $\psi_1,\ldots,\psi_k$ is a basis for $V$. By the induction hypothesis, there are operators $\Delta_1^*,\ldots,\Delta_t^*$ of orders $\le 0,1,\ldots,t-1$, respectively, with

$$W_1 = \det(\Delta_i^*\psi_j) \neq 0 \qquad (1 \le i,j \le t) \quad .$$

If $c_{t+1},\ldots,c_k$ are real numbers, not all of which are $0$, then

$$\frac{\partial}{\partial X_1}(c_{t+1}\psi_{t+1}+\cdots+c_k\psi_k) \neq 0 \quad :$$

This is true because the subspace spanned by $\psi_{t+1},\ldots,\psi_k$ has intersection $\underline{\underline{0}}$ with $W$. In other words, the rational functions

$$\frac{\partial}{\partial X_1}\psi_{t+1},\ldots,\frac{\partial}{\partial X_1}\psi_k$$

are linearly independent over the reals. By induction, there are operators $\Delta_{t+1}^*,\ldots,\Delta_k^*$ of orders $\le 0,1,\ldots,k-t-1$, respectively, with

$$W_2 = \det\left(\Delta_i^*\frac{\partial}{\partial X_1}\psi_j\right) \neq 0 \qquad (t < i,j \le k) \quad .$$

We define operators $\Delta_i$ $(1 \le i \le k)$ as follows:

$$\Delta_i = \begin{cases} \Delta_i^* & , \text{ if } 1 \le i \le t \\ \Delta_i^*\dfrac{\partial}{\partial X_1} & , \text{ if } t < i \le k \end{cases} \quad .$$

Note that each $\Delta_i$ has order $\le i-1$. We have

$$\det(\Delta_i \psi_j) = \det \begin{pmatrix} \Delta_i^* \psi_j & \Delta_i^* \psi_j \\ 0 & \Delta_i^* \frac{\partial}{\partial X_1} \psi_j \end{pmatrix} \begin{matrix} 1 \le i \le t \\ t < i \le k \end{matrix}$$

(columns: $1 \le j \le t$, $t < j \le k$)

$$= W_1 W_2 \ne 0 \quad .$$

Because $\psi_1, \ldots, \psi_k$ is a basis of the vector space spanned by $\varphi_1, \ldots, \varphi_k$, it follows that

$$\det(\Delta_i \varphi_j) \ne 0 \quad .$$

This completes the proof.

## §10. Roth's Lemma.

THEOREM 10A. (Roth (1955a)). *Suppose that*

(10.1) $$0 < \varepsilon < \frac{1}{12} \quad .$$

*Let* $m$ *be a fixed positive integer. Put*

(10.2) $$\omega = \omega(m,\varepsilon) = 24 \cdot 2^{-m} \left(\frac{\varepsilon}{12}\right)^{2^{m-1}} \quad .$$

*Let* $r_1, \ldots, r_m$ *be positive integers with*

(10.3) $$\omega r_h \ge r_{h+1} \qquad (1 \le h < m) \; .$$

*Suppose that* $0 < \gamma \le 1$ , *and let* $(p_1, q_1), \ldots, (p_m, q_m)$ *be pairs of coprime integers with* $q_h > 0$ $(1 \le h \le m)$ *and with*

(10.4) $$q_h^{r_h} \ge q_1^{\gamma r_1} \qquad (1 \le h \le m) \; ,$$

(10.5) $$q_h^{\omega\gamma} \ge 2^{3m} \qquad (1 \le h \le m) \; .$$

*Further, suppose that* $P(X_1, \ldots, X_m) \ne 0$ *is a polynomial of degree* $\le r_h$

<u>in</u> $X_h$ $(1 \le h \le m)$ <u>with rational integer coefficients and with</u>

(10.6) $$\overline{|P|} \le q_1^{\omega\gamma r_1} .$$

<u>Then the index of</u> P <u>with respect to</u> $\left(\frac{p_1}{q_1},\ldots,\frac{p_m}{q_m}; r_1,\ldots,r_m\right)$ <u>is</u> $\le \varepsilon$ .

<u>Remark</u> This theorem is usually stated with $\gamma = 1$ . The presence of the parameter $\gamma$ is not necessary for the proof of Roth's Theorem, but will be useful in later generalizations.

<u>Proof</u>. We use induction on m .

<u>m = 1</u> . We may write

$$P(X) = \left(X - \frac{p_1}{q_1}\right)^{\ell} M(X) ,$$

where M(X) is a polynomial with rational coefficients having $M\left(\frac{p_1}{q_1}\right) \ne 0$ . Thus

(10.7) $$P(X) = (q_1X - p_1)^{\ell} R(X) ,$$

where $R(X) = q_1^{-\ell} M(X)$ . It is a consequence of Gauss' Lemma that R(X) has integer coefficients.

It now follows from (10.7) that the leading coefficient of P(X) is divisible by $q_1^{\ell}$ . Thus

$$q_1^{\ell} \le \overline{|P|} \le q_1^{\omega r_1} \qquad \text{(by (10.6))}$$

$$= q_1^{\varepsilon r_1} \qquad \text{(by (10.2) and since } m = 1) ,$$

so that $\frac{\ell}{r_1} \le \varepsilon$ , since $q_1 > 1$ by (10.5) . But $\frac{\ell}{r_1}$ is the index of P with respect to $\left(\frac{p_1}{q_1}; r_1\right)$ , and the theorem is true if m = 1 .

<u>Inductive Step m-1 ⇒ m</u> . Consider decompositions

$$P(X_1,\ldots,X_m) = \sum_{j=1}^{k} \varphi_j(X_1,\ldots,X_{m-1})\psi_j(X_m) \quad , \tag{10.8}$$

where $\varphi_1,\ldots,\varphi_k$ and $\psi_1,\ldots,\psi_k$ are polynomials with rational coefficients. For example, take $k = r_m + 1$ and take $\psi_1,\ldots,\psi_k$ equal to $1, X_m, X_m^2, \ldots, X_m^{r_m}$ , respectively. We now choose a decomposition where $k$ is minimal. Then in particular,

$$k \le r_m + 1 \quad .$$

Since this decomposition is one in which $k$ is minimal, <u>the functions $\varphi_1,\ldots,\varphi_k$ are linearly independent over the reals</u>. For otherwise, there would exist real numbers $c_1,\ldots,c_k$ , not all $0$ , such that

$$c_1\varphi_1 + \cdots + c_k\varphi_k = 0 \quad .$$

Since $\varphi_1,\ldots,\varphi_k$ have <u>rational</u> coefficients, there would, in fact, exist <u>rational</u> numbers $c_1,\ldots,c_k$ with these properties. But then if, say, $c_k \ne 0$ , we would have

$$P = \sum_{j=1}^{k-1} \varphi_j\left(\psi_j - \frac{c_j}{c_k}\psi_k\right) \quad ,$$

contradicting the minimality of $k$ . Similarly, $\psi_1,\ldots,\psi_k$ are linearly independent over the reals.

Let us write

$$U(X_m) = \det\left(\frac{1}{(i-1)!}\frac{\partial^{i-1}}{\partial X_m^{i-1}}\psi_j(X_m)\right) \qquad (1 \le i,j \le k).$$

By Lemma 9A and the remark preceding it,

$$U(X_m) \ne 0 \quad .$$

Also by Lemma 9A , there exist operators

$$\Delta_i' = \frac{1}{i_1!\cdots i_{m-1}!}\,\frac{\partial^{i_1+\cdots+i_{m-1}}}{\partial X_1^{i_1}\cdots\partial X_{m-1}^{i_{m-1}}} \qquad (1 \le i \le k)$$

with orders

(10.9) $$i_1+\cdots+i_{m-1} \le i-1 \le k-1 \le r_m$$

having

$$V(X_1,\ldots,X_{m-1}) \overset{\text{def}}{=} \det_{1\le i,j\le k}(\Delta_i'\varphi_j) \ne 0 \quad .$$

Put

$$W(X_1,\ldots,X_m) = \det\left(\frac{1}{(j-1)!}\,\frac{\partial^{j-1}}{\partial X_m^{j-1}}\,\Delta_i' P\right) \qquad (1 \le i,j \le k).$$

Then

$$W(X_1,\ldots,X_m) = \det\left(\sum_{r=1}^{k}(\Delta_i'\varphi_r)\left(\frac{1}{(j-1)!)}\,\frac{\partial^{j-1}}{\partial X_m^{j-1}}\,\psi_r\right)\right) \qquad \text{(by (10.8))}$$

$$= V(X_1,\ldots,X_{m-1})U(X_m) \ne 0 \quad .$$

The entries in the determinant defining $W$ are of the type $P_{i_1\cdots i_{m-1}j-1}$, and hence these entries have rational integer coefficients. Thus $W$ is a polynomial with rational integer coefficients. Before proceeding further we need

LEMMA 10B. *Let* $\Theta$ *be the index of* $W$ *with respect to* $\left(\frac{p_1}{q_1},\ldots,\frac{p_m}{q_m};r_1,\ldots,r_m\right)$. *Then*

$$\Theta \le \frac{k\varepsilon^2}{6} \quad .$$

*Proof of the Lemma*. The polynomials $U, V$ do not necessarily have integer coefficients. But clearly there is a factorization

$$W(X_1,\ldots,X_m) = V^*(X_1,\ldots,X_{m-1})U^*(X_m) \quad ,$$

where $U^*, V^*$ have rational integer coefficients.

Next, we obtain estimates for the heights of $U^*$ and $V^*$. First, we have

$$\overline{|P_{i_1\ldots i_{m-1}j-1}|} \le 2^{r_1+\cdots+r_m}\overline{|P|} \qquad \text{(by Lemma 5a)}$$

$$\le 2^{r_1+\cdots+r_m} q_1^{\omega\gamma r_1} \qquad \text{(by (10.6)).}$$

Furthermore, the number of terms in $P_{i_1\ldots i_{m-1}j-1}$ is $\le 2^{r_1+\cdots+r_m}$, and the number of summands in the determinant expansion for $W$ is $k! \le k^{k-1} \le k^{r_m} \le 2^{kr_m}$. It follows that

$$\overline{|W|} \le 2^{kr_m}\left(2^{r_1+\cdots+r_m}2^{r_1+\cdots+r_m}q_1^{\omega r_1\gamma}\right)^k \le \left(2^{3mr_1}q_1^{\omega r_1\gamma}\right)^k ,$$

since $r_1 \ge r_2 \ge \cdots \ge r_m$ by (10.2) and (10.3). From (10.5) we obtain

$$\overline{|W|} \le \left(q_1^{2\omega r_1\gamma}\right)^k = q_1^{2\omega\gamma r_1 k} .$$

This yields the estimates

$$\overline{|U^*|} \le q_1^{2\omega\gamma r_1 k} \le q_m^{2\omega r_m k}, \quad \overline{|V^*|} \le q_1^{2\omega\gamma r_1 k} . \tag{10.10}$$

We now apply the inductive hypothesis. More precisely, we are going to apply Theorem 10A with $m-1$ for $m$, with $kr_1,\ldots,kr_{m-1}$ for $r_1,\ldots,r_m$, with $\frac{\varepsilon^2}{12}$ for $\varepsilon$, and with $V^*(X_1,\ldots,X_{m-1})$ for $P(X_1,\ldots,X_m)$. Namely, (10.3) and (10.5) are true with $\omega = \omega(m,\varepsilon)$ by our assumption, hence are true _a fortiori_ with $\omega(m-1,\frac{\varepsilon^2}{12}) = 2\omega(m,\varepsilon)$. It is clear that (10.4) holds and that (10.1) holds with $\frac{\varepsilon^2}{12}$ for $\varepsilon$. The analogue of (10.6) holds by (10.10), since

$$\boxed{V^*} \le q_1^{\omega\left(m-1,\frac{\varepsilon^2}{12}\right)\gamma(kr_1)} .$$

Hence the index of $V^*$ with respect to $\left(\frac{p_1}{q_1},\ldots,\frac{p_{m-1}}{q_{m-1}}; kr_1,\ldots,kr_{m-1}\right)$ is $\le \frac{\varepsilon^2}{12}$, and therefore the index of $V^*$ with respect to $\left(\frac{p_1}{q_1},\ldots,\frac{p_{m-1}}{q_{m-1}}; r_1,\ldots,r_{m-1}\right)$ is $\le \frac{k\varepsilon^2}{12}$. If we now consider $V^*$ as a polynomial in $X_1,\ldots,X_m$, then $V^*$ still has index $\le \frac{k\varepsilon^2}{12}$ with respect to $\left(\frac{p_1}{q_1},\ldots,\frac{p_m}{q_m}; r_1,\ldots,r_m\right)$.

It is easily checked that the hypotheses of Roth's Lemma are satisfied with $\gamma = 1$, with $m = 1$, with $kr_m$ for $r_1,\ldots,r_m$, with $\frac{\varepsilon^2}{12}$ for $\varepsilon$, and with $U^* = U^*(X_m)$ for $P(X_1,\ldots,X_m)$. (Note that $\omega\left(1,\frac{\varepsilon^2}{12}\right) \ge 2\omega(m,\varepsilon)$.) Since the case $m = 1$ of Roth's Lemma has been established, $U^*$ has index $\le \frac{k\varepsilon^2}{12}$ with respect to $\left(\frac{p_1}{q_1},\ldots,\frac{p_m}{q_m}; r_1,\ldots,r_m\right)$. Since $W = U^*V^*$, it follows from part (iii) of Lemma 6A that

$$\Theta \le \frac{k\varepsilon^2}{12} + \frac{k\varepsilon^2}{12} = \frac{k\varepsilon^2}{6} .$$

Lemma 10B is established.

<u>Completion of the proof of Theorem 10A</u>. Let $\theta$ denote the index of $P$ with respect to $\left(\frac{p_1}{q_1},\ldots,\frac{p_m}{q_m}; r_1,\ldots,r_m\right)$. Then

$$\begin{aligned}
\operatorname{Ind} P_{i_1\ldots i_{m-1}j-1} &\ge \theta - \frac{i_1}{r_1} - \ldots - \frac{i_{m-1}}{r_{m-1}} - \frac{j-1}{r_m} && \text{(by Lemma 6A)}\\
&\ge \theta - \frac{i_1+\ldots+i_{m-1}}{r_{m-1}} - \frac{j-1}{r_m} && (\text{since } r_1 \ge \cdots \ge r_{m-1})\\
&\ge \theta - \frac{r_m}{r_{m-1}} - \frac{j-1}{r_m} && \text{(by (10.9))}\\
&\ge \theta - \omega - \frac{j-1}{r_m} && \text{(by (10.3))}\\
&\ge \theta - \frac{\varepsilon^2}{24} - \frac{j-1}{r_m} && (\text{by (10.2), since } m \ge 2).
\end{aligned}$$

As noted earlier, each entry in column $j$ in the determinant defining $W$ is of the type $P_{i_1 \dots i_{m-1} j-1}$. Recall the identity

$$\mathrm{Ind}\, P^{(1)}P^{(2)} = \mathrm{Ind}\, P^{(1)} + \mathrm{Ind}\, P^{(2)}$$

and the inequality $\mathrm{Ind}(P^{(1)} + P^{(2)}) \geqq \min(\mathrm{Ind}\, P^{(1)}, \mathrm{Ind}\, P^{(2)})$ from Lemma 6A. Since $W$ is a sum of products of $k$ elements, one from each column, we see that

$$\begin{aligned} \Theta = \mathrm{Ind}\, W &\geq \sum_{j=1}^{k} \max\left(\theta - \frac{\varepsilon^2}{24} - \frac{j-1}{r_m}, 0\right) \\ &\geq -\frac{k\varepsilon^2}{24} + \sum_{i=0}^{k-1} \max\left(\theta - \frac{i}{r_m}, 0\right) . \end{aligned}$$

Thus

$$\sum_{i=0}^{k-1} \max\left(\theta - \frac{i}{r_m}, 0\right) \leq \Theta + \frac{k\varepsilon^2}{24} \leq \frac{k\varepsilon^2}{6} + \frac{k\varepsilon^2}{24} < \frac{k\varepsilon^2}{4} . \tag{10.11}$$

We consider two cases:

<u>Case I</u>. $\theta > \frac{k-1}{r_m}$. Then (10.11) becomes

$$\frac{1}{2}k\left(\theta + \theta - \frac{k-1}{r_m}\right) < k\,\frac{\varepsilon^2}{4} ,$$

which is equivalent to

$$\theta + \left(\theta - \frac{k-1}{r_m}\right) < \frac{\varepsilon^2}{2} .$$

But $\theta - \frac{k-1}{r_m} > 0$, so $\theta < \frac{\varepsilon^2}{2} < \varepsilon$.

<u>Case II</u>. $\theta \leq \frac{k-1}{r_m}$. Then (10.11) becomes

$$\sum_{i=0}^{[\theta r_m]} \left(\theta - \frac{i}{r_m}\right) < \frac{k\varepsilon^2}{4} ,$$

which yields

$$\frac{1}{2}\theta([\theta r_m]+1) < \frac{k\varepsilon^2}{4} \quad ,$$

whence

$$\frac{1}{2}\theta^2 r_m < \frac{k\varepsilon^2}{4} \quad .$$

But $k \le r_m + 1 \le 2r_m$ , so $\frac{1}{2}\theta^2 r_m < \frac{1}{2}\varepsilon^2 r_m$ , and therefore $\theta < \varepsilon$ .

In either case, we see that the index $\theta$ of $P$ with respect to $\left(\frac{p_1}{q_1},\dots,\frac{p_m}{q_m}; r_1,\dots,r_m\right)$ is $< \varepsilon$ . This completes the proof of Theorem 10A.

## §11. Conclusion of the proof of Roth's Theorem.

As was pointed out in §5 , we may restrict ourselves to algebraic integers. Suppose then that there exists a $\delta > 0$ such that

$$\left|\alpha - \frac{p}{q}\right| < q^{-2-\delta} \tag{11.1}$$

has infinitely many rational solutions $\frac{p}{q}$ , where $\alpha$ is an algebraic integer of degree $d$ , $d \ge 2$ . We proceed as follows.

(i) Assume, without loss of generality, that $0 < \delta < 1$ .

(ii) Choose $\varepsilon$ with $0 < \varepsilon < \frac{\delta}{36}$ . This is (8.1) . It implies that $0 < \varepsilon < \frac{1}{12}$ , which is (10.1) .

(iii) Choose an integer $m$ with $m > 16\varepsilon^{-2}\log 4d$ . Thus (7.1) holds. Define $\omega = \omega(m,\varepsilon)$ by (10.2) .

(iv) Let $\frac{p_1}{q_1}$ be a solution of (11.1) with $(p_1,q_1) = 1$ , $q_1 > 0$ , such that $q_1^{\omega} > B^m$ , where $B = B(\alpha)$ is the quantity of Theorem 7A , and such that (8.3) , (10.5) hold with $\gamma = 1$ and $h = 1$ .

(v) Choose $\frac{p_2}{q_2},\dots,\frac{p_m}{q_m}$ successively with (11.1) , with $(p_h,q_h) = 1$ , $q_h > 0$ $(2 \le h \le m)$ , and so that

$$\omega \log q_{h+1} \geq 2 \log q_h \qquad (1 \leq h \leq m-1).$$

This implies that $q_1 < q_2 < \cdots < q_m$ , and hence (8.3) and (10.5) with $\gamma = 1$ hold for $h = 1,2,\ldots,m$ .

(vi) Let $r_1$ be an integer so large that $\varepsilon r_1 \log q_1 \geq \log q_m$ .

(vii) For $2 \leq h \leq m$ , put

$$r_h = \left[\frac{r_1 \log q_1}{\log q_h}\right] + 1 .$$

Then for $2 \leq h \leq m$ we have

$$\begin{aligned} r_1 \log q_1 &< r_h \log q_h && \text{(by (vii))} \\ &\leq r_1 \log q_1 + \log q_h && \text{(by (vii))} \\ &\leq (1+\varepsilon) r_1 \log q_1 && \text{(by (vi))} . \end{aligned}$$

This gives (8.4) , (10.4) with $\gamma = 1$ . From this sequence of inequalities, it follows that

$$r_{h+1} \log q_{h+1} \leq (1+\varepsilon) r_h \log q_h \qquad (1 \leq h \leq m-1) .$$

Consequently,

$$\begin{aligned} \omega r_h &\geq \omega \frac{r_{h+1} \log q_{h+1}}{(1+\varepsilon) \log q_h} && (1 \leq h \leq m-1) \\ &\geq \frac{2}{1+\varepsilon} r_{h+1} \end{aligned}$$

by (v) , whence

$$\omega r_h \geq r_{h+1} \qquad (1 \leq h \leq m-1) ,$$

which is (10.3) .

The conditions of Theorem 7A (Index Theorem) are satisfied, since

(7.1) holds. Let $P(X_1,\ldots,X_m)$ be a polynomial satisfying the conclusions of the Index Theorem. The hypotheses of Theorem 8A , (i.e., (8.1), (8.2), (8.3), (8.4)) hold. Hence the index of $P$ with respect to $\left(\frac{p_1}{q_1},\ldots,\frac{p_m}{q_m};r_1,\ldots,r_m\right)$ is

$$\geq \varepsilon m \quad . \tag{11.2}$$

On the other hand, the hypotheses of Theorem 10A , (i.e., (10.1), (10.2), (10.3), (10.4), (10.5), (10.6)) hold with $\gamma = 1$ : Namely, (10.6) holds with $\gamma = 1$ because

$$\begin{aligned} \overline{|P|} &\leq B^{r_1+\cdots+r_m} && \text{(by Theorem 6A)} \\ &\leq B^{mr_1} \leq q_1^{\omega r_1} && \text{(by (iv))} \ . \end{aligned}$$

The conclusion of Theorem 10A is that the index of $P$ with respect to $\left(\frac{p_1}{q_1},\ldots,\frac{p_m}{q_m};r_1,\ldots,r_m\right)$ is $\leq \varepsilon$ . But this contradicts (11.2) . This is the desired contradiction, and Roth's Theorem follows.

## VI. Simultaneous Approximation to Algebraic Numbers

References: Schmidt (1970, 1971b,c)

§1. Basic Results. The only easy result here is as follows. Suppose[†)] $1,\alpha_1,\ldots,\alpha_v$ are algebraic and linearly independent over the rationals. Let $d$ be the degree of the number field generated by $\alpha_1,\ldots,\alpha_v$ and let $1,\alpha_1,\ldots,\alpha_v,\ldots,\alpha_{d-1}$ be a basis of this field. We saw in Theorem 4A of Chapter II that $\alpha_1,\ldots,\alpha_{d-1}$ are badly approximable, so that

$$|\alpha_1 q_1+\cdots+\alpha_{d-1}q_{d-1} - p| > c_1 q^{-d+1}$$

where $q_1,\ldots,q_{d-1}, p$ are rational integers and where $q = \max(|q_1|,\ldots,|q_{d-1}|) \neq 0$. Taking $q_{v+1} = \cdots = q_{d-1} = 0$ , we have

LEMMA 1A. *Suppose* $1,\alpha_1,\ldots,\alpha_v$ *are linearly independent over* $\mathbb{Q}$ , *and they generate an algebraic number field of degree* $d$ . *Then*

$$|\alpha_1 q_1+\cdots+\alpha_v q_v - p| > c_1 q^{-d+1}$$

*for arbitrary integers* $q_1,\ldots,q_v, p$ *having* $q = \max(|q_1|,\ldots,|q_v|) > 0$ .

When $v = 1$ we obtain Liouville's result. The exponent is clearly best possible when $d = v + 1$ , but will be improved in Corollary 1E below if $d > v + 1$ . We now turn to more difficult "Roth-type" results:

THEOREM 1B. *Suppose* $\alpha_1,\ldots,\alpha_u$ *are real algebraic numbers such that* $1,\alpha_1,\ldots,\alpha_u$ *are linearly independent over the rationals, and suppose* $\delta > 0$ . *Then there are only finitely many positive integers* $q$ *with*

$$q^{1+\delta}\,\|\alpha_1 q\|\cdots\|\alpha_u q\| < 1 \quad . \tag{1.1}$$

---

†) We ask the reader to forgive us for using $u$ or $v$ in this section instead of the more customary letters $n$ or $m$ , which will be needed later.

The inequalities

$$|\alpha_i - \frac{p_i}{q}| < q^{-1-(1/u)-\delta} \qquad (i = 1,\ldots,u) \tag{1.2}$$

imply that $\|\alpha_i q\| < q^{-(1/u)-\delta}$, and hence they imply (1.1). Therefore (1.2) has only finitely many solutions, and we obtain

<u>COROLLARY 1C</u>. <u>Suppose</u> $\alpha_1,\ldots,\alpha_u$ <u>and</u> $\delta$ <u>are as in Theorem</u> 1B. <u>Then there are only finitely many rational</u> u-<u>tuples</u> $(\frac{p_1}{q},\ldots,\frac{p_u}{q})$ <u>with</u> (1.2).

<u>THEOREM 1D</u>. <u>Assume that</u> $1,\alpha_1,\ldots,\alpha_v$ <u>are real algebraic and linearly independent over the rationals, and suppose</u> $\delta > 0$. <u>Then there are only finitely many</u> v-<u>tuples of nonzero integers</u> $q_1,\ldots,q_v$ <u>with</u>

$$|q_1 q_2 \cdots q_v|^{1+\delta} \|\alpha_1 q_1 + \cdots + \alpha_v q_v\| < 1 . \tag{1.3}$$

By applying Theorem 1D to all the non-empty subsets of $\alpha_1,\ldots,\alpha_v$ one deduces

<u>COROLLARY 1E</u>. <u>If</u> $\alpha_1,\ldots,\alpha_v$ <u>are as in Theorem</u> 1B, <u>then there are only finitely many</u> (v+1)-<u>tuples of integers</u> $q_1,\ldots,q_v,p$ <u>with</u> $q = \max(|q_1|,\ldots,|q_v|) > 0$ <u>and with</u>

$$|\alpha_1 q_1 + \cdots + \alpha_v q_v - p| < q^{-v-\delta} . \tag{1.4}$$

The exponents in Corollaries 1C and 1E are essentially best possible by Dirichlet's theorems on simultaneous approximation (see Ch. II). In view of Khintchine's transference principle (see the end of §5 of Chapter IV), these two corollaries are equivalent†), and they say that an u-tuple such as given in Theorem 1B is not very well approximable. The case u = 1 is of course Roth's Theorem. Theorems 1B and 1D were

---

†) Also, Theorems 1B and 1D are equivalent. See Wang, Yu and Zhu (1979).

proved by Schmidt (1970), after he had derived weaker results in (1965b, 1967). Baker (1967), Feldman (1970) and Osgood (1970) proved weaker but effective versions of Corollary 1E for special algebraic numbers $\alpha_1,\dots,\alpha_v$ .

THEOREM 1F. (Subspace Theorem) (Schmidt (1972)). *Suppose $L_1(\underline{x}),\dots,L_n(\underline{x})$ are linearly independent forms in $\underline{x} = (x_1,\dots,x_n)$ with real or complex algebraic coefficients. Given $\delta > 0$, there are finitely many proper subspaces $T_1,\dots,T_w$ of $\mathbb{R}^n$ such that every integer point $\underline{x} \neq \underline{0}$ with*

$$|L_1(\underline{x}) \cdots L_n(\underline{x})| < \overline{|\underline{x}|}^{-\delta} \tag{1.5}$$

*lies in one of these subspaces.*

The integer points in a subspace $T$ span a rational linear subspace, i.e. one defined by linear equations with rational coefficients, and hence $T_1,\dots,T_w$ may be taken to be rational subspaces.

We now proceed to derive Theorem 1B from the Subspace Theorem. Given $q$ with (1.1) choose $p_1,\dots,p_u$ with $\|\alpha_i q\| = |\alpha_i q - p_i|$ $(i = 1,\dots,u)$ . Then with $n = u+1$ and

$$\underline{x} = (x_1,\dots,x_n) = (p_1,\dots,p_u,q)$$

it is clear that $|\underline{x}| \ll q$ . Introduce linear forms

$$(1.6)\qquad \begin{aligned} L_i(\underline{X}) &= \alpha_i X_n - X_i \qquad (1 \le i \le u) , \\ L_n(\underline{X}) &= X_n \quad . \end{aligned}$$

Then (1.1) implies that

$$|L_1(\underline{x}) \cdots L_n(\underline{x})| < |\underline{x}|^{-\delta/2} \quad , \tag{1.7}$$

at least when $q$ is large. By the Subspace Theorem, the solutions to (1.7) lie in finitely many proper rational subspaces. If a typical such subspace $T$ is defined by

$$c_1x_1 + \cdots + c_ux_u + c_nx_n = 0 \quad , \tag{1.8}$$

then for $\underline{\underline{x}}$ in $T$ we have

$$c_1(\alpha_1 q - p_1) + \cdots + c_u(\alpha_u q - p_u) = (c_1\alpha_1 + \cdots + c_u\alpha_u + c_n)q \ ,$$

and thus

$$|c_1| \, \|\alpha_1 q\| + \cdots + |c_u| \, \|\alpha_u q\| \geq \gamma q \tag{1.9}$$

where $\gamma = |c_1\alpha_1 + \cdots + c_u\alpha_u + c_n| > 0$ in view of our linear independence condition. But (1.9) shows that $q$ is bounded.

Theorem 1D also is a consequence of the Subspace Theorem. Given $q_1,\ldots,q_v$ with (1.3) choose $p$ with $\|\alpha_1q_1+\ldots+\alpha_vq_v\| = |\alpha_1q_1+\ldots+\alpha_vq_v - p$ Put $n = v+1$ and write

$$\underline{\underline{x}} = (x_1,\ldots,x_n) = (q_1,\ldots,q_v,p) \quad ;$$

then $\overline{|\underline{\underline{x}}|} \ll q$ . In view of (1.3) we have (1.7) , at least when $q$ is large, with linear forms

$$\begin{aligned} L_i(\underline{\underline{X}}) &= X_i \qquad\qquad (1 \leq i \leq v) \ , \\ L_n(\underline{\underline{X}}) &= \alpha_1X_1 + \cdots + \alpha_vX_v - X_n \quad . \end{aligned} \tag{1.10}$$

In view of the Subspace Theorem, the solutions lie in a finite number of rational subspaces; let a typical such subspace be given by $c_1x_1 + \cdots + c_vx_v + c_nx_n = 0$ .

Now if $c_v \neq 0$ , we note that for $\underline{\underline{x}}$ satisfying this equation,

$$c_v\|\alpha_1 q_1 + \cdots + \alpha_v q_v\| = c_v|\alpha_1 q_1 + \cdots + \alpha_v q_v - p|$$

$$= |(c_v\alpha_1 - c_1\alpha_v)q_1 + \cdots + (c_v\alpha_{v-1} - c_{v-1}\alpha_v)q_{v-1} - (c_v + c_n\alpha_v)p|$$

$$= |c_v + c_n\alpha_v|\,|\alpha'_1 q_1 + \cdots + \alpha'_{v-1}q_{v-1} - p|$$

where $\alpha'_i = (c_v\alpha_i - c_i\alpha_v)/(c_v + c_n\alpha_v)$ $(i = 1,\ldots,v)$. For $v = 1$, we get $c_1\|\alpha_1 q_1\| \gg |p| \geq 1$ if $q$ is large, so that (1.3) has only finitely many solutions. For $v > 1$ we obtain

$$c_v\|\alpha_1 q_1 + \cdots + \alpha_v q_v\| \gg \|\alpha'_1 q_1 + \cdots + \alpha'_{v-1}q_{v-1}\| \quad ,$$

and hence

$$|q_1 q_2 \cdots q_{v-1}|^{1+(\delta/2)}\|\alpha'_1 q_1 + \cdots + \alpha'_{v-1}q_{v-1}\| < 1$$

from (1.3). Since $1,\alpha'_1,\ldots,\alpha'_{v-1}$ are linearly independent over $\mathbb{Q}$, this has only finitely many solutions by induction.

The situation is the same if some $c_j \neq 0$ where $1 \leq j \leq v$, and it is similar if $c_1 = \cdots = c_v = 0$ but $c_n \neq 0$. For another result on linear forms see §VII.11.

## §2. Roth Systems.

This section may be omitted at first reading.

Suppose $n$, $u$, $v$ are positive integers with

$$n = u + v \quad . \tag{2.1}$$

Let $L_1(\underline{X}),\ldots,L_u(\underline{X})$ be linear forms in $\underline{X} = (X_1,\ldots,X_n)$ with real coefficients. The rank of these forms is the maximum number of linearly independent forms among them. Suppose at the moment that the rank is $u$. By reordering the variables we may assume that the $n$ linear forms

$X_1,\dots,X_v$ , $L_1,\dots,L_u$ are linearly independent. By Minkowski's Theorem on Linear Forms (see Theorem 2C of Chapter II) there is for every $Q > 0$ an integer point $\underline{\underline{x}} \neq 0$ with

$$|x_i| \ll Q^u \quad (1 \leqq i \leqq v) \quad \text{and} \quad |L_j(\underline{\underline{x}})| \leqq Q^{-v} \quad (1 \leqq j \leqq u) \quad ,$$

and with the constant in $\ll$ depending only on $L_1,\dots,L_u$ . Writing $|\overline{\underline{\underline{x}}}| = \max(|x_1|,\dots,|x_n|)$ we obtain

$$(2.2) \qquad |\underline{\underline{x}}| \ll Q^u \quad \text{and} \quad |L_j(\underline{\underline{x}})| \leqq Q^{-v} \quad (1 \leqq j \leqq u) \quad ,$$

and hence

$$(2.3) \qquad |L_j(\underline{\underline{x}})| \ll |\overline{\underline{\underline{x}}}|^{-v/u} \qquad (1 \leqq j \leqq u) \; .$$

Since $Q$ in (2.2) may be taken arbitrarily large, the inequalities (2.3) have infinitely many solutions in integer points $\underline{\underline{x}} \neq \underline{\underline{0}}$ . (The discerning reader will note that (2.3) is related to Corollary 1F in Chapter II. If there is no integer point $\underline{\underline{x}}_0 \neq \underline{\underline{0}}$ with $L_1(\underline{\underline{x}}_0) = \cdots = L_u(\underline{\underline{x}}_0)$ then we get infinitely many solutions to (2.3) in integer points with coprime coordinates. If there is such an $\underline{\underline{x}}_0$ , then the multiples of $\underline{\underline{x}}_0$ all satisfy (2.3) .)

It is easy to see that (2.2) and (2.3) still have solutions if $L_1,\dots,L_u$ have rank less than $u$ . More precisely, if $L_1,\dots,L_u$ have rank $r$ , then we may replace (2.3) by

$$(2.4) \qquad |L_j(\underline{\underline{x}})| \ll |\underline{\underline{x}}|^{-(n-r)/r} \qquad (1 \leq j \leq u) \; .$$

When $r < u$ , this is stronger than (2.3) .

Suppose now that $L_1,\dots,L_u$ are linear forms with real <u>algebraic</u> coefficients. We shall call $L_1,\dots,L_u$ a <u>Roth system</u> if for every $\delta > 0$

the inequalities

$$|L_j(\underline{x})| < |\underline{x}|^{-(v/u)-\delta} \qquad (1 \le j \le u) \tag{2.5}$$

have only finitely many solutions in integer points $\underline{x}$. Roth's theorem says precisely that for $u = v = 1$, $n = 2$, the single form $L(\underline{X}) = \alpha X_1 - X_2$ is a Roth system if $\alpha$ is an algebraic irrationality. If $v = 1$ and $L_i(\underline{X}) = \alpha_i X_n - X_i$ $(1 \leq i \leq u)$, and if $1, \alpha_1, \ldots, \alpha_u$ are linearly independent, then $L_1, \ldots, L_u$ is a Roth System by Corollary 1C. If $u = 1$ and $L_1(\underline{X}) = \alpha_1 X_1 + \cdots + \alpha_v X_v - X_n$ with linearly independent $1, \alpha_1, \ldots, \alpha_v$, then $L_1$ is a Roth System by Corollary 1E.

We shall say that $L_1, \ldots, L_u$ have rank $r$ on a subspace $S$ of $\mathbb{R}^n$ if the restrictions of $L_1, \ldots, L_u$ to $S$ have rank $r$.

THEOREM 2A. *Suppose $L_1, \ldots, L_u$ have real algebraic coefficients. Necessary and sufficient for $L_1, \ldots, L_u$ to be a Roth system is that on every rational subspace $S^d$ of dimensions $d$ with $1 \le d \le n$, the forms $L_1, \ldots, L_u$ have rank $r$ satisfying*

$$r \ge du/n \,. \tag{2.6}$$

Both Corollary 1C and 1E can be derived from this theorem as an exercise.

The necessity of the condition is easy to see: the integer points in $S^d$ form a point lattice (see Corollary 8B of Ch. IV), and the restrictions of $L_1, \ldots, L_u$ to this point lattice behave like linear forms defined on the integer points of $d$-dimensional space. Hence by (2.4) there are infinitely many integer points $\underline{x} \neq \underline{0}$ in $S^d$ having

$$|L_j(\underline{x})| \ll |\underline{x}|^{-(d-r)/r} = |\underline{x}|^{1-(d/r)} \qquad (j = 1, \ldots, u) \,. \tag{2.7}$$

Now if $r < du/n$ , say $r = dun^{-1}(1+\delta)^{-1}$ , then

$$|L_j(\underline{x})| \ll \overline{|\underline{x}|}^{1-(n/u)(1+\delta)} \ll \overline{|\underline{x}|}^{-(v/u)-\delta} \qquad (j = 1,\dots,u) ,$$

and $L_1,\dots,L_u$ is not a Roth System.

We now turn to a generalization. Let $L_1,\dots,L_n$ be linear forms in $\underline{X} = (X_1,\dots,X_n)$ with real coefficients. Let $c_1,\dots,c_n$ be reals with

$$c_1 + \dots + c_n = 0 \quad . \tag{2.8}$$

By Minkowski's theorem on linear forms, there is for every $Q > 0$ an integer point $\underline{x} \neq 0$ with

$$|L_j(\underline{x})| \ll Q^{c_j} \qquad (1 \le j \le n) , \tag{2.9}$$

with the constant in $\ll$ depending only on the determinant of the forms $L_1,\dots,L_n$ .

We shall say that $(L_1,\dots,L_n ; c_1,\dots,c_n)$ is a <u>general Roth system</u> if $L_1,\dots,L_n$ have algebraic coefficients and if for every $\delta > 0$ there is a $Q_1 = Q_1(\delta ; L_1,\dots,L_n ; c_1,\dots,c_n)$ such that the inequalities

$$|L_j(\underline{x})| \le Q^{c_j - \delta} \qquad (1 \le j \le n) \tag{2.10}$$

have no integer solution $\underline{x} \neq \underline{0}$ if $Q > Q_1$ . Again by Roth's Theorem, $(L_1 = X_1 , L_2 = \alpha X_1 - X_2 ; c_1 = 1 , c_2 = -1)$ is a general Roth system if $\alpha$ is an algebraic irrationality.

From now on we shall assume that

$$c_1 \le c_2 \le \dots \le c_n \quad . \tag{2.11}$$

Now let $S^d$ be a subspace of $\mathbb{R}^n$ of dimension $d > 0$ , and suppose that

$L_1,\dots,L_n$ have rank $r$ on $S^d$. If $r = d$, let $t_1$ be the smallest integer such that $L_{t_1} \neq 0$ on $S^d$, i.e., that $L_{t_1}$ has rank 1 on $S^d$. Let $t_2$ be the smallest integer such that $L_{t_1}, L_{t_2}$ have rank 2 on $S^d$; and so on. In this way we obtain $d$ integers $t_1, t_2, \dots, t_d$. Put

$$c(S^d) = c_{t_1} + \cdots + c_{t_d} . \tag{2.12}$$

All this was under the assumption that $r = d$. When $r < d$, put $c(S^d) = +\infty$.

THEOREM 2B. *Suppose $L_1,\dots,L_n$ are linear forms with real algebraic coefficients, and $c_1,\dots,c_n$ are constants subject to (2.8) and (2.11). Necessary and sufficient for $(L_1,\dots,L_n\,;\,c_1,\dots,c_n)$ to be a general Roth system is that*

$$c(S^d) \leq 0 \tag{2.13}$$

*for every rational subspace $S^d \neq \underline{0}$.*

The necessity of the condition (2.13) is easy to show. Observe again that the restrictions of $L_1,\dots,L_n$ to the integer points of a rational subspace $S^d$ behave like linear forms on the integer points of $\mathbb{R}^d$. Suppose that $c(S^d) > 0$ for some rational subspace $S^d \neq \underline{0}$. If $L_1,\dots,L_n$ have rank $r < d$ on $S^d$, then for every $\varepsilon > 0$ there is an integer point $\underline{x} \neq \underline{0}$ on $S^d$ with $|L_j(\underline{x})| < \varepsilon$ $(1 \leq j \leq n)$, and hence for arbitrary $Q > 0$, $\delta > 0$ there is an $\underline{x} \neq \underline{0}$ with (2.10). Thus in this case we do not have a general Roth system. Assume, therefore, that $L_1,\dots,L_n$ have rank $r = d$ on $S^d$. Let $\delta$ be the positive number with $2d\delta = c(S^d)$. By Minkowski's theorem on linear forms, applied to the lattice of integer points on $S^d$ , there is for every $Q > 0$ an integer point $\underline{\underline{x}} \neq \underline{\underline{0}}$ in $S^d$ with

$$|L_{t_i}(\underline{\underline{x}})| \ll Q^{c_{t_i}-2\delta} \qquad (1 \le i \le d) .$$

By (2.11) , by our choice of $t_1,\ldots,t_d$ , and since $\underline{\underline{x}} \in S^d$ , this implies that

$$|L_j(\underline{\underline{x}})| \ll Q^{c_j-2\delta} \qquad (1 \le j \le n) ,$$

and hence it implies (2.10) if $Q$ is sufficiently large. Thus again $(L_1,\ldots,L_n ; c_1,\ldots,c_n)$ is not a general Roth system.

We now shall derive Theorem 2A from Theorem 2B . By what we said above, it will suffice to prove the <u>sufficiency</u> of (2.6) . This inequality, with $d = n$ , implies that $L_1,\ldots,L_u$ are independent. By renaming the variables, if necessary, we may assume that the $n$ forms $L_1,\ldots,L_u$ , $X_1,\ldots,X_v$ are independent. Let us take a look at the system

$$(2.14) \qquad (L_1,\ldots,L_u,X_1,\ldots,X_v;\ c_1 = -v,\ldots,c_u = -v\ ,\ c_{u+1} = u,\ldots,c_n = u) .$$

Since $L_1,\ldots,L_u,X_1,\ldots,X_v$ have rank $n$ , they have rank $d$ on every subspace of dimension $d$ . Let $S^d$ be a rational subspace of dimension $d$ , and let $r$ be the rank of $L_1,\ldots,L_u$ on $S^d$ . Then $t_1 < t_2 < \cdots < t_r \le u < t_{r+1} < \cdots < t_d$ and

$$\begin{aligned} c(S^d) &= c_{t_1} + \cdots + c_{t_r} + c_{t_{r+1}} + \cdots + c_{t_d} = r(-v) + (d-r)u \\ &= ud - nr \ . \end{aligned}$$

The condition (2.6) implies that $c(S^d) \le 0$ , and hence that (2.14) is a general Roth system by Theorem 2B . Thus for any $\delta > 0$ , the inequalities

$$|L_j(\underline{\underline{x}})| \le Q^{-v-\delta} \quad (1 \le j \le u) \;, \quad |x_i| \le Q^{u-\delta} \quad (1 \le i \le v)$$

have no integer solution $\underline{\underline{x}} \neq \underline{\underline{0}}$ if $Q$ is large. This implies that $L_1,\ldots,L_u$ is a Roth system.

We end this section with a deduction of Theorem 2B from the Subspace Theorem. By what we said above, it will suffice to prove the sufficiency of (2.13) .

Suppose we did not have a General Roth System. Then for some $\delta > 0$ and some arbitrarily large values of $Q$ , the relations (2.10) have a nontrivial solution $\underline{\underline{x}}$ . Choose a subspace $S$ which is minimal in the sense that (2.10) has for some arbitrarily large values of $Q$ a non-zero integer solution $\underline{\underline{x}} \in S$ , but that for every proper subspace $S'$ of $S$ the relations (2.10) have no such solution in $S'$ if $Q > Q_0(S')$ .

Suppose $\dim S = d$ and write $S = S^d$ . In view of (2.13) , the forms $L_1,\ldots,L_n$ have rank $d$ on $S$ . Pick $t_1,\ldots,t_d$ as in the definition of $c(S^d)$ . Then conditions (2.10) for large $Q$ imply that $\overline{|\underline{\underline{x}}|} \leqq Q^2$ , hence that

$$(2.15) \qquad |L_{t_1}(\underline{\underline{x}})\ldots L_{t_d}(\underline{\underline{x}})| \leqq Q^{c(S^d)-d\delta} \leqq Q^{-\delta} \leqq \overline{|\underline{\underline{x}}|}^{-\delta/2} .$$

The integer points $\underline{\underline{x}}$ of $S^d$ form a point lattice $\Lambda$ . (See §IV.8). There is a nonsingular linear transformation $T$ from $\mathbb{R}^d$ onto $S^d$ which provides a 1-1 map from the lattice of integer points in $\mathbb{R}^d$ onto $\Lambda$ . If $\underline{\underline{x}} = T(\underline{\underline{y}})$ and if $\overline{|\underline{\underline{x}}|}$ is large, then

$$(2.16) \qquad \overline{|\underline{\underline{y}}|} \leqq \overline{|\underline{\underline{x}}|}^2 .$$

The forms

$$\tilde{L}_i(\underline{\underline{y}}) = L_{t_i}(T(\underline{\underline{y}})) \qquad (i = 1,\ldots,d)$$

are $d$ linearly independent linear forms on $\mathbb{R}^d$ with algebraic coefficients. If $\mathbf{x}$ is a nonzero integer point in $S^d$ with (2.10) and if $\mathbf{x} = T(\mathbf{y})$, then (2.15), (2.16) yield

$$(2.17) \qquad |\tilde{L}_1(\mathbf{y}) \ldots \tilde{L}_d(\mathbf{y})| \leq \overline{|\mathbf{y}|}^{-\delta/4} .$$

By the Subspace Theorem there are finitely many subspaces $V_1, \ldots, V_w$ of $\mathbb{R}^d$ such that each solution $\mathbf{y}$ of (2.17) lies in one of these subspaces. But then $\mathbf{x} = T(\mathbf{y})$ lies in one of the spaces $W_1 = T(V_1), \ldots, W_w = T(V_w)$, which are proper subspaces of $S^d$. So for one of these subspaces $W_j$, the relations (2.10) have solutions $\mathbf{x}$ in $W_j$ for arbitrarily large values of $Q$. This contradicts the minimality of $S^d$.

Generalizations of the results of this section may be found in Vehmanen (1976).

## §3. The Strong Subspace Theorem.

Everything in this chapter will follow from the Strong Subspace Theorem. The interdependence of results is then as follows.

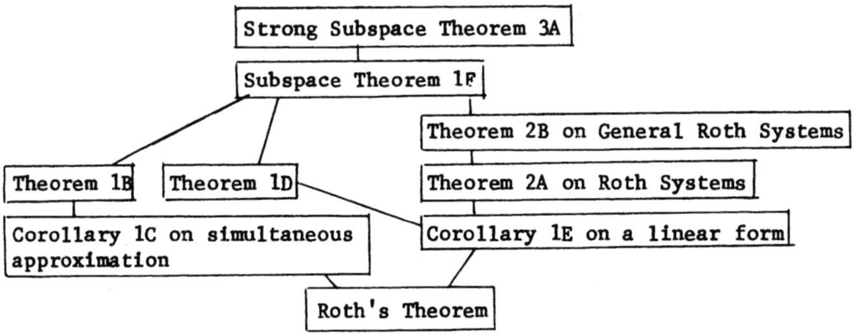


Moreover, by our transference results, the two corollaries are equivalent, and according to Wang, Yu and Zhu (1979), Theorems 1B and 1D are equivalent.

So what is the Strong Subspace Theorem? Here it is:

THEOREM 3A. *Let* $L_1,\ldots,L_n$ *be independent linear forms in* $\underline{X} = (X_1,\ldots,X_n)$ *with real algebraic coefficients, and let* $c_1,\ldots,c_n$ *be constants with* (2.8) . *For every* $Q > 0$ *the inequalities*

$$|L_j(\underline{x})| \leq Q^{c_j} \qquad (1 \leq j \leq n) \tag{3.1}$$

*define a parallelepiped* $\Pi = \Pi(Q)$ . *Denote the successive minima (in the sense of the Geometry of Numbers) of* $\Pi$ *by* $\lambda_1 = \lambda_1(Q),\ldots,\lambda_n = \lambda_n(Q)$ . *Suppose there is a* $\delta > 0$ , *a number* $d$ *with* $1 \leq d \leq n-1$ *and an unbounded set* $\mathfrak{N}$ *of positive numbers such that*

$$\lambda_d < \lambda_{d+1} Q^{-\delta} \tag{3.2}$$

*for every* $Q$ *in* $\mathfrak{N}$ .

*Then there is a fixed rational subspace* $S^d$ *of dimension* $d$ *and an unbounded subset* $\mathfrak{N}'$ *of* $\mathfrak{N}$ *such that for every* $Q$ *in* $\mathfrak{N}'$ *the first* $d$ *successive minima of* $\Pi(Q)$ *are assumed by points* $\underline{g}_1,\ldots,\underline{g}_d$ *in* $S^d$ .

We now proceed to derive the Subspace Theorem.

LEMMA 3B. *Let* $L_1,\ldots,L_n,c_1,\ldots,c_n$ *be as in Theorem 3A . Then there are finitely many proper rational subspaces* $T_1,\ldots,T_w$ *of* $\mathbb{R}^n$ *such that every integer point* $\underline{x} \neq \underline{0}$ *with*

$$|L_i(\underline{x})| \leq |\underline{x}|^{c_i-\delta} \qquad (i = 1,\ldots,n) \tag{3.3}$$

*lies in one of these subspaces.*

*Proof.* Suppose this were wrong. We could then construct a sequence $\underline{x}_1,\underline{x}_2,\ldots$ of solutions of (3.3) with any $n$ elements of this sequence linearly independent. For every $Q > 0$ consider the parallelepiped $\Pi$ defined by (3.1), choose independent integer points $\underline{g}_1,\dots,\underline{g}_n$ with $\underline{g}_i \in \lambda_i\Pi$ $(i = 1,\dots,n)$, and let $S_i$ be the subspace spanned by $\underline{g}_1,\dots,\underline{g}_i$. Now it is clear that $\lambda_1 \leq Q^{-\delta}$ if $Q = |\underline{x}_j|$ for some $j$, and hence (since $\lambda_1\cdots\lambda_n \gg 1$) $\lambda_n > 1$, at least when $Q = |\underline{x}_j|$ is large. Thus $\underline{x}_j$ will certainly lie in $S_{n-1}$; let $k$ be the smallest integer such that $\underline{x}_j$ lies in $S_k$. Then $\lambda_k \leq Q^{-\delta}$, and there is a $d$ in $k \leq d \leq n-1$ with

$$\lambda_d < \lambda_{d+1} Q^{-\delta/n} \tag{3.4}$$

There is a subsequence $\underline{x}_{j_1},\underline{x}_{j_2},\dots$ and a fixed $d$ such that (3.4) holds for $Q = |\underline{x}_{j_u}|$. Now if $\mathfrak{N}$ is the set of values $|\underline{x}_{j_u}|$ $(u = 1,2,\dots)$, then by Theorem 3A there is an infinite subsequence whose absolute values make up a set $\mathfrak{N}'$ such that for $Q \in \mathfrak{N}'$, the points $\underline{g}_1,\dots,\underline{g}_d$ lie in a fixed subspace $S^d$. Then $\underline{x}_{j_u} \in S_k \subseteq S_d = S^d$ for elements of this subsequence, contradicting the linear independence condition on our sequence.

Now as for Theorem 1F, the points $\underline{x}$ having $L_j(\underline{x}) = 0$ for some $j$ clearly lie in finitely many proper subspaces. We may therefore restrict our attention to points $\underline{x}$ with $L_1(\underline{x})\cdots L_n(\underline{x}) \neq 0$. Using norms, one sees that for these integer points one has $|\underline{x}|^{-A} \ll |L_j(\underline{x})| \ll |\underline{x}|$ $(j = 1,\dots,n)$ with some $A > 0$, hence $|\underline{x}|^{-2A} \leq |L_j(\underline{x})| < |\underline{x}|^2$ if $|\underline{x}|$ is large. We divide the interval $-2A \leq \xi < 2$ into finitely many subintervals $c' \leq \xi < c$ of length less than $\delta/2n$ . If $c_1' \leq \xi < c_1'', \ldots, c_n' \leq \xi < c_n''$ are any $n$ such intervals, it will suffice to show that the solutions $\underline{\underline{x}}$ of (1.5) with

$$(3.5) \qquad |\underline{\underline{x}}|^{c_j'} \leq |L_j(\underline{\underline{x}})| < |\underline{\underline{x}}|^{c_j''} \qquad (j = 1,\ldots,n)$$

lie in finitely many subspaces. Now (1.5) is compatible with (3.5) only if $c_1' + \cdots + c_n' < -\delta$ , so that $c_1'' + \cdots + c_n'' < -\delta/2$ . Setting $c_j = c_j'' - n^{-1}(c_1'' + \cdots + c_n'')$ , we have (2.8) and $c_j'' < c_j - (\delta/2n)$ , whence

$$|L_j(\underline{\underline{x}})| < |\underline{\underline{x}}|^{c_j - (\delta/2n)} \qquad (j = 1,\ldots,n) ,$$

and by Lemma 3B the solutions to this system lie in a finite number of subspaces.

We are finished with the proof of Theorem 1F , except that so far all our forms had real coefficients. To prove the general case, we use induction on the number of linear forms whose coefficients are not real. Suppose, say, that $L_n(\underline{\underline{x}})$ has some non-real coefficients, and that $L_n(\underline{\underline{x}}) = R(\underline{\underline{x}}) + i\, I(\underline{\underline{x}})$ where $R, I$ have real coefficients. If $L_1, \ldots, L_{n-1}, R$ are linearly independent, put $L_n' = R$ ; otherwise $L_1, \ldots, L_{n-1}, I$ are linearly independent, and set $L_n' = I$ . The forms $L_1, \ldots, L_{n-1}, L_n'$ again satisfy the conditions of Theorem 1F , and fewer of them are non-real. The truth of Theorem 1F is now evident.

The remainder of this chapter will be devoted to a proof of the Strong Subspace Theorem. We will imitate Ch. V as much as possible. The reader should observe that the basic ideas of the (perhaps somewhat intricate) proof are geometrical.

## §4. The Index of a Polynomial.

Suppose that $n$ and $m$ are positive integers. Let $\mathfrak{R}$ denote the ring of polynomials

$$P(X_{11},\dots,X_{1n};X_{21},\dots,X_{2n};\dots;X_{m1},\dots,X_{mn})$$

with real coefficients. Let $L_1,\dots,L_m$ be linear forms, none identically 0 , where $L_h$ is a linear form in the variables $X_{h1},\dots,X_{hn}$ : thus,

$$L_h = \alpha_{h1}X_{h1}+\dots+\alpha_{hn}X_{hn} \qquad (1 \le h \le m) ,$$

where for every $h$ at least one $\alpha_{hk}$ is different from $0$. Let $r_1,\dots,r_m$ be positive integers. For each $c \geq 0$, let $I(c)$ be the ideal in $\mathfrak{R}$ generated by the polynomials $L_1^{i_1} L_2^{i_2} \cdots L_m^{i_m}$ with

$$\frac{i_1}{r_1} + \frac{i_2}{r_2} + \cdots + \frac{i_m}{r_m} \geq c \quad .$$

Clearly, $I(0) = \mathfrak{R}$. Also, as $c$ increases, $I(c)$ remains constant or decreases.

Definition. Suppose that $P \in \mathfrak{R}$, $P \neq 0$. The index of $P$ with respect to $(L_1,\dots,L_m;r_1,\dots,r_m)$ is the largest number $c$ such that $P \in I(c)$. (The existence of such a $c$ is easily seen.) We define the index of the zero polynomial to be $+\infty$. We denote the index of $P$ by $\operatorname{Ind} P$.

LEMMA 4A. *Suppose that $L_1,\dots,L_m;r_1,\dots,r_m$ are given as above. With respect to these parameters, we have:*

(i) $\operatorname{Ind}(P+Q) \geq \min(\operatorname{Ind} P, \operatorname{Ind} Q)$,

(ii) $\operatorname{Ind}(PQ) = \operatorname{Ind} P + \operatorname{Ind} Q$.

Proof. (i). $P \in I(\operatorname{Ind} P)$ and $Q \in I(\operatorname{Ind} Q)$, so $P,Q \in I(\min(\operatorname{Ind}P,\operatorname{Ind}Q))$. Then $P + Q \in I(\min(\operatorname{Ind} P,\operatorname{Ind} Q))$, so

$$\operatorname{Ind}(P+Q) \geq \min(\operatorname{Ind} P, \operatorname{Ind} Q) \quad .$$

(ii). Put

$$L_h = \alpha_{h1}X_{h1} + \cdots + \alpha_{hn}X_{hn} \qquad (1 \leq h \leq m),$$

and assume without loss of generality that $\alpha_{h1} \neq 0$ for each $h$. Then every polynomial in $X_{h1},\dots,X_{hn}$ is a polynomial in $L_h,X_{h2},\dots,X_{hn}$.

We may write (uniquely)

$$P = \sum c(j_1, a_{12}, \ldots, a_{1n}; \ldots; j_m, a_{m2}, \ldots, a_{mn}) \cdot$$
(4.1)
$$L_1^{j_1} X_{12}^{a_{12}} \cdots X_{1n}^{a_{1n}} \cdots L_m^{j_m} X_{m2}^{a_{m2}} \cdots X_{mn}^{a_{mn}} \quad .$$

Put

$$\mu = \min\left(\frac{j_1}{r_1} + \frac{j_2}{r_2} + \ldots + \frac{j_m}{r_m}\right) \quad ,$$

where the minimum is taken over all m-tuples $j_1, j_2, \ldots, j_m$ of nonnegative integers such that $c(j_1, a_{12}, \ldots) \neq 0$ for some choice of the $a_{ij}$ -s .

We claim that $\mu = \text{Ind } P$ . Namely, it is clear that $P \in I(\mu)$ , whence that $\text{Ind } P \geq \mu$ . But since $P \in I(\text{Ind } P)$ , $P$ is a linear combination (with polynomial coefficients) of $L_1^{j_1} L_2^{j_2} \cdots L_m^{j_m}$ with

(4.2)
$$\frac{j_1}{r_1} + \frac{j_2}{r_2} + \ldots + \frac{j_m}{r_m} \geq \text{Ind } P \quad .$$

By the uniqueness of the representation (4.1) for $P$ , each term in (4.1) with nonzero coefficient satisfies (4.2) . Hence $\mu \geq \text{Ind } P$ , and the claim is true.

Now let $P_1$ and $P_2$ be the sum of those terms in (4.1) for which

$$\sum_{h=1}^{m} \frac{j_h}{r_h} = \text{Ind } P \quad \text{and} \quad \sum_{h=1}^{m} \frac{j_h}{r_h} > \text{Ind } P \quad ,$$

respectively. Similarly define $Q_1$ and $Q_2$ . Then

$$PQ = P_1Q_1 + P_1Q_2 + P_2Q_1 + P_2Q_2 \quad .$$

Here $P_1Q_1$ is non-zero and has index $\text{Ind } P + \text{Ind } Q$ , while each of $P_1Q_2$ , $P_2Q_1$ , $P_2Q_2$ has an index $> \text{Ind } P + \text{Ind } Q$ . It follows that $\text{Ind}(PQ) = \text{Ind } P + \text{Ind } Q$ .

Remark. Let $P(X_1,\dots,X_m)$ be a polynomial of degree $\le r_h$ in $X_h$ $(1 \le h \le m)$. Let

$$\hat{P}(X_{11},X_{12};\dots;X_{m1},X_{m2})$$

denote the polynomial obtained from $P$ by replacing $X_h^{i_h}$ by $X_{h1}^{i_h} X_{h2}^{r_h-i_h}$ $(0 \le i_h \le r_h \ ; \ 1 \le h \le m)$. Then $\hat{P}$ is homogeneous in $X_{h1}$, $X_{h2}$ of degree $r_h$ $(1 \le h \le m)$.

Now let $(\alpha_1,\dots,\alpha_m) \in \mathbb{R}^m$. By Taylor's formula,

$$(4.3) \qquad P(X_1,\dots,X_m) = \sum c(j_1,\dots,j_m)(X_1-\alpha_1)^{j_1} \cdots (X_m-\alpha_m)^{j_m} ,$$

where the sum is over all $m$-tuples of nonnegative integers $j_1,\dots,j_m$ with $0 \le j_h \le r_h$ $(1 \le h \le m)$. In the notation of Chapter V, §5,

$$c(j_1,\dots,j_m) = P_{j_1\dots j_m}(\alpha_1,\dots,\alpha_m) .$$

Therefore the Roth index of $P$ with respect to $(\alpha_1,\dots,\alpha_m;r_1,\dots,r_m)$ is

$$\mu = \min\left(\frac{j_1}{r_1} + \cdots + \frac{j_m}{r_m}\right) ,$$

where the minimum is taken over all $m$-tuples $j_1,\dots,j_m$ for which $c(j_1,\dots,j_m) \ne 0$ in (4.3).

We have

$$\begin{aligned}\hat{P}&(X_{11},X_{12};\dots;X_{m1},X_{m2})\\ &= \sum c(j_1,\dots,j_m)X_{12}^{r_1-j_1}\dots X_{m2}^{r_m-j_m}(X_{11}-\alpha_1X_{12})^{j_1}\dots(X_{m1}-\alpha_mX_{m2})^{j_m}\\ &= \sum c(j_1,\dots,j_m)L_1^{j_1}X_{12}^{r_1-j_1}L_2^{j_2}X_{22}^{r_2-j_2}\cdots L_m^{j_m}X_{m2}^{r_m-j_m} ,\end{aligned}$$

where

$$L_h = X_{h1} - \alpha_h X_{h2} \qquad (1 \le h \le m) .$$

By the characterization of the index of a polynomial established in the proof of part (ii) of Lemma 4A, the index of $\hat{P}$ with respect to $(L_1,\dots,L_m;r_1,\dots,r_m)$ also is $\mu$. Thus, the Roth index of $P$ with respect to $(\alpha_1,\dots,\alpha_m;r_1,\dots,r_m)$ equals the index of $\hat{P}$ with respect to $(L_1,\dots,L_m;r_1,\dots,r_m)$.

We now introduce the following notation. If $r_1,\dots,r_m$ is an $m$-tuple of positive integers, we write $\underline{\underline{r}} = (r_1,\dots,r_m)$. $\mathfrak{J}$ will denote an $nm$-tuple of nonnegative integers

$$(i_{11},\dots,i_{1n};i_{21},\dots,i_{2n};\dots;i_{m1},\dots,i_{mn}) .$$

Given $\underline{\underline{r}}$ and $\mathfrak{J}$, we put

$$(\mathfrak{J}/\underline{\underline{r}}) = \sum_{h=1}^{m} \frac{i_{h1}+i_{h2}+\dots+i_{hn}}{r_h} .$$

Finally, if $P \in \mathfrak{R}$, we put

$$P^{\mathfrak{J}} = \frac{1}{i_{11}!\dots i_{mn}!} \frac{\partial^{i_{11}+\dots+i_{mn}}}{\partial X_{11}^{i_{11}} \dots \partial X_{mn}^{i_{mn}}} P .$$

<u>LEMMA 4B</u>. <u>Suppose that</u> $L_1,\dots,L_m$ <u>are linear forms as above and that</u> $\underline{\underline{r}} = (r_1,\dots,r_m)$ <u>has positive integer components</u>. <u>Then for any</u> $P \in \mathfrak{R}$ <u>and any</u> $\mathfrak{J}$,

$$\operatorname{Ind} P^{\mathfrak{J}} \ge \operatorname{Ind} P - (\mathfrak{J}/\underline{\underline{r}}) .$$

<u>Further, if</u> $(\mathfrak{J}/\underline{\underline{r}}) < \operatorname{Ind} P$, <u>then</u> $P^{\mathfrak{J}} = 0$ <u>on the</u> $(nm-m)$-<u>dimensional subspace</u> $T$ <u>of</u> $\mathbb{R}^{mn}$ <u>defined by</u> $L_1 = L_2 = \dots = L_m = 0$.

<u>Proof</u>. The first assertion follows from the representation of $P$ in (4.1) and from the characterization of the index in terms of $\mu$.

If $(\mathfrak{J}/\underline{\underline{r}}) < \text{Ind } P$, then $\text{Ind } P^{\mathfrak{J}} > 0$ from what has already been shown. Let us write

$$(4.4) \qquad P^{\mathfrak{J}} = \sum \overset{\mathfrak{J}}{c}(j_1,a_{12},\ldots,a_{1n};\ldots;j_m,a_{m2},\ldots,a_{mn}) L_1^{j_1} X_{12}^{a_{12}} \ldots X_{1n}^{a_{1n}} \ldots L_m^{j_m} X_{m2}^{a_{m2}} \ldots X_{mn}^{a_{mn}} .$$

It was shown in the proof of part (ii) of Lemma 4A that

$$\text{Ind } P^{\mathfrak{J}} = \mu^{\mathfrak{J}} = \min\left(\frac{j_1}{r_1} + \cdots + \frac{j_m}{r_m}\right) ,$$

where the minimum is taken over all m-tuples $j_1,\ldots,j_m$ of nonnegative integers such that $\overset{\mathfrak{J}}{c}(j_1,a_{12},\ldots) \neq 0$ for some choice of the $a_{ij}$-s. Since $\text{Ind } P^{\mathfrak{J}} > 0$, each m-tuple $(j_1,\ldots,j_m)$ for which $c(j_1,a_{12},\ldots) \neq 0$ in (4.4) is different from $\underline{\underline{0}}$, and therefore each term in (4.4) is divisible by at least one of $L_1,\ldots,L_m$. Hence $P^{\mathfrak{J}}$ vanishes on T.

LEMMA 4C. *Suppose that $L_1,\ldots,L_m$ and $\underline{\underline{r}}$ are as in Lemma 4B. Let T be the subspace of $\mathbb{R}^{mn}$ defined by $L_1 = \cdots = L_m = 0$. Suppose that $P \in \mathfrak{R}$, $P \neq 0$. Then there is an $\mathfrak{J}$ with $(\mathfrak{J}/\underline{\underline{r}}) = \text{Ind } P$ such that $P^{\mathfrak{J}} \neq 0$ on T. Moreover, if in*

$$L_h = \alpha_{h1}X_{h1} + \cdots + \alpha_{hn}X_{hn} \qquad (1 \leq h \leq m)$$

*we have $\alpha_{h1} \neq 0$ for each h, then there is such an $\mathfrak{J}$ of the type*

$$\mathfrak{J} = (i_1,0,\ldots,0;i_2,0,\ldots,0;\ldots;i_m,0,\ldots,0) .$$

*Proof.* We may assume that $\alpha_{h1} \neq 0$ $(1 \leq h \leq m)$ by relabeling if necessary. Then P may be written (uniquely) as

$$P = \sum P(j_1,\ldots,j_m|X_{12},\ldots,X_{1n};\ldots;X_{m2},\ldots,X_{mn}) \cdot L_1^{j_1} L_2^{j_2} \ldots L_m^{j_m} ,$$

where, for each fixed $j_1,\ldots,j_m$, $P(j_1,\ldots,j_m|X_{12},\ldots,X_{mn})$ is a polynomial in the mn-m variables $X_{12},\ldots,X_{1n};\ldots;X_{m2},\ldots,X_{mn}$. From the definition of the index of $P$ with respect to $(L_1,\ldots,L_m;r_1,\ldots,r_m)$, we have that $P(j_1,\ldots,j_m|X_{12},\ldots,X_{mn}) = 0$ if

$$\frac{j_1}{r_1} + \cdots + \frac{j_m}{r_m} < \text{Ind } P \quad ,$$

but $P(\bar{j}_1,\ldots,\bar{j}_m|X_{12},\ldots,X_{mn}) \neq 0$ for some $\bar{j}_1,\ldots,\bar{j}_m$ with

$$\frac{\bar{j}_1}{r_1} + \cdots + \frac{\bar{j}_m}{r_m} = \text{Ind } P \quad .$$

It is clear that with

$$\mathfrak{J} = (\bar{j}_1,0,\ldots,0;\bar{j}_2,0,\ldots,0;\bar{j}_m,0,\ldots,0) \quad ,$$

we have $P^{\mathfrak{J}} \neq 0$ on $T$.

## §5. Some Auxiliary Lemmas.

If $P \in \mathfrak{R}$, $P = \sum c(j_{11},\ldots,j_{mn})X_{11}^{j_{11}}\cdots X_{mn}^{j_{mn}}$, put

$$\lceil P \rceil = \max_{j_{11},\ldots,j_{mn}} (|c(j_{11},\ldots,j_{mn})|) \quad .$$

LEMMA 5A. *Suppose* $P$ *has rational integer coefficients. Then so does* $P^{\mathfrak{J}}$. *Further suppose that* $P$ *is homogeneous in* $X_{h1},\ldots,X_{hn}$ *of degree* $r_h$ $(1 \le h \le m)$. *Then for any* $\mathfrak{J}$,

$$\lceil P^{\mathfrak{J}} \rceil \le 2^{r_1+\cdots+r_m} \lceil P \rceil \quad . \tag{5.1}$$

*Proof*. The first assertion follows as in Chapter V. It will suffice to prove the estimate (5.1) for monomials. Now

$$\left(X_{11}^{j_{11}}\cdots X_{mn}^{j_{mn}}\right)^{\mathfrak{J}} = \binom{j_{11}}{i_{11}}\cdots\binom{j_{mn}}{i_{mn}} X_{11}^{j_{11}-i_{11}}\cdots X_{mn}^{j_{mn}-i_{mn}} \quad .$$

Since $\binom{j_{11}}{i_{11}}\cdots\binom{j_{mn}}{i_{mn}} \le 2^{j_{11}+\cdots+j_{mn}} \le 2^{r_1+\cdots+r_m}$, everything follows.

We now suppose that $\alpha_1,\dots,\alpha_n$ are fixed algebraic integers. Denote the degree of the field $K = \mathbb{Q}(\alpha_1,\dots,\alpha_n)$ by $\Delta$, and let

$$\beta_1,\dots,\beta_\Delta$$

be an integer basis of $K$. There are relations

$$\beta_i\beta_j = \sum_{k=1}^{\Delta} c_{ijk}\beta_k \qquad (1 \le i,j \le \Delta)$$

with rational integer coefficients $c_{ijk}$. Put $A = \max_{i,j,k}(|c_{ijk}|)$. An element $\gamma \in K$ may uniquely be written as

$$\gamma = \sum_{j=1}^{\Delta} r_j\beta_j \quad \text{with rational } r_1,\dots,r_\Delta .$$

Put

$$\lceil\gamma\rceil = A\Delta^2 \max_j |r_j| .$$

Then for $\gamma,\delta \in K$,

$$\lceil\gamma+\delta\rceil \le \lceil\gamma\rceil + \lceil\delta\rceil . \tag{5.2}$$

If, say $\delta = \sum_{j=1}^{\Delta} s_j\beta_j$ with rational $s_1,\dots,s_\Delta$, then

$$\gamma\delta = \sum_i \sum_j r_i s_j \beta_i\beta_j = \sum_i \sum_j \sum_k r_i s_j c_{ijk}\beta_k .$$

Here the coefficient of $\beta_k$ is rational and has absolute value

$$\le A\Delta^2 \max(|r_i|)\max(|s_j|) = (A\Delta^2)^{-1}\lceil\gamma\rceil\lceil\delta\rceil ,$$

and this proves that

$$\lceil\gamma\delta\rceil \le \lceil\gamma\rceil\lceil\delta\rceil . \tag{5.3}$$

Put

$$B = \max(\overline{|\alpha_1|}, \dots, \overline{|\alpha_n|}) \quad ,$$

where $\alpha_1, \dots, \alpha_n$ were the given integers.

LEMMA 5B. *Suppose* $\alpha_1, \dots, \alpha_n$ *are algebraic integers as above. Consider the polynomials*

$$P(X_{11}, \dots, X_{1n}; \dots; X_{m1}, \dots, X_{mn}) = \sum c(j_{11}, \dots, j_{mn}) X_{11}^{j_{11}} \dots X_{mn}^{j_{mn}}$$

*in* $\mathfrak{R}$ *which are homogeneous of degree* $r_h$ *in the variables* $X_{h1}, \dots, X_{hn}$ $(1 \le h \le m)$. *We construct new polynomials* $P^*$ *in the* $nm - m$ *variables*

$$X_{12}, \dots, X_{1n}; \dots; X_{m2}, \dots, X_{mn}$$

*by putting*

$$P^* = P^{\mathfrak{J}}(-\alpha_2 X_{12} - \dots - \alpha_n X_{1n}, \alpha_1 X_{12}, \dots, \alpha_1 X_{1n}; \dots$$

$$; -\alpha_2 X_{m2} - \dots - \alpha_n X_{mn}, \alpha_1 X_{m2}, \dots, \alpha_1 X_{mn}).$$

*where* $\mathfrak{J}$ *is of the type*

$$\mathfrak{J} = (j_1, 0, \dots, 0; \dots; j_m, 0, \dots, 0) \quad .$$

*Then every coefficient of* $P^*$ *is a linear form in the coefficients* $c(j_{11}, \dots, j_{mn})$ *of* $P$. *The coefficients* $\gamma$ *of these linear forms are integers of the field* $K = Q(\alpha_1, \dots, \alpha_n)$ *with*

$$\overline{|\gamma|} \le (4^n B)^{r_1 + \dots + r_m} \quad .$$

*Proof*. Only the upper bound for $\overline{|\gamma|}$ is to be proved. By Lemma 4B of Chapter V , $P^*$ is a sum of at most

$$\binom{r_1+n-1}{n-1} \dots \binom{r_m+n-1}{n-1} \le 2^{(r_1+n-1)+\dots+(r_m+n-1)} \le 2^{r_1 n + \dots + r_m n} = 2^{n(r_1+\dots+r_m)}$$

summands $S$ of the type

$$S = \pm c(\dots)\binom{j_{11}}{j_1}\dots\binom{j_{m1}}{j_m}\left(\alpha_2 X_{12}+\dots+\alpha_n X_{1n}\right)^{j_{11}-j_1}(\alpha_1 X_{12})^{j_{12}}\dots(\alpha_1 X_{1n})^{j_{1n}}$$

$$\dots(\alpha_2 X_{m2}+\dots+\alpha_n X_{mn})^{j_{m1}-j_m}(\alpha_1 X_{m2})^{j_{m2}}\dots(\alpha_1 X_{mn})^{j_{mn}}$$

$$= \pm c(\dots)\binom{j_{11}}{j_1}\dots\binom{j_{m1}}{j_m} S_1 S_2 \dots S_m \quad ,$$

where for $1 \le h \le m$ ,

$$S_h = (\alpha_2 X_{h2}+\dots+\alpha_n X_{hn})^{j_{h1}-j_h}(\alpha_1 X_{h2})^{j_{h2}}\dots(\alpha_1 X_{hn})^{j_{hn}}$$

$$= \sum_{\substack{c_2,\dots,c_n \\ c_2+\dots+c_n = j_{h1}-j_h}} \frac{(j_{h1}-j_h)!}{c_2!\dots c_n!}\alpha_2^{c_2}\dots\alpha_n^{c_n}\alpha_1^{j_{h2}+\dots+j_{hn}} X_{h2}^{c_2+j_{h2}}\dots X_{hn}^{c_n+j_{hn}} \quad .$$

Here the sum of the numerical coefficients in front is $(n-1)^{j_{h1}-j_h} < n^{r_h}$ , and we note that

$$\left|\overline{\alpha_2^{c_2}\dots\alpha_n^{c_n}\alpha_1^{j_{h2}+\dots+j_{hn}}}\right| \le B^{c_2+\dots+c_n+j_{h2}+\dots+j_{hn}} = B^{r_h-j_h} \le B^{r_h} \quad .$$

Thus every coefficient $\alpha$ of $S_h$ is an integer of $K$ with $\left|\overline{\alpha}\right| \le (nB)^{r_h}$ . Thus every coefficient of $S$ is of the type $c(\dots)\beta$ where $\beta \in K$ and

$$\left|\overline{\beta}\right| \le \binom{j_{11}}{j_1}\dots\binom{j_{m1}}{j_m}(nB)^{r_1+\dots+r_m} \le (2nB)^{r_1+\dots+r_m} \quad .$$

Now since, as pointed out above, $P^*$ is a sum of at most $2^{(r_1+\dots+r_m)n}$ polynomials of the type $S$ , each coefficient of $P^*$ is a linear form in the $c(\dots)$ with coefficients $\gamma$ having

$$\left|\overline{\gamma}\right| \le 2^{n(r_1+\dots+r_m)}(2nB)^{r_1+\dots+r_m} \le (2^{2n}B)^{r_1+\dots+r_m} \quad .$$

§6. The Index Theorem.

Let $L = \alpha_1 X_1 + \cdots + \alpha_n X_n$ be a non-zero linear form in $X_1, \ldots, X_n$. We make $m$ polynomials out of it by putting $L_h = \alpha_1 X_{h1} + \cdots + \alpha_n X_{hn}$ $(1 \le h \le m)$. When we speak of the index of a polynomial $P \in \mathfrak{R}$ with respect to

$$(L, L, \ldots, L; r_1, \ldots, r_m) ,$$

we really mean the index of $P$ with respect to $(L_1, \ldots, L_m; r_1, \ldots, r_m)$.

[This is like considering in Chapter V the index of a polynomial with respect to $(\alpha, \ldots, \alpha; r_1, \ldots, r_m)$.]

THEOREM 6A. (Index Theorem). *Let $n, t$ be positive integers and $L^{(1)}, \ldots, L^{(t)}$ linear forms in $X_1, \ldots, X_n$, none identically zero. Suppose that*

$$L^{(i)} = \alpha_{i1} X_1 + \cdots + \alpha_{in} X_n \qquad (1 \le i \le t),$$

*where the $\alpha_{ij}$ are algebraic integers. We put $\Delta_i$ for the degree of the field $K_i = \mathbb{Q}(\alpha_{i1}, \ldots, \alpha_{in})$, and $\Delta = \max(\Delta_1, \ldots, \Delta_t)$.*

*Suppose $\varepsilon > 0$, and $m$ is an integer so large that*

$$m \ge 4\varepsilon^{-2} \log(2t\Delta) . \tag{6.1}$$

*Let $r_1, \ldots, r_m$ be positive integers.*

*Then there exists a non-zero polynomial $P \in \mathfrak{R}$ with rational integer coefficients such that*

(i) *$P$ is homogeneous in $X_{h1}, \ldots, X_{hn}$ of degree $r_h$* $(1 \le h \le m)$,

(ii) *$P$ has index $\ge (\frac{1}{n} - \varepsilon)m$ with respect to*

$$(L^{(i)}, \ldots, L^{(i)}; r_1, \ldots, r_m) \qquad (1 \le i \le t),$$

(iii) $\overline{|P|} \le D^{r_1+\cdots+r_m}$

with a constant $D$ depending only on the $\alpha_{ij}$-s .

Proof. We write $P$ in the form

$$P = \sum_{\substack{j_{h1}+\cdots+j_{hn}=r_h \\ (1\le h\le m)}} c(j_{11},\ldots,j_{mn})X_{11}^{j_{11}}\ldots X_{mn}^{j_{mn}} \quad ,$$

and we shall check what the conditions (ii) imply for the coefficients $c(j_{11},\ldots,j_{mn})$ . What for instance does it mean that the index with respect to $(L^{(1)},\ldots,L^{(1)};r_1,\ldots,r_m)$ should be $\ge (\frac{1}{n}-\varepsilon)m$ ?

Without loss of generality we may assume that $\alpha_{11} \ne 0$ . By Lemma 4C , the index satisfies the desired inequality if each polynomial $P^{\mathfrak{J}}$ , where

$$\mathfrak{J} = (i_1,0,\ldots,0;\ldots;i_m,0,\ldots,0)$$

and where

$$\sum_{h=1}^{m} \frac{i_h}{r_h} < (\frac{1}{n}-\varepsilon)m \quad \text{and} \quad 0 \le i_h \le r_h ,$$

vanishes on the subspace $T$ defined by $L_1^{(1)} = L_2^{(1)} = \cdots = L_m^{(1)} = 0$ . (For the definition of $L_1^{(1)},\ldots,L_m^{(1)}$ see the beginning of this section.)

Keep $i_1,\ldots,i_m$ fixed at the moment. Our condition then is that

$$P^* = P^{\mathfrak{J}}(-\alpha_{12}X_{12}-\cdots-\alpha_{1n}X_{1n},\alpha_{11}X_{12},\ldots,\alpha_{11}X_{1n};\ldots$$
$$;-\alpha_{12}X_{m2}-\cdots-\alpha_{1n}X_{mn},\alpha_{11}X_{m2},\ldots,\alpha_{11}X_{mn})$$

is identically zero. In general $P^*$ is homogeneous in $X_{h2},\ldots,X_{hn}$ of degree $r_h - i_h$ $(1 \le h \le m)$ , hence has at most

$$\binom{r_1-i_1+n-2}{n-2}\cdots\binom{r_m-i_m+n-2}{n-2} = f_1(i_1)\ldots f_m(i_m)$$

coefficients. By Lemma 5B, each coefficient is a linear form in the $c(\dots)$, with the coefficients of these linear forms being integers of $K_1 = \mathbb{Q}(\alpha_{11},\dots,\alpha_{1n})$. We thus obtain

$$f_1(i_1)\dots f_m(i_m)$$

linear homogeneous equations in the numbers $c(\dots)$, and the coefficients of these linear equations are algebraic integers $\gamma$ of $K$ with $\overline{|\gamma|} \le (4^n B_1)^{r_1+\dots+r_m}$. Here $B_1$ is the constant $B$ of the last section with respect to $\alpha_{11},\dots,\alpha_{1n}$. Each of these linear equations follows from $\Delta_1$ linear homogeneous equations with rational integer coefficients $k$ having

$$|k| \le (4^n B_1)^{r_1+\dots+r_m} . \tag{6.2}$$

By taking the sum over all m-tuples $i_1,\dots,i_m$ with $\sum_{h=1}^{m} \frac{i_h}{r_h} < (\frac{1}{n} - \varepsilon)m$, one sees: In order that (ii) is true with $(L^{(1)},\dots,L^{(1)};r_1,\dots,r_m)$, the $c(\dots)$ have to satisfy at most

$$\Delta_1 \Sigma f_1(i_1)\dots f_m(i_m) \tag{6.3}$$

linear homogeneous equations with rational integer coefficients $k$ satisfying (6.2), where the sume in (6.3) is extended over all $i_1,\dots,i_m$ with

$$\sum_{h=1}^{m} \frac{i_h}{r_h} < (\frac{1}{n} - \varepsilon)m \quad \text{and} \quad 0 \le i_h \le r_h .$$

Now by the formula (4.4) in Chapter V the expression in (6.3) equals

$$\Delta_1 M_- ,$$

since $f_j(i_j) = \binom{r_j - i_j + n-2}{n-2}$ is the number of (n-1)-tuples $i_{j2},\dots,i_{jn}$ with $i_{j2}+\dots+i_{jn} = r_j - i_j$. By (4.3) of Chapter V we have

$$M_- \le \binom{r_1+n-1}{r_1}\cdots\binom{r_m+n-1}{r_m} e^{-\varepsilon^2 m/4} ,$$

and the total number of equations to satisfy (ii) with $i = 1$ is

$$\le \Delta_1 \binom{r_1+n-1}{r_1}\cdots\binom{r_m+n-1}{r_m}\frac{1}{2t\Delta} \qquad \text{by (6.1)} ,$$

and hence is

$$\le \frac{N}{2t}$$

where

$$N = \binom{r_1+n-1}{r_1}\cdots\binom{r_m+n-1}{r_m} .$$

Thus the total number $M$ of homogeneous linear equations to be satisfied by the $N$ unknowns $c(j_{11},\dots,j_{mn})$ to obtain (ii) for $i = 1,\dots,t$ satisfies

$$M \le \frac{N}{2} .$$

The absolute values of the rational integer coefficients of these $M$ equations are at most

$$A = (4^n B)^{r_1+\dots+r_m} , \quad \text{where } B = \max(B_1,\dots,B_t) .$$

By Lemma 5B of Chapter V, this system of homogeneous linear equations has rational integer solutions $c(\dots)$, not all zero, with

$$|c(\dots)| \le (NA)^{M/(N-M)} \le NA = \binom{r_1+n-1}{r_1}\cdots\binom{r_m+n-1}{r_m}A$$

$$\le 2^{n(r_1+\dots+r_m)}A = (8^n B)^{r_1+\dots+r_m} = D^{r_1+\dots+r_m} , \text{ say.}$$

So we obtain a polynomial $P \neq 0$ with (i), (ii) and (iii).

## §7. The Polynomial Theorem.

THEOREM 7A. (The Polynomial Theorem). *Let*

$$L^{(i)} = \alpha_{i1}X_1 + \cdots + \alpha_{in}X_n \qquad (1 \le i \le n)$$

*be* $n$ *linear forms with algebraic integer coefficients* $\alpha_{ij}$ *of non-vanishing determinant. Put*

$$L_h^{(i)} = \alpha_{i1}X_{h1} + \cdots + \alpha_{in}X_{hn} \qquad (1 \le h \le m,\ 1 \le i \le n)$$

*Suppose that* $\varepsilon > 0$ *and that* $m$ *satisfies*

$$m > 4\varepsilon^{-2}\log(2n\Delta) \quad , \tag{7.1}$$

*where* $\Delta$ *is defined as in the Index Theorem* (Theorem 6A). *Also, let* $r_1, r_2, \ldots, r_m$ *be positive integers.*

*Then there is a* $P \in \Re$, $P \neq 0$, *with rational integer coefficients, which has the following properties:*

(i) $P$ *is homogeneous in* $X_{h1}, \ldots, X_{hn}$ *of degree* $r_h$ $(1 \le h \le m)$.

(ii) $\overline{|P|} \le D^{r_1 + \cdots + r_m}$, *where* $D$ *depends only on the* $\alpha_{ij}$-s.

(iii) *If we write* (*uniquely*)

$$P^{\mathfrak{J}} = \sum d^{\mathfrak{J}}(j_{11}, \ldots, j_{mn}) L_1^{(1)j_{11}} \cdots L_1^{(n)j_{1n}} \cdots L_m^{(1)j_{m1}} \cdots L_m^{(n)j_{mn}} \quad , \tag{7.2}$$

*then*

$$|d^{\mathfrak{J}}(j_{11}, \ldots, j_{mn})| \le E^{r_1 + \cdots + r_m}$$

*for any* $\mathfrak{J}$ *and any* $(j_{11}, \ldots, j_{mn})$, *where* $E$ *depends only on the* $\alpha_{ij}$-s.

(iv) *If* $(\mathfrak{J}/\underline{\underline{r}}) \le 2\varepsilon m$, *then* $d^{\mathfrak{J}}(j_{11}, \ldots, j_{mn}) = 0$ *in* (7.2), *unless*

$$\left|\left(\sum_{h=1}^{m} \frac{j_{hk}}{r_h}\right) - \frac{m}{n}\right| \leq 3nm\varepsilon \qquad (1 \leq k \leq n) .$$

Proof. We may apply the Index Theorem with $t = n$. Let $P \neq 0$ be a polynomial in $\mathfrak{R}$ with rational integer coefficients satisfying the conditions (i), (ii), (iii) of the Index Theorem. Then the new conditions (i) and (ii) hold. We shall show that also (iii) and (iv) hold.

Verification of (iii). Since $\det(L^{(1)},\dots,L^{(n)}) \neq 0$, one can go from $X_1,\dots,X_n$ to $L^{(1)},\dots,L^{(n)}$ by a nonsingular linear transformation. Thus, every polynomial in $X_1,\dots,X_n$ is a polynomial in $L^{(1)},\dots,L^{(n)}$, and therefore every polynomial in $X_{h1},\dots,X_{hn}$ is a polynomial in $L_h^{(1)},\dots,L_h^{(n)}$ $(1 \leq h \leq m)$. This makes it clear that $P^{\mathfrak{J}}$ may be uniquely written as (7.2).

If $(\beta_{ij})$ is the inverse of the matrix $(\alpha_{ij})$, then

$$X_i = \sum_{k=1}^{n} \beta_{ik} L^{(k)} , \qquad (1 \leq i \leq n)$$

whence

$$\text{(7.3)} \qquad X_{hi} = \sum_{k=1}^{n} \beta_{ik} L_h^{(k)} \qquad (1 \leq h \leq m,\ 1 \leq i \leq n) .$$

Let us write

$$\text{(7.4)} \qquad P^{\mathfrak{J}} = \sum c^{\mathfrak{J}}(j_{11},\dots,j_{mn}) X_{11}^{j_{11}} \cdots X_{mn}^{j_{mn}} .$$

Now $\lceil P^{\mathfrak{J}} \rceil \leq 2^{r_1+\dots+r_m} \lceil P \rceil \leq (2D)^{r_1+\dots+r_m}$ by (ii), so

$$|c^{\mathfrak{J}}(j_{11},\dots,j_{mn})| \leq (2D)^{r_1+\dots+r_m}$$

for any $\mathfrak{J}$ and any $(j_{11},\dots,j_{mn})$.

One gets from (7.4) to (7.2) by substituting the right hand side of (7.3) for each $X_{hi}$. A typical product $X_{11}^{j_{11}} \cdots X_{mn}^{j_{mn}}$ in (7.4) then becomes

$$\left(\sum_{k=1}^{n} \beta_{1k}L_1^{(k)}\right)^{j_{11}} \cdots \left(\sum_{k=1}^{n} \beta_{nk}L_m^{(k)}\right)^{j_{mn}} . \tag{7.5}$$

We put $G = \max(1,|\beta_{11}|,\dots,|\beta_{mn}|)$ . Then (7.5) is a polynomial in $L_h^{(i)}$ $(1 \le h \le m,\ 1 \le i \le n)$ , whose coefficients have absolute values

$$\le (nG)^{j_{11}+\cdots+j_{mn}} \le (nG)^{r_1+\cdots+r_m} .$$

So $c^{\mathfrak{J}}(j_{11},\dots,j_{mn})X_{11}^{j_{11}}\cdots X_{mn}^{j_{mn}}$ as a polynomial in the $L_h^{(i)}$ has coefficients with absolute value

$$\le (2DnG)^{r_1+\cdots+r_m} .$$

By Lemma 4B of Chapter V , the total number of summands in (7.4) is

$$\le \binom{r_1+n-1}{r_1}\cdots\binom{r_m+n-1}{r_m} \le 2^{(r_1+n-1)+\cdots+(r_m+n-1)}$$

$$\le 2^{r_1n+\cdots+r_mn} = 2^{n(r_1+\cdots+r_m)} .$$

Hence $P^{\mathfrak{J}}$ as a polynomial in $L_h^{(i)}$ $(1 \le h \le m,\ 1 \le i \le n)$ has coefficients of absolute value

$$\le (2^n \cdot 2DnG)^{r_1+\cdots+r_m} = E^{r_1+\cdots+r_m} , \text{ say.}$$

Verification of (iv). By the Index Theorem, the index of $P$ with respect to

$$(L_1^{(k)},\dots,L_m^{(k)};r_1,\dots,r_m) \text{ is } \ge \left(\frac{1}{n}-\varepsilon\right)m \qquad (1 \le k \le n) .$$

Now if $(\mathfrak{J}/\underline{r}) \le 2\varepsilon m$ , then it follows from Lemma 4B that $P^{\mathfrak{J}}$ has index $\ge \left(\frac{1}{n}-\varepsilon\right)m - 2\varepsilon m = \left(\frac{1}{n}-3\varepsilon\right)m$ with respect to

$$(L_1^{(k)},L_2^{(k)},\dots,L_m^{(k)};r_1,r_2,\dots,r_m) \qquad (1 \le k \le n) .$$

Therefore if $d^{\mathfrak{I}}(j_{11},\ldots,j_{mn}) \neq 0$ in (7.2), then

$$\sum_{h=1}^{m} \frac{j_{hk}}{r_h} \geq \left(\frac{1}{n} - 3\varepsilon\right)m \qquad (1 \leq k \leq n) .$$

This inequality may be rewritten as

$$(7.6) \qquad \left(\sum_{h=1}^{m} \frac{j_{hk}}{r_h}\right) - \frac{m}{n} \geq -3\varepsilon m \qquad (1 \leq k \leq n),$$

which gives half of the desired inequality.

P is homogeneous in $X_{h1},\ldots,X_{hn}$ of degree $r_h$, hence $P^{\mathfrak{I}}$ is of total degree $\leq r_h$ in the block of variables $X_{h1},\ldots,X_{hn}$ $(1 \leq h \leq m)$. Therefore

$$\sum_{k=1}^{n} \frac{j_{hk}}{r_h} \leq 1 \qquad (1 \leq h \leq m),$$

whence

$$\sum_{k=1}^{n} \sum_{h=1}^{m} \frac{j_{hk}}{r_h} \leq m \quad ,$$

or

$$\sum_{k=1}^{n}\left(\left(\sum_{h=1}^{m} \frac{j_{hk}}{r_h}\right) - \frac{m}{n}\right) \leq 0 \quad .$$

Together with (7.6) this gives

$$\left(\sum_{h=1}^{m} \frac{j_{hk}}{r_h}\right) - \frac{m}{n} \leq 3m\varepsilon(n-1) \qquad (1 \leq k \leq n) \quad .$$

Thus the other half of the desired inequality in (iv) is established, and the theorem is proved.

## §8. Grids.

Let $n$ be an integer with $n \geq 2$. Put $q = n - 1$, and let $H^q$ be a linear subspace of $\mathbb{R}^n$ of dimension $q$.

Definition. Let $\underline{\underline{w}}_1,\ldots,\underline{\underline{w}}_q$ be $q$ linearly independent points of $H^q$, and let $s$ be a positive integer. Denote by

$$\Gamma = \Gamma(s;\underline{\underline{w}}_1,\ldots,\underline{\underline{w}}_q)$$

the set of all points $\underline{\underline{w}} = h_1\underline{\underline{w}}_1+\cdots+h_q\underline{\underline{w}}_q$ where $h_1,\ldots,h_q$ are integers with $1 \le h_i \le s$ $(1 \le i \le q)$. (Thus $\Gamma$ consists of $s^q$ points.) We shall say that $\Gamma$ is a grid on $H^q$ of size $s$.

In what follows, we will study polynomials $P(X_1,\ldots,X_n)$ which will be regarded as functions on $\mathbb{R}^n$.

LEMMA 8A. *Let $P(X_1,\ldots,X_n)$ be a polynomial with real coefficients, and of total degree $\le r$. Let $s$ be a positive integer, and let $\Gamma$ be a grid of size $s$ on some subspace $H^q$ of $\mathbb{R}^n$. Finally let $t$ be an integer with*

$$s(t+1) > r \quad , \tag{8.1}$$

*and assume that $P$ and the partial derivatives*

$$\frac{\partial^{t_1+\cdots+t_n}}{\partial X_1^{t_1}\cdots\partial X_n^{t_n}} P \quad \text{with} \quad t_1+\cdots+t_n \le t$$

*vanish on $\Gamma$.*

*Then $P$ vanishes identically on $H^q$.*

Proof. After a linear transformation, we may assume that the grid $\Gamma$ is spanned by $\underline{\underline{w}}_1 = (1,0,\ldots,0),\ldots,\underline{\underline{w}}_q = (0,\ldots,0,1,0)$, and hence that $H^q$ is the coordinate plane $x_n = 0$.

Put $Q(X_1,\ldots,X_q) = P(X_1,\ldots,X_q,0)$. We have to show that *a polynomial $Q(X_1,\ldots,X_q)$ of total degree $\le r$ is identically zero if $Q$ and its partial derivatives of order $\le t$ vanish on the $s^q$ integer* points $(h_1,\ldots,h_q)$ where $1 \le h_i \le s$ $(1 \le i \le q)$. We shall prove this by induction on $q$.

Suppose that $q = 1$. We have now a polynomial $Q(X_1)$ in a single variable. If $Q, Q', Q'', \ldots, Q^{(t)}$ each vanish at $X_1 = 1,2,\ldots,s$, then the number of zeros of $Q$, counted with multiplicities, is $\ge s(t+1)$. Since $Q$ is of degree $\le r < s(t+1)$, we have $Q = 0$.

Assume now that the above statement is true for $q-1$ in place of $q$, and that $q \ge 2$. To prove the statement with $q$, it will suffice to show that $(X_1 - h)^{t+1}$ divides $Q$ for $h = 1,2,\ldots,s$. [For then the polynomial $K = (X_1 - 1)^{t+1}(X_1 - 2)^{t+1}\ldots(X_1 - s)^{t+1}$ divides $Q$. But $K$ has degree $s(t+1) > r \ge \deg Q$, whence $Q = 0$.] To this end, let $e_h$ be the largest exponent so that $(X_1 - h)^{e_h}$ divides $Q$ $(1 \le h \le s)$. To complete the proof it is enough to show that $e \ge t+1$, where $e = \min(e_1,\ldots,e_s)$.

Suppose, to the contrary, that $e \le t$. Suppose, say, that $e = e_1 \le t$. We write

$$(8.2) \qquad Q(X_1,\ldots,X_q) = (X_1 - 1)^{e_1}(X_1 - 2)^{e_2}\ldots(X_1 - s)^{e_s}R(X_1,\ldots,X_q) \quad .$$

Here $R$ has total degree $\le r - e_1 - e_2 - \cdots - e_s \le r - se$. Take the $e = e_1$ - th derivative with respect to $X_1$, then put $X_1 = 1$. The right hand side of (8.2) becomes

$$e!(1-2)^{e_2}\ldots(1-s)^{e_s}R(1,X_2,\ldots,X_q) \quad .$$

$R(1,X_2,\ldots,X_q)$ is a polynomial in $q-1$ variables of total degree $\le r - se$. Because $Q$ and its partial derivatives of order $\le t$ vanish on the $s^q$ integer points $(h_1,h_2,\ldots,h_q)$, where $1 \le h_i \le s$ $(1 \le i \le q)$, the polynomial $R(1,X_2,\ldots,X_q)$ and its partial derivatives of order $\le t-e$ vanish on the $s^{q-1}$ integer points $(h_2,\ldots,h_q)$ where $1 \le h_i \le s$. Also, inequality (8.1) gives

$$s(t-e+1) > r-se \quad .$$

By the inductive hypothesis, we conclude that $R(1,X_2,\ldots,X_q)=0$. But then $(X_1-1)$ divides $R(1,X_2,\ldots,X_q)$, and therefore $(X_1-1)^{e_1+1}$ divides $Q(X_1,X_2,\ldots,X_q)$. Since this contradicts the definition of $e_1$, we conclude that $e \geq t+1$.

LEMMA 8B. *Let $P$ be a polynomial in $\mathfrak{R}$ of total degree $\leq r_h$ in the block of variables $X_{h1},\ldots,X_{hn}$ $(1 \leq h \leq m)$. We may write $P = P(\mathfrak{X}_1,\ldots,\mathfrak{X}_m)$, where $\mathfrak{X}_h \in \mathbb{R}^n$ $(1 \leq h \leq m)$. Thus $P$ may be regarded as a function on the product space*

$$\underbrace{\mathbb{R}^n \times \mathbb{R}^n \times\cdots\times \mathbb{R}^n}_{m \text{ factors}} \quad .$$

*Let $H_1^q,\ldots,H_m^q$ be subspaces of $\mathbb{R}^n$, each of dimension $q=n-1$, and let $T = H_1^q \times H_2^q \times\cdots\times H_m^q$ consist of all points $(\mathfrak{X}_1,\ldots,\mathfrak{X}_m)$ with $\mathfrak{X}_h \in H_h^q$ $(1 \leq h \leq m)$. (Thus, $T$ is a subspace of $\mathbb{R}^{nm}$ of dimension $mn-m$.) For $h=1,2,\ldots,m$, let $\Gamma_h$ be a grid on $H_h^q$ of size $s_h$, and let $t_1,t_2,\ldots,t_m$ be integers with*

$$s_h(t_h+1) > r_h \qquad (1 \leq h \leq m) \quad . \tag{8.3}$$

*In an obvious notation, let $\Gamma^* = \Gamma_1 \times \Gamma_2 \times\cdots\times \Gamma_m$. Finally, suppose that $P$ and all polynomials $P^{\mathfrak{T}}$ with $\mathfrak{T} = (t_{11},\ldots,t_{1n};\ldots;t_{m1},\ldots,t_{mn})$ satisfying*

$$t_{h1}+t_{h2}+\cdots+t_{hn} \leq t_h \qquad (1 \leq h \leq m)$$

*vanish on $\Gamma^*$.*

Then $P = 0$ on $T$ .

Proof. Apply Lemma 8A and induction on $m$ .

## §9. The Index of $P$ with Respect to certain Rational Linear Forms.

Suppose that $n \geq 2$ is an integer. Put

$$q = n-1 \quad . \tag{9.1}$$

Suppose that $\underline{\underline{w}}_1,\ldots,\underline{\underline{w}}_q$ are linearly independent integer points in $\mathbb{R}^n$ . Then there exists a linear form

$$M(\underline{\underline{X}}) = m_1X_1 + \cdots + m_nX_n \quad ,$$

not identically zero, with coprime rational integer coefficients and such that

$$M(\underline{\underline{w}}_t) = 0 \qquad (1 \leq t \leq q) .$$

It is clear that $M$ is unique up to a factor $\pm 1$ . We shall put

$$M = M\{\underline{\underline{w}}_1,\ldots,\underline{\underline{w}}_q\} \quad .$$

THEOREM 9A. (Corresponds to Theorem 8A of Chapter V.) Let $c_1,\ldots,c_n$ be real numbers with

$$c_1 + \cdots + c_n = 0 \quad \text{and} \quad |c_i| \leq 1 \qquad (1 \leq i \leq n) . \tag{9.2}$$

Suppose that $\varepsilon > 0$ , $0 < \delta < 1$ , and

$$16n^2\varepsilon < \delta \quad . \tag{9.3}$$

Let $L^{(1)},\ldots,L^{(n)}$ be linear forms and $m;r_1,\ldots,r_m$ positive integers satisfying the hypotheses of Theorem 7A . Let $E$ be the constant of *part (iii) of Theorem 7A, and let P be the polynomial described in that theorem.*

*Let $Q_1, \ldots, Q_m$ be real numbers with*

$$Q_h^{\varepsilon} > 2^n E \, , \quad Q_h^{\varepsilon} > n(\varepsilon^{-1}+1) \qquad (1 \le h \le m) \, , \tag{9.4}$$

$$r_1 \log Q_1 \le r_h \log Q_h \le (1+\varepsilon) r_1 \log Q_1 \qquad (1 \le h \le m) \, . \tag{9.5}$$

*For each h, $1 \le h \le m$, let $\underline{\underline{g}}_{h1}, \ldots, \underline{\underline{g}}_{hq}$ be q linearly independent integer points in $\mathbb{R}^n$ with*

$$|L^{(k)}(\underline{\underline{g}}_{ht})| \le Q_h^{c_k - \delta} \qquad (1 \le k \le n \, , \, 1 \le h \le m \, , \, 1 \le t \le q) \, . \tag{9.6}$$

*Then P has index $\ge m\varepsilon$ with respect to $(M_1, \ldots, M_m; r_1, \ldots, r_m)$, where each $M_h$ is the linear form in $X_{h1}, \ldots, X_{hn}$ given by $M_h = M_h\{\underline{\underline{g}}_{h1}, \ldots, \underline{\underline{g}}_{hq}\}$.*

*Proof.* Let T be the subspace of $\mathbb{R}^{mn}$ where $M_1 = M_2 = \cdots = M_m = 0$. By Lemma 4C, it will suffice to show that $P^{\mathfrak{J}} = 0$ on T whenever $(\mathfrak{J}/\underline{\underline{r}}) < \varepsilon m$. Let $\Gamma_h$ be the grid

$$\Gamma_h = \Gamma([\varepsilon^{-1}]+1; \underline{\underline{g}}_{h1}, \ldots, \underline{\underline{g}}_{hq}) \qquad (1 \le h \le m) \, .$$

By Lemma 8B, it will suffice to show that

$$(P^{\mathfrak{J}})^{\mathfrak{T}}(\underline{\underline{v}}_1, \ldots, \underline{\underline{v}}_m) = 0$$

for $\underline{\underline{v}}_h \in \Gamma_h$ $(1 \le h \le m)$ and $\mathfrak{T} = (t_{11}, \ldots, t_{1n}; \ldots; t_{m1}, \ldots, t_{mn})$ satisfying

$$t_{h1} + \cdots + t_{hn} \le [r_h \varepsilon] \qquad (1 \le h \le m) \, .$$

Namely, $s_h = [\varepsilon^{-1}]+1$, $t_h = [r_h \varepsilon]$ satisfy $s_h(t_h+1) > \varepsilon^{-1} r_h \varepsilon = r_h$, i.e.

(8.3). Now $(\mathfrak{J}/\underline{\underline{r}}) < \varepsilon m$ and $(\mathfrak{I}/\underline{\underline{r}}) \le \frac{[r_1\varepsilon]}{r_1} + \dots + \frac{[r_m\varepsilon]}{r_m} \le \varepsilon m$, so $(P^{\mathfrak{J}})^{\mathfrak{I}}$ (except for a constant factor) is of type $P^{\mathfrak{J}'}$ with $(\mathfrak{J}'/\underline{\underline{r}}) < 2\varepsilon m$.

Hence it will suffice to show that

$$P^{\mathfrak{J}}(\underline{\underline{v}}_1, \dots, \underline{\underline{v}}_m) = 0 \tag{9.7}$$

when $(\mathfrak{J}/\underline{\underline{r}}) < 2\varepsilon m$ and $\underline{\underline{v}}_h \in \Gamma_h$ $(1 \le h \le m)$.

The left hand side of (9.7) may be written as

$$\sum d^{\mathfrak{J}}(j_{11}, \dots, j_{mn}) L^{(1)}(\underline{\underline{v}}_1)^{j_{11}} \dots L^{(n)}(\underline{\underline{v}}_1)^{j_{1n}} \dots L^{(1)}(\underline{\underline{v}}_m)^{j_{m1}} \dots L^{(n)}(\underline{\underline{v}}_m)^{j_{mn}}. \tag{9.8}$$

By (9.3), (9.4) and (9.6), we have

$$|L^{(k)}(\underline{\underline{v}}_h)| < n\left(\frac{1}{\varepsilon}+1\right) Q_h^{c_k-\delta} < Q_h^{c_k-\delta+\varepsilon} \le Q_h^{c_k-15n^2\varepsilon} \tag{9.9}$$

$$(1 \le k \le n, \ 1 \le h \le m),$$

whence

$$|L^{(k)}(\underline{\underline{v}}_1)^{j_{1k}} \dots L^{(k)}(\underline{\underline{v}}_m)^{j_{mk}}| \le \exp\left((c_k - 15n^2\varepsilon) \sum_{h=1}^{m} j_{hk} \log Q_h\right) \qquad (1 \le k \le n).$$

By (9.5) and by (iv) of Theorem 7A, we have $d^{\mathfrak{J}}(j_{11}, \dots, j_{mn}) = 0$ unless

$$\sum_{h=1}^{m} j_{hk} \log Q_h \ge r_1 \log Q_1 \cdot \sum_{h=1}^{m} \frac{j_{hk}}{r_h} \ge r_1 \log Q_1 \cdot \left(\frac{1}{n} - 3n\varepsilon\right) m$$

and

$$\begin{aligned} \sum_{h=1}^{m} j_{hk} \log Q_h &\le (1+\varepsilon) r_1 \log Q_1 \cdot \sum_{h=1}^{m} \frac{j_{hk}}{r_h} \\ &\le r_1 \log Q_1 \cdot (1+\varepsilon)\left(\frac{1}{n} + 3n\varepsilon\right) m \\ &\le r_1 \log Q_1 \cdot \left(\frac{1}{n} + 7n\varepsilon\right) m. \end{aligned}$$

Both inequalities together give

$$\left| \sum_{h=1}^{m} j_{hk} \log Q_h - (r_1 \log Q_1) \cdot \frac{m}{n} \right| \le (r_1 \log Q_1) \cdot 7n\varepsilon m \quad .$$

This estimate together with $|c_k - 15n^2\varepsilon| \le 2$ (which follows from (9.2) , (9.3)) yields

$$|L^{(k)}(\underline{v}_1)^{j_{1k}} \cdots L^{(k)}(\underline{v}_m)^{j_{mk}}| \le Q_1^{r_1 \frac{m}{n}(c_k - 15n^2\varepsilon) + 2r_1 \cdot 7n\varepsilon m}$$
$$= Q_1^{r_1(\frac{m}{n}c_k - nm\varepsilon)} \quad .$$

Altogether each summand of (9.8) has absolute value

$$\le E^{r_1+\cdots+r_m} Q_1^{r_1 \frac{m}{n}(c_1+\cdots+c_n) - r_1 n^2 m\varepsilon} = E^{r_1+\cdots+r_m} Q_1^{-r_1 n^2 m\varepsilon} \qquad \text{(by (9.2))}$$

$$\le E^{r_1+\cdots+r_m}\left(Q_1^{-r_1\varepsilon} \cdots Q_m^{-r_m\varepsilon}\right)^{\frac{n^2}{1+\varepsilon}} \qquad \text{(by (9.5))}$$

$$\le E^{r_1+\cdots+r_m} Q_1^{-r_1\varepsilon} \cdots Q_m^{-r_m\varepsilon} \quad .$$

The number of summands in (9.8) is $\le 2^{n(r_1+\cdots+r_m)}$ . Therefore (9.8) , which is the left hand side of (9.7) , has absolute value

$$\le \prod_{h=1}^{m} \left(2^n E Q_h^{-\varepsilon}\right)^{r_h} < 1$$

by (9.4) . But the left side of (9.7) is necessarily a rational integer, so it must be zero. This completes the proof of Theorem 9A .

## §10. An Analogue of Roth's Lemma.

LEMMA 10A. *Suppose that* $n \ge 2$ *is an integer. Let* $m_1, \ldots, m_n$ *be coprime rational integers,* $m_1 \ne 0$ . *Then there is a* $j$ *with* $2 \le j \le n$ *such that*

$$\text{g.c.d.}(m_1, m_j) \le |m_1|^{\frac{n-2}{n-1}} \quad .$$

*Proof*. Put $d_j = \text{g.c.d.}(m_1, m_j)$ for $j = 2,\ldots,n$. Then $\text{g.c.d.}(d_2,\ldots,d_n) = 1$. Each $d_j$ divides $m_1$, hence $d_2 d_3 \ldots d_n$ divides $m_1^{n-2}$, and therefore

$$d_2 d_3 \ldots d_n \leq |m_1|^{n-2} .$$

The Lemma follows at once from this inequality.

THEOREM 10B. *Suppose that* $0 < \varepsilon < \frac{1}{12}$, *that* $m$ *is a positive integer, and*

$$\omega = 24 \cdot 2^{-m} (\varepsilon/12)^{2^{m-1}} . \tag{10.1}$$

*Let* $r_1,\ldots,r_m$ *be positive integers with*

$$\omega r_h \geq r_{h+1} \qquad (1 \leq h \leq m) . \tag{10.2}$$

*Suppose that* $n \geq 2$ *is an integer, and put* $q = n-1$. *Let* $M_1,\ldots,M_m$ *be nonzero linear forms in* $n$ *variables with coprime rational integer coefficients. Suppose that* $0 < \Gamma \leq q$ *and that*

$$\overline{|M_h|}^{\,r_h} \geq \overline{|M_1|}^{\,r_1 \Gamma} \qquad (1 \leq h \leq m) , \tag{10.3}$$

$$\overline{|M_h|}^{\,\omega \Gamma} \geq 2^{3mq^2} \qquad (1 \leq h \leq m) . \tag{10.4}$$

*Let* $P \in \mathfrak{R}$, $P \neq 0$, *with rational integer coefficients, homogeneous in* $X_{h1},\ldots,X_{hn}$ *of degree* $r_h$ $(1 \leq h \leq m)$. *Further suppose that*

$$\overline{|P|}^{\,q^2} \leq \overline{|M_1|}^{\,\omega r_1 \Gamma} . \tag{10.5}$$

*Then the index of* $P$ *with respect to* $(M_1,\ldots,M_m; r_1,\ldots,r_m)$ *is*

$$\leq \varepsilon .$$

*Proof*. Without loss of generality, we may suppose that $\overline{|M_h|} = |m_{h1}|$

for each $h$ , where

$$M_h = m_{h1}X_{h1}+\dots+m_{hn}X_{hn} \qquad (1 \le h \le m)^{\dagger)}.$$

Further, by Lemma 10A , we may assume that

$$\text{(10.6)} \qquad \text{g.c.d.}(m_{h1},m_{h2}) \le |m_{h1}|^{(n-2)/(n-1)} = |m_{h1}|^{(q-1)/q} \qquad (1 \le h \le m) .$$

Let $\eta$ be the index of $P$ with respect to $(M_1,\dots,M_m;r_1,\dots,r_m)$ . Then $P$ lies in the ideal $I(\eta)$ generated by polynomials $M_1^{i_1}M_2^{i_2}\dots M_m^{i_m}$ with

$$\text{(10.7)} \qquad \sum_{h=1}^{m} \frac{i_h}{r_h} \ge \eta \quad .$$

We construct a new polynomial $\hat{P}(X_{11},X_{12};X_{21},X_{22};\dots;X_{m1},X_{m2})$ from $P$ as follows: If $n = 2$ , put $\hat{P} = P$ . Otherwise, if $n > 2$ , we get rid of the $p = (n-2)m$ variables $X_{13},\dots,X_{1n};\dots;X_{m3},\dots,X_{mn}$ in the following manner. $P$ may be written (uniquely) as a polynomial in $M_1,X_{12},X_{13},\dots,X_{1n};\dots;M_m,X_{m2},X_{m3},\dots,X_{mn}$ . Let $a$ be the largest non-negative integer $u$ such that $X_{13}^u$ divides $P$ . Put $\bar{P} = X_{13}^{-a}P$ . Then $\bar{P} \in \Re$ , $\boxed{\bar{P}} = \boxed{P}$ and $\bar{P} \in I(\eta)$ . Also, $\bar{P}$ is homogeneous in the block of variables $X_{h1},\dots,X_{hn}$ of some degree $\le r_h$ $(1 \le h \le m)$ .

Put

$$P^{(1)} = \bar{P}(X_{11},X_{12},0,X_{14},\dots,X_{1n};\dots;X_{m1},X_{m2},X_{m3},\dots,X_{mn}) \quad .$$

Then $P^{(1)} \ne 0$ , $\boxed{P^{(1)}} \le \boxed{\bar{P}}$ , further $P^{(1)}$ is homogeneous in $X_{11},X_{12},X_{14},\dots,X_{1n}$ of degree $\le r_1$ and homogeneous in $X_{h1},X_{h2},\dots,X_{hn}$ of degree $r_h$ $(2 \le h \le m)$ , and $P^{(1)}$ lies in the ring of polynomials in $mn-1$ variables in the ideal $I^{(1)}(\eta)$ generated by polynomials

†) Even though the letter $m$ is used both for the coefficients of $M_h$ and fo[r] the number of indices $h$ , there should be no confusion.

$$M_1(X_{11},X_{12},0,X_{14},\ldots,X_{1n})^{i_1} M_2^{i_2}\ldots M_m^{i_m}$$

with (10.7) .

Carrying out this procedure $p$ times we obtain a polynomial $P^{(p)}$ . Put $\hat{P} = P^{(p)}$ . Then $\hat{P}$ is a polynomial in $X_{11},X_{12};X_{21},X_{22};\ldots;X_{m1},X_{m2}$ , homogeneous in $X_{h1},X_{h2}$ of degree $\le r_h$ $(1 \le h \le m)$ , with rational integer coefficients, $\hat{P} \neq 0$ , $\lceil \hat{P} \rceil \le \lceil P \rceil$ , and $\hat{P}$ lies in the ideal $\hat{I}(\eta)$ generated by polynomials

$$(m_{11}X_{11} + m_{12}X_{12})^{i_1}\ldots(m_{m1}X_{m1} + m_{m2}X_{m2})^{i_m}$$

with (10.7) . Put

$$\tilde{P}(X_1,\ldots,X_m) = \hat{P}(X_1,1;X_2,1;\ldots;X_m,1) \quad .$$

Then $\tilde{P} \neq 0$ since $\hat{P} \neq 0$ and since $\hat{P}$ was homogeneous in each block of variables. Also, $\lceil \tilde{P} \rceil = \lceil \hat{P} \rceil \le \lceil P \rceil$ . Further $\tilde{P}$ lies in the ideal $\tilde{I}(\eta)$ generated by polynomials

(10.8) $$\left(X_1 + \frac{m_{12}}{m_{11}}\right)^{i_1} \cdots \left(X_m + \frac{m_{m2}}{m_{m1}}\right)^{i_m}$$

with (10.7) .

Now put

$$q_h = \frac{|m_{h1}|}{(m_{h1},m_{h2})} \quad , \quad p_h = -\frac{m_{h2}}{(m_{h1},m_{h2})} \cdot \frac{m_{h1}}{|m_{h1}|} \qquad (1 \le h \le m) \ .$$

Clearly $q_h > 0$ and $(p_h,q_h) = 1$ . The polynomials (10.8) may be written as

$$\left(X_1 - \frac{p_1}{q_1}\right)^{i_1} \cdots \left(X_m - \frac{p_m}{q_m}\right)^{i_m}$$

with (10.7). For $j_1,\ldots,j_m$ with $\sum_{h=1}^{m} \frac{j_h}{r_h} < \eta$ , we have

$$\tilde{P}_{j_1 \cdots j_m}\left(\frac{p_1}{q_1}, \ldots, \frac{p_m}{q_m}\right) = 0 \quad .$$

Therefore $\tilde{P}$ has Roth index $\geq \eta$ with respect to $\left(\frac{p_1}{q_1}, \ldots, \frac{p_m}{q_m}; r_1, \ldots, r_m\right)$. This will give the desired relation $\eta \leq \varepsilon$ if we can show that $\tilde{P}$ and $\frac{p_1}{q_1}, \ldots, \frac{p_m}{q_m}$ satisfy the hypotheses of Roth's Lemma (Theorem 10A of Chapter V).

First of all, $\tilde{P}$ has degree $\leqq r_h$ in $X_h$. It is clear that conditions (10.1), (10.2) and (10.3) of Roth's Lemma hold. Put $\gamma = \Gamma/q$ (so that $0 < \gamma \leq 1$) and recall that $(p_h, q_h) = 1$ $(1 \leq h \leq m)$. We have

$$\lceil M_h \rceil^{\frac{1}{q}} = |m_{h1}|^{\frac{1}{q}} \leq \frac{|m_{h1}|}{(m_{h1}, m_{h2})} \qquad \text{(by (10.6))}$$

$$= q_h \leq \lceil M_h \rceil \qquad (1 \leq h \leq m) \; .$$

Thus

$$q_h^{r_h} \geq \lceil M_h \rceil^{r_h/q} \geq \lceil M_1 \rceil^{r_1 \Gamma/q} \qquad \text{(by (10.3))}$$

$$= \lceil M_1 \rceil^{r_1 \gamma} \geq q_1^{r_1 \gamma} .$$

But this is inequality (10.4) of Chapter V. Also,

$$q_h^{\omega\gamma} = q_h^{\omega\Gamma/q} \geq \lceil M_h \rceil^{\omega\Gamma/q^2} \geq 2^{3m} \qquad (1 \geq h \geq m)$$

by our (10.4), which gives (10.5) of Chapter V. Finally, our (10.5) yields

$$\lceil \tilde{P} \rceil \leq \lceil P \rceil \leq \lceil M_1 \rceil^{\omega r_1 \Gamma/q^2} = \lceil M_1 \rceil^{\omega r_1 \gamma/q} \leq q_1^{\omega\gamma r_1} \; .$$

This is (10.6) of Chapter V, and the desired conclusion follows.

## §11. The Size of $\underline{g}_n^*$ .

Let again $L_i(\underline{X}) = \underline{a}_i\underline{X}$ $(i = 1,\dots,n)$ be independent linear forms with real algebraic coefficients, and let $\underline{a}_1^*,\dots,\underline{a}_n^*$ be reciprocal to $\underline{a}_1,\dots,\underline{a}_n$ . Let $c_1,\dots,c_n$ be constants with (9.2) , let $\delta > 0$ , and let $\mathfrak{S}$ be the subset of $\{1,\dots,n\}$ of numbers $i$ with

$$c_i + (\delta/2) \geq 0 \quad . \tag{11.1}$$

Clearly $\mathfrak{S}$ is nonempty.

LEMMA 11A. *Suppose* $Q > 1$ , *and let* $\Pi$ *be the parallelepiped*

$$|L_i(\underline{x})| = |\underline{a}_i\underline{x}| \leq Q^{c_i} \qquad (1 \leq i \leq n) . \tag{11.2}$$

*Let* $\lambda_1,\dots,\lambda_n$ *be the successive minima of* $\Pi$ , *let* $\underline{g}_1,\dots,\underline{g}_n$ *be independent integer points with* $\underline{g}_j \in \lambda_j\Pi$ $(1 \leq j \leq n)$ , *and let* $\underline{g}_1^*,\dots,\underline{g}_n^*$ *be the reciprocal points. Suppose we have*

$$\lambda_{n-1} < Q^{-\delta} \quad , \tag{11.3}$$

*and suppose there is an* $i \in \mathfrak{S}$ *having*

$$\underline{a}_i^*\underline{g}_n^* \neq 0 \quad . \tag{11.4}$$

*Then*

$$Q^{C_1} \leq |\underline{g}_n^*| \leq Q^{C_2} \tag{11.5}$$

*if* $Q \geq C_3$ . *Here* $C_i = C_i(\delta;L_1,\dots,L_n;c_1,\dots,c_n) > 0$ .

*Proof.* By Mahler's theory of reciprocal parallelepipeds (Theorem 4A of Chapter IV), applied to the reciprocal bases

$$Q^{-c_i}\underline{a}_i \quad (1 \leq i \leq n) \quad \text{and} \quad Q^{c_i}\underline{a}_i^* \quad (1 \leq i \leq n) \quad ,$$

we have

$$(11.6) \qquad |\underline{\underline{a}}_i^* \underline{\underline{g}}_n^*| \ll \lambda_n^{-1} Q^{-c_i} \ll Q^{-\delta(n-1)-c_i} \qquad (1 \leq i \leq n) ,$$

since $\lambda_n^{-1} \ll \lambda_1 \cdots \lambda_{n-1} < Q^{-\delta(n-1)}$ by (11.3) . (The constants in $\ll$ here and in the rest of this section depend only on $L_1,\ldots,L_n$) . Since $|c_i| \leq 1$ and since $\underline{\underline{a}}_1^*,\ldots,\underline{\underline{a}}_n^*$ are independent, it follows that $|\underline{\underline{g}}_n^*| \ll Q^{1-\delta(n-1)}$ . In particular we have $|\underline{\underline{g}}_n^*| \leq Q^{C_2}$ if $Q$ is large. As for the lower bound, we note that in view of (11.1) and (11.6) we have

$$(11.7) \qquad |\underline{\underline{a}}_i^* \underline{\underline{g}}_n^*| \ll Q^{-\delta(n-1)+(\delta/2)} \leq Q^{-\delta/2}$$

for $i \in \mathfrak{S}$ . In particular, this is true for the integer $i$ in (11.4) . The components of $\underline{\underline{a}}_i^*$ are algebraic and generate a number field, of degree $\Delta_i^*$ say. Since $\underline{\underline{g}}_n^*$ has rational coordinates with denominator $E \leq (n!)^2$ (an easy consequence of (1.9) of Chapter IV), the norm of $\underline{\underline{a}}_i^* \underline{\underline{g}}_n^*$ satisfies $|\mathfrak{N}(\underline{\underline{a}}_i^* \underline{\underline{g}}_n^*)| \gg 1$ . The conjugates of $\underline{\underline{a}}_i^* \underline{\underline{g}}_n^*$ have absolute values $\ll |\underline{\underline{g}}_n^*|$ , and we obtain

$$|\underline{\underline{a}}_i^* \underline{\underline{g}}_n^*| \gg |\underline{\underline{g}}_n^*|^{1-\Delta_i^*} \geq |\underline{\underline{g}}_n^*|^{1-\Delta^*} ,$$

where $\Delta^*$ is the maximum of $\Delta_1^*,\ldots,\Delta_n^*$ . In conjunction with (11.7) this shows that $\Delta^* > 1$ and that $|\underline{\underline{g}}_n^*|^{\Delta^*-1} \gg Q^{\delta/2}$ . Thus $|\underline{\underline{g}}_n^*| \geq Q^{C_1}$ with $C_1 = \frac{1}{3}\delta/(\Delta^*-1)$ if $Q$ is large, and the proof of Lemma 11A is complete.

Let $M$ be the linear form $M = M\{\underline{\underline{g}}_1,\ldots,\underline{\underline{g}}_{n-1}\}$ . Then $M(\underline{\underline{X}}) = \underline{\underline{m}}\underline{\underline{X}}$ where $\underline{\underline{m}}$ has coprime integer coordinates, and $\underline{\underline{g}}_n^* = (t/E)\underline{\underline{m}}$ with integral $t$ . In fact, $t$ is a divisor of $E$ since $\underline{\underline{g}}_n \underline{\underline{g}}_n^* = (t/E)\underline{\underline{g}}_n \underline{\underline{m}} = 1$ . Putting $F = E/t$ we have

(11.8) $$\underline{\underline{g}}_n^* = (1/F)\underline{\underline{m}} \quad \text{with} \quad 1 \leq |F| \leq (n!)^2 \quad .$$

It follows that $\lceil \underline{\underline{m}} \rceil \ll \lceil \underline{\underline{g}}_n^* \rceil \leq \lceil \underline{\underline{m}} \rceil$ , and $\lceil M \rceil \gg \ll \lceil \underline{\underline{g}}_n^* \rceil$ , so that

(11.9) $$Q^{c_4} \leq \lceil M \rceil \leq Q^{c_5}$$

provided $Q \geq C_6$ , where $C_i = C_i(\delta; L_1,\ldots,L_n; c_1,\ldots,c_n) > 0 \quad (i = 4,5,6)$ .

## §12. The Next to Last Minimum.

THEOREM 12A. (Theorem on the Next to Last Minimum). *Suppose* $L_1,\ldots,L_n, \underline{\underline{a}}_1,\ldots,\underline{\underline{a}}_n, \underline{\underline{a}}_1^*,\ldots,\underline{\underline{a}}_n^*, c_1,\ldots,c_n$ *and* $\delta$ *and* $\mathfrak{S}$ *are as in* §11. *Again let* $Q > 1$ *and* $\Pi$ *the parallelepiped* (11.2) , *with minima* $\lambda_1,\ldots,\lambda_n$ *having* (11.3) . *Let* $\underline{\underline{g}}_1,\ldots,\underline{\underline{g}}_n$ *and* $\underline{\underline{g}}_1^*,\ldots,\underline{\underline{g}}_n^*$ *be as above. Then if* $Q > Q_1(\delta; L_1,\ldots,L_n; c_1,\ldots,c_n)$ , *we have*

(12.1) $$\underline{\underline{a}}_i^* \underline{\underline{g}}_n^* = 0 \quad \textit{for} \quad i \in \mathfrak{S} \quad .$$

This theorem *appears* to be weak. Firstly, a very good rational approximation makes the *first* minimum $\lambda_1$ of some parallelepiped small, so that the hypothesis (11.3) seems to be far too strong. Secondly, the conclusion (12.1) appears to be weak, since ideally we would want to say that "for $Q > Q_1$ this is impossible." However, it will turn out that Theorem 12A gives us everything we need. So here is a siutation where a result which looks temporary gives in fact the final answer.

Let us briefly discuss a special case pertaining to Corollary 1C. Set $n = u+1$ and let $L_1,\ldots,L_n$ be given by (1.6) , and $c_1 = \cdots = c_u = -1/u$ , but $c_n = 1$ . Here $\mathfrak{S} = \{n\}$ and $\underline{\underline{a}}_n^* = (\alpha_1,\ldots,\alpha_u,1)$ . By the linear independence of $1,\alpha_1,\ldots,\alpha_u$ , the relation (12.1) is impossible (since $\underline{\underline{g}}_n^*$ has rational components), so that in fact we do get the desired conclusion.

<u>Proof of Theorem 12A</u>. Let $\mathfrak{Q}$ be the set of $Q > 1$ such that the parallelepiped $\Pi = \Pi(Q)$ given by (11.2) has (11.3) but not (12.1). We have to show that $\mathfrak{Q}$ is bounded. We shall assume indirectly that $\mathfrak{Q}$ is unbounded, and we shall derive a contradiction.

A short reflection shows that we may suppose without loss of generality that the coefficients of $L_1,\dots,L_n$ are real algebraic <u>integers</u>. Again write

$$q = n-1 \quad .$$

(i) Pick $\delta_1$ with $0 < \delta_1 < 1$ and $\delta_1 \leqq \delta$ .

(ii) Pick $\varepsilon > 0$ small enough to satisfy

$$16n^2\varepsilon < \delta_1 \quad . \tag{12.2}$$

(iii) Define $\Delta$ as in §7 and choose an integer $m > 4\varepsilon^{-2}\log(2n\Delta)$ , so that (7.1) holds. Define $\omega = \omega(m,\varepsilon)$ by (10.1).

(iv) Let $Q_1 \in \mathfrak{Q}$ be so large that (9.4) holds for $h = 1$ , and that $Q_1 \geqq C_6$ and $Q_1^{\omega C_4^2/C_5} \geqq 2^{3mq^2}$ and $Q_1^{\omega C_4^2/C_5} \geqq D^{mq^2}$ , where $C_4, C_4, C_6$ are the constants of §11, and $D$ is described in §7.

(v) Successively choose $Q_2,\dots,Q_m$ in $\mathfrak{Q}$ with

$$\omega \log Q_{h+1} \geqq 2 \log Q_h \qquad (h = 1,\dots,m-1) \; .$$

Then $Q_1 < \dots < Q_m$ , so that (9.4) holds for each $h$ , and also

$$Q_h \geqq C_6 \quad \text{and} \quad Q_h^{\omega C_4^2/C_5} \geqq 2^{3mq^2} \qquad (h = 1,\dots,m) \; . \tag{12.3}$$

(vi) Let $r_1$ be an integer so large that $\varepsilon r_1 \log Q_1 \geqq \log Q_m$ .

(viii) For $h = 2,\dots,m$ , put

$$r_h = \left[\frac{r_1 \log Q_1}{\log Q_h}\right] + 1 \quad .$$

We then have for $h = 2,\ldots,m$ ,

$$r_1 \log Q_1 \leqq r_h \log Q_h$$
$$\leqq r_1 \log Q_1 + \log Q_h$$
$$\leqq (1+\varepsilon) r_1 \log Q_1 \quad ,$$

which is (9.5) .

The linear forms $L_1,\ldots,L_n$ as well as $\varepsilon, m$ and $r_1,\ldots,r_m$ satisfy the hypotheses of the Polynomial Theorem (Theorem 7A). Let $P(X_{11},\ldots,X_{1n};\ldots;X_{m1},\ldots,X_{mn})$ be the polynomial described there. Now also the hypotheses of Theorem 9A are satisfied, with $\delta_1$ in place of $\delta$ . In particular, if for $1 \leqq h \leqq m$ the integer points $\underline{g}_{h1},\ldots,\underline{g}_{hn}$ are linearly independent with $\underline{g}_{hi} \in \lambda_i(Q_h)\Pi(Q_h)$ , then (9.6) holds since $q = n-1$ and since $\lambda_q(Q_h) = \lambda_{n-1}(Q_h) < Q_h^{-\delta}$ . Let $M_1,\ldots,M_m$ be the linear forms described in Theorem 9A . The conclusion is that

P <u>has index at least</u> $m\varepsilon$ <u>with respect to</u> $(M_1,\ldots,M_m;r_1,\ldots,r_m)$ .

From (9.5) we have $r_{h+1} \log Q_{h+1} \leqq (1+\varepsilon) r_h \log Q_h$ , so that

$$\omega r_h \geqq \omega \frac{r_{h+1} \log Q_{h+1}}{(1+\varepsilon)\log Q_h} \qquad (1 \leqq h \leqq m-1)$$
$$\geqq \frac{2}{1+\varepsilon} r_{h+1}$$

from (v) , and therefore $\omega r_h \geqq r_{h+1}$ , which is (10.2) .

Since (12.1) is not true for $Q_1,\ldots,Q_m$ , the inequality (11.9) applies:

$$Q_h^{C_4} \leqq \overline{|M_h|} \leqq Q_h^{C_5} \qquad (1 \leqq h \leqq m) \quad .$$

Therefore

$$\overline{|M_h|}^{r_h} \geqq Q_h^{r_h C_4} \geqq Q_1^{r_1 C_4} \geqq \overline{|M_1|}^{r_1 C_4/C_5} \quad ,$$

whence

$$\overline{|M_h|}^{\,r_h} \geqq \overline{|M_1|}^{\,r_1\Gamma} \qquad (1 \leqq h \leqq m)$$

with $\Gamma = C_4/C_5$ , which is (10.3) . Furthermore, by (12.3) ,

$$\overline{|M_h|}^{\,\omega\Gamma} \geqq Q_h^{\omega\Gamma C_4} \geqq 2^{3mq^2} \qquad (1 \leqq h \leqq m) ,$$

which is (10.4) . By Theorem 7A we have

$$\overline{|P|} \leqq D^{r_1+\cdots+r_m} \leqq D^{mr_1} ,$$

which by (iv) gives

$$\overline{|P|}^{\,q^2} \leqq D^{mq^2r_1} \leqq Q_1^{\omega r_1 C_4^2/C_5} \leqq \overline{|M_1|}^{\,\omega r_1 C_4/C_5} = \overline{|M_1|}^{\,\omega r_1\Gamma} ,$$

i.e. (10.5) . All the hypotheses of Theorem 10B are satisfied, and we may conclude that

> $P$ *has index at most* $\varepsilon$ *with respect to* $(M_1,\ldots,M_m,r_1,\ldots,r_m)$ . This contradicts the lower bound for the index given above.

## §13. The Constancy of $\mathbf{g}_n^*$ .

LEMMA 13A. *Suppose* $L_1,\ldots,L_n,\mathbf{a}_1,\ldots,\mathbf{a}_n,c_1,\ldots,c_n$ *and* $\delta$ *are as in* §11,12. *For* $Q > 1$ *let* $\Pi$ *be the parallelepiped* (11.2) , *with minima* $\lambda_1,\ldots,\lambda_n$ . *Let* $\mathbf{g}_1,\ldots,\mathbf{g}_n$ *and* $\mathbf{g}_1^*,\ldots,\mathbf{g}_n^*$ *be as above. Let* $\mathfrak{Q}$ *be an unbounded set of reals* $> 1$ *such that* (11.3) *holds for* $Q \in \mathfrak{Q}$ . *Then there is a fixed point* $\mathbf{h}$ *and an unbounded subset* $\mathfrak{Q}'$ *of* $\mathfrak{Q}$ *such that*

$$\mathbf{g}_n^* = \mathbf{h}$$

*for every* $Q \in \mathfrak{Q}'$ .

Proof. Let $\bar{Q}$ be a fixed large element of $\mathfrak{Q}$. By Theorem 12A we have $\underline{a}_i^* \bar{\underline{g}}_n^* = 0$ for $i \in \mathfrak{S}$, where $\bar{\underline{g}}_n^* = \underline{g}_n^*(\bar{Q})$. Thus there is an integer point $\bar{\underline{h}} \neq \underline{0}$ with coprime components such that

$$\underline{a}_i^* \bar{\underline{h}} = 0 \qquad \text{for } i \in \mathfrak{S} . \tag{13.1}$$

This point $\bar{\underline{h}}$ will be fixed in the sequel. Since $\bar{\underline{h}}$ is fixed we have $|\underline{a}_i^* \bar{\underline{h}}| \ll 1$ $(1 \leq i \leq n)$. For $i \notin \mathfrak{S}$ we have $c_i + (\delta/2) < 0$, so that

$$|\underline{a}_i^* \bar{\underline{h}}| \ll 1 \leq Q^{-c_i - (\delta/2)} \qquad (i \notin \mathfrak{S}) .$$

So $\bar{\underline{h}}$ lies in the parallelepiped $\Pi^* = \Pi^*(Q)$ reciprocal to $\Pi = \Pi(Q)$. This holds for every large $Q$, in particular for $Q \in \mathfrak{Q}$. For $Q \in \mathfrak{Q}$ we have $\lambda_{n-1} < Q^{-\delta}$, whence $\lambda_2^* \gg Q^{\delta}$ by Theorem 4A of Chapter IV. So $\lambda_2^* > 1$ if $Q \in \mathfrak{Q}$ is large, and therefore every integer point in $\Pi^*$ is a multiple of $\bar{\underline{h}}$.

Again let $Q \in \mathfrak{Q}$ be large, and let $\underline{g}_n^* = \underline{g}_n^*(Q)$. By the argument leading to (11.8) we have $\underline{g}_n^* = (1/F)\underline{m}$ where $1 \leq |F| \leq (n!)^2$ and where $\underline{m}$ has coprime integer coordinates. A priori $F$ and $\underline{m}$ may vary with $Q$. But now by (11.6), $F\underline{g}_n^*$ lies in $\Pi^*$, and so is proportional to $\bar{\underline{h}}$. It follows that $\underline{m} = \pm\bar{\underline{h}}$ and $\underline{g}_n^* = \pm(1/F)\bar{\underline{h}}$. The number of possibilities for $F$ and hence for $\underline{g}_n^*$ is finite.

A slightly more general result will be needed:

LEMMA 13B. *Suppose* $L_i(\underline{X}) = \underline{a}_i \underline{X}$ $(i = 1,\dots,n)$ *are independent linear forms with real algebraic coefficients, and let* $\delta > 0$. *Let* $\mathfrak{P}$ *be an unbounded set of reals* $Q > 1$, *and suppose that with each* $Q \in \mathfrak{P}$ *we are given positive* $A_1,\dots,A_n$ *satisfying*

(13.1) $A_1A_2\ldots A_n = 1$ and $\max(A_1,\ldots,A_n,A_1^{-1},\ldots,A_n^{-1}) \le Q$ .

*Let* $\Pi(Q)$ *be the parallelepiped given by*

$$|L_i(\underline{x})| \le A_i \qquad (i = 1,\ldots,n) .$$

*Let* $\lambda_1,\ldots,\lambda_n$ *be the minima of* $\Pi = \Pi(Q)$ , *and define* $\underline{g}_1,\ldots,\underline{g}_n$, $\underline{g}_1^*,\ldots,\underline{g}_n^*$ *as before. Suppose that* (11.3) *holds for* $Q \in \mathfrak{P}$ . *Then there is a fixed point* $\underline{h}$ *and an unbounded subset* $\mathfrak{P}'$ *of* $\mathfrak{P}$ *such that* $\underline{g}_n^* = \underline{h}$ *for* $Q \in \mathfrak{P}'$ .

*Proof.* Set $A_i = Q^{c_i}$ $(i = 1,\ldots,n)$ and use a compactness argument, such as was used in the deduction of Theorem 1F from Lemma 3B . The reader is urged to do the details as an *exercise*.

§14. The Last Two Minima. Since the last minimum $\lambda_n$ has $\lambda_n \gg 1$ , it is clear that the hypothesis

(14.1) $$\lambda_{n-1} < \lambda_n Q^{-\delta}$$

is weaker than (11.3) . In general, an inequality (14.1) does not imply an inequality of the type (11.3) . But we will see that if (14.1) holds for a parallelepiped $\Pi$ , then something like (11.3) holds for a related parallelepiped $\Pi'$ .

LEMMA 14A. *The conclusion of Lemma* 13A *remains valid if* (11.3) *is replaced by the weaker hypothesis* (14.1).

*Proof.* Given $Q \in \mathfrak{Q}$ put $\rho_0 = (\lambda_1\lambda_2 \cdots \lambda_{n-2}\lambda_{n-1}^2)^{1/n}$ and

$$\rho_1 = \rho_0/\lambda_1,\ldots,\rho_{n-1} = \rho_0/\lambda_{n-1}, \text{ but } \rho_n = \rho_0/\lambda_{n-1} .$$

Then equations (3.1) , (3.2) , (3.3) of Chapter IV hold. By Davenport's

Lemma (Theorem 3A of Chapter IV) there is a permutation $t_1,\dots,t_n$ of $1,\dots,n$ such that the successive minima $\lambda'_1,\dots,\lambda'_n$ of the parallelepiped $\Pi'$ given by

$$|L_i(\underline{\underline{x}})| \leqq Q^{c_i}\rho_{t_i}^{-1} \quad (= A_i\ ,\ \text{say}) \qquad (1 \leqq i \leqq n)$$

satisfy

$$\lambda_j\rho_j \ll \lambda'_j \ll \lambda_j\rho_j \qquad (1 \leqq j \leqq n)\ .$$

In particular, we have

$$(14.2) \qquad \lambda'_{n-1} \ll \rho_{n-1}\lambda_{n-1} = \rho_0 \ll (\lambda_{n-1}/\lambda_n)^{1/n} \ll Q^{-\delta/n}$$

by (14.1) .

We observe that

$$\max(Q^{-c_1}|L_1(\underline{\underline{x}})|,\dots,Q^{-c_n}|L_n(\underline{\underline{x}})|) \geq Q^{-1}\max(|L_1(\underline{\underline{x}})|,\dots,|L_n(\underline{\underline{x}})|) \gg Q^{-1}|\overline{\underline{\underline{x}}}| \geq Q^{-1}$$

for every integer point $\underline{\underline{x}} \neq \underline{\underline{0}}$ , and therefore $\lambda_1 \gg Q^{-1}$ . On the other hand, the basis vectors $\underline{\underline{e}}_1,\dots,\underline{\underline{e}}_n$ satisfy $|L_i(\underline{\underline{e}}_j)| \ll 1 \ll Q^{c_i+1}$ $(1 \leqq i,j \leqq n)$ , so that $\lambda_n \ll Q$ . We have

$$\rho_1 = \lambda_1^{-1}\rho_0 \ll \lambda_1^{-1}(\lambda_{n-1}/\lambda_n)^{1/n} \leqq \lambda_1^{-1} \ll Q$$

and

$$\rho_n = \lambda_{n-1}^{-1}\rho_0 \gg \lambda_{n-1}^{-1}(\lambda_{n-1}/\lambda_n)^{1/n} \geqq \lambda_n^{-1} \gg Q^{-1}\ .$$

Hence we obtain

$$Q^{-1} \ll \rho_n \leqq \rho_{n-1} \leqq \cdots \leqq \rho_1 \ll Q\ ,$$

and the quotients $A_i = Q^{c_i}\rho_{t_i}^{-1}$ satisfy $Q^{-2} \ll A_i \ll Q^2$ . Since

$c_1+\cdots+c_n = 0$ , and in view of $\rho_1\rho_2\cdots\rho_n = 1$ , we have for large $Q$ ,

$$A_1 \cdots A_n = 1 \quad \text{and} \quad \max(A_1,\ldots,A_n,A_1^{-1},\ldots,A_n^{-1}) \leq Q^3 \quad .$$

Further, for every large $Q$ in $\Omega$ we have

$$\lambda'_{n-1} < Q^{-(3\delta/4n)} = (Q^3)^{-\delta_2}$$

by (14.2) , with $\delta_2 = \delta/(4n)$ . The hypotheses of Lemma 13B are thus satisfied with $\delta_2$ in place of $\delta$ , and with $\mathfrak{P}$ consisting of large $Q^3$ with $Q \in \Omega$ . Let $\underline{\underline{g}}'_1,\ldots,\underline{\underline{g}}'_n$ be independent integer points with $\underline{\underline{g}}'_j \in \lambda'_j\Pi'$ $(1 \leq j \leq n)$ , and let $\underline{\underline{g}}'^*_1,\ldots,\underline{\underline{g}}'^*_n$ be their reciprocals. By Lemma 13B there is an unbounded subset $\mathfrak{P}'$ of $\mathfrak{P}$ such that for $Q^3 \in \mathfrak{P}'$ ,

$$\underline{\underline{g}}'^*_n = \underline{\underline{h}}' \quad , \tag{14.3}$$

where $\underline{\underline{h}}'$ is fixed.

Let $T^{n-1}$ be the subspace spanned by $\underline{\underline{g}}_1,\ldots,\underline{\underline{g}}_{n-1}$ . By the last assertion of Theorem 3A of Chapter IV (Davenport's Lemma), we see that for every integer point $\underline{\underline{g}} \notin T^{n-1}$ ,

$$\max_{1\leq i\leq n} (|L_i(\underline{\underline{g}})|\rho_{t_i} Q^{-c_i}) = \max_{1\leq i\leq n} (|L_i(\underline{\underline{g}})|A_i^{-1}) \gg \lambda'_n \gg 1 \quad .$$

On the other hand, we have $\lambda'_{n-1} \ll Q^{-\delta/n}$ by (14.2) , so that $\underline{\underline{g}}'_1,\ldots,\underline{\underline{g}}'_{n-1}$ must lie in $T^{n-1}$ . Thus $\underline{\underline{g}}_1,\ldots,\underline{\underline{g}}_{n-1}$ and $\underline{\underline{g}}'_1,\ldots,\underline{\underline{g}}'_{n-1}$ span the same space, and $\underline{\underline{g}}^*_n$ and $\underline{\underline{g}}'^*_n$ are proportional. By the argument leading to (11.8) we see that $\underline{\underline{g}}^*_n = F^{-1}\underline{\underline{m}}$ , $\underline{\underline{g}}'^*_n = F'^{-1}\underline{\underline{m}}$ where $\underline{\underline{m}}$ has coprime integer components and where $1 \leq |F|,|F'| \leq (n!)^2$ . Since for $Q^3 \in \mathfrak{P}'$ the vector $\underline{\underline{g}}'^*_n$ is fixed by (14.3) , we get only finitely many possibilities for $\underline{\underline{g}}^*_n$ , so that indeed $\underline{\underline{g}}^*_n$ is fixed for an unbounded subset $\mathfrak{Q}'$ of $\mathfrak{Q}$ .

§15. Proof of the Strong Subspace Theorem. Our arguments will depend heavily on Mahler's Theory of compound parallelepipeds developed in §7 of Chapter IV. We clearly may suppose that $L_1(\underline{X}) = \underline{a}_1\underline{X},\ldots,L_n(\underline{X}) = \underline{a}_n\underline{X}$ have determinant 1 . Further, for reasons of homogeneity, we may suppose that $c_1,\ldots,c_n$ satisfy not only (2.8) , i.e., $c_1+\cdots+c_n = 0$ , but also $|c_i| \leq 1/n$ $(i = 1,\ldots,n)$ . Suppose that (3.2) holds for some arbitrarily large values of $Q$ , say for $Q$ in some unbounded set $\mathfrak{N}$ . Put

$$p = n-d \quad . \tag{15.1}$$

For $\sigma = \{i_1 < \cdots < i_p\}$ in $C(n,p)$ , define $\underline{A}_\sigma$ by (6.10) of Chapter IV , and put

$$c_\sigma = \sum_{i\in\sigma} c_i \quad . \tag{15.2}$$

Then

$$\sum_{\sigma\in C(n,p)} c_\sigma = 0 \quad \text{and} \quad |c_\sigma| \leq 1 \quad \text{for} \quad \sigma \in C(n,p) \quad . \tag{15.3}$$

We now apply the theory of §7 of Chapter IV to the points $Q^{-c_1}\underline{a}_1,\ldots,Q^{-c_n}\underline{a}_n$ . The parallelepiped $\Pi$ given by $|\underline{a}_i\underline{x}| \leq Q^{c_i}$ $(i = 1,\ldots,n)$ has a p-th pseudocompound $\Pi^{(p)}$ given by

$$|\underline{A}_\sigma\underline{X}^{(p)}| \leq Q^{c_\sigma} \qquad (\sigma \in C(n,p)) \ . \tag{15.4}$$

Denote the successive minima of $\Pi^{(p)}$ by $\nu_1,\ldots,\nu_\ell$ where $\ell = \binom{n}{d} = \binom{n}{p}$ . It is clear that in (7.2) of Chapter IV we may take $\tau_\ell = \{n-p+1,n-p+2,\ldots,n\} = \{d+1,d+2,\ldots,n\}$ and

$\tau_{\ell-1} = \{n-p, n-p+2, \ldots, n\} = \{d, d+2, \ldots, n\}$ . By Theorem 7A of Chapter IV we have

$$\nu_\ell \ll \lambda_{d+1}\lambda_{d+2} \cdots \lambda_n \ll \nu_\ell$$

and

$$\nu_{\ell-1} \ll \lambda_d \lambda_{d+2} \cdots \lambda_n \ll \nu_{\ell-1} \quad ,$$

and (3.2) yields $\nu_{\ell-1} \ll \nu_\ell Q^{-\delta}$ . In particular, we have

$$\nu_{\ell-1} < \nu_\ell Q^{-\delta/2} \tag{15.5}$$

if $Q$ in $\mathfrak{N}$ is large.

Next we shall apply Lemma 14A to the vectors $\underline{\underline{A}}_\sigma$ and constants $c_\sigma$ and to the parallelepiped $\Pi^{(p)}$ . The conditions (15.3) replace (2.8) . The inequality (15.5) replaces (14.1) . Let $\underline{\underline{V}}_1^{(p)}, \ldots, \underline{\underline{V}}_\ell^{(p)}$ be independent integer points in $\mathbb{R}_p^n$ with $\underline{\underline{V}}_j^{(p)} \in \nu_j \Pi^{(p)}$ $(1 \le j \le 1)$ , and let $\underline{\underline{V}}_1^{(p)*}, \ldots, \underline{\underline{V}}_\ell^{(p)*}$ be their reciprocals. The conclusion is that there is a fixed $\underline{\underline{H}}^{(p)}$ such that

$$\underline{\underline{V}}_\ell^{(p)*} = \underline{\underline{V}}_\ell^{(p)*}(Q) = \underline{\underline{H}}^{(p)} \tag{15.6}$$

for every $Q$ in some unbounded subset $\mathfrak{N}'$ of $\mathfrak{N}$ .

Let $\underline{\underline{g}}_1, \ldots, \underline{\underline{g}}_n$ be independent integer points of $\mathbb{R}^n$ with $\underline{\underline{g}}_j \in \lambda_j \Pi$ $(1 \le j \le n)$ , and for $\tau = \{j_1 < \cdots < j_p\}$ in $C(n,p)$ put $\underline{\underline{G}}_\tau = \underline{\underline{g}}_{j_1} \wedge \cdots \wedge \underline{\underline{g}}_{j_p}$ . By Theorem 7A of Chapter IV we see that

$$|\underline{\underline{A}}_\sigma \underline{\underline{G}}_\tau| \ll \nu_{\ell-1} Q^{c_\sigma}$$

for $\sigma, \tau$ in $C(n,p)$ with $\tau \ne \tau_\ell = \{d+1, d+2, \ldots, n\}$ . Since $\nu_{\ell-1} \ll \nu_\ell Q^{-\delta}$ ., this implies that the $\ell - 1$ vectors $\underline{\underline{G}}_\tau$ with $\tau \ne \tau_\ell$

span the same subspace as $\underline{\underline{V}}_1^{(p)},\ldots,\underline{\underline{V}}_{\ell-1}^{(p)}$ . Hence the vectors $(\underline{\underline{G}}_{\tau_\ell})^*$ and $\underline{\underline{V}}_\ell^{(p)*}$ are proportional, and in view of (6.12) of Chapter IV the vectors $(\underline{\underline{G}}^*)_{\tau_\ell}$ and $\underline{\underline{V}}_\ell^{(p)*}$ are proportional. Thus for every Q in $\mathfrak{N}'$ we have

$$(\underline{\underline{G}}^*)_{\tau_\ell} = \underline{\underline{g}}_{d+1}^* \wedge \cdots \wedge \underline{\underline{g}}_n^* = \lambda \underline{\underline{H}}^{(p)} \quad .$$

By Lemma 6C of Chapter IV this implies that there is a fixed subspace $S^*$ of dimension $p = n-d$ such that $\underline{\underline{g}}_{d+1}^*,\ldots,\underline{\underline{g}}_n^*$ lies in $S^*$ when $Q \in \mathfrak{N}'$ . Let $S^d$ be the orthogonal complement to $S^*$ . Whenever $\underline{\underline{g}}_{d+1}^*,\ldots,\underline{\underline{g}}_n^*$ lie in $S^*$ , the points $\underline{\underline{g}}_1,\ldots,\underline{\underline{g}}_d$ lie in $S^d$ . Hence for every Q in $\mathfrak{N}'$ , the points $\underline{\underline{g}}_1,\ldots,\underline{\underline{g}}_d$ lie in $S^d$ .

# VII. Norm Form Equations.

References: Borevich and Shafarevich (1966), Schmidt (1971c, 1972).

§1. Norm Form Equations. In this chapter we shall require some knowledge about algebraic number fields. Throughout, $K$ will be an algebraic number field of degree $k$, and $\mathfrak{N}$ will denote the norm from $K$ to the field $\mathbb{Q}$ of rationals. There are $k$ isomorphic embeddings $\varphi_1, \dots, \varphi_k$ of $K$ into the complex numbers; denote the image of an element $\alpha$ of $K$ under $\varphi_i$ by $\alpha^{(i)}$. We will always tacitly assume that $\varphi_1$ is the identity map, so that $\alpha^{(1)} = \alpha$, and we shall say that $\alpha^{(1)} = \alpha, \alpha^{(2)}, \dots, \alpha^{(k)}$ are the conjugates of $\alpha$. In this notation $\mathfrak{N}(\alpha) = \alpha^{(1)} \cdots \alpha^{(k)}$. Given a linear form

$$M(\underline{X}) = M(X_1, \dots, X_n) = \alpha_1 X_1 + \cdots + \alpha_n X_n \quad , \tag{1.1}$$

with coefficients in $K$ we write

$$\mathfrak{N}(M(\underline{X})) = \prod_{i=1}^{k} M^{(i)}(X_1, \dots, X_n) = \prod_{i=1}^{k} (\alpha_1^{(i)} X_1 + \cdots + \alpha_n^{(i)} X_n) \quad .$$

A form $N(\underline{X})$ of the type $N(\underline{X}) = \mathfrak{N}(M(\underline{X}))$ will be called a <u>norm form</u>. For example, if $\vartheta = \sqrt[4]{2}$ and $K = \mathbb{Q}(\vartheta)$, and if $M(X_1, X_2, X_3) = X_1 + \vartheta X_2 + \vartheta^2 X_3$, then

$$\mathfrak{N}(M(\underline{X})) = X_1^4 - 2X_2^4 + 4X_3^4 - 4X_1^2 X_3^2 + 8X_1 X_2^2 X_3 \quad .$$

A norm form is a form of degree $k$ with rational coefficients. These coefficients will be rational integers if $\alpha_1, \dots, \alpha_n$ are integers in $K$.

A <u>norm form equation</u> is an equation of the type

$$\mathfrak{N}(M(\underline{x})) = a \quad , \tag{1.2}$$

where $a \neq 0$ is a given rational number, and one is interested in solutions $\underline{\underline{x}} = (x_1,\ldots,x_n)$ with rational integer components. As $\underline{\underline{x}}$ runs through integer points, $M(\underline{\underline{x}})$ will run through a subset $\mathfrak{M}$ of $K$. This subset $\mathfrak{M}$ is clearly a <u>module</u> in $K$, i.e. it is a nonempty subset which is closed under addition and subtraction, it is closed under multiplication by elements of $\mathbb{Z}$, the rational integers, <u>and</u> there is a fixed rational integer $z \neq 0$ such that $z\mu$ is an algebraic integer for every $\mu \in \mathfrak{M}$. The equation (1.2) may be written in the form

$$\mathfrak{N}(\mu) = a \quad , \tag{1.3}$$

and we are seeking solutions $\mu \in \mathfrak{M}$.

<u>LEMMA 1A</u>. <u>A module $\mathfrak{M}$ in $K$ has a basis. That is, there is an $n$ in $1 \leq n \leq k$ and there are elements $\beta_1,\ldots,\beta_n$ in $\mathfrak{M}$, linearly independent over $\mathbb{Q}$, such that every $\mu \in \mathfrak{M}$ may uniquely be written as</u>

$$\mu = \beta_1 y_1 + \cdots + \beta_n y_n$$

<u>with $y_1,\ldots,y_n$ in $\mathbb{Z}$</u>.

For a proof, see e.g. the Corollary on p. 85 of Borevich and Shafarevich (1966). The number $n$ depends only on $\mathfrak{M}$ and is called the <u>rank</u> of $\mathfrak{M}$. In view of the lemma we may restrict ourselves to forms $M$ whose coefficients $\alpha_1,\ldots,\alpha_n$ are linearly independent over $\mathbb{Q}$.

<u>LEMMA 1B</u>. <u>The norm form</u>

$$\mathfrak{N}(\alpha_1 X_1 + \cdots + \alpha_n X_n)$$

<u>is irreducible over the rationals if and only if</u> $K = \mathbb{Q}(\alpha_2/\alpha_1,\ldots,\alpha_n/\alpha_1)$.

<u>Proof</u>. Since the form equals $\mathfrak{N}(\alpha_1)\mathfrak{N}(X_1 + (\alpha_2/\alpha_1)X_2 + \cdots + (\alpha_n/\alpha_1)X_n)$,

we may suppose without loss of generality that $\alpha_1 = 1$ . Now if $L = \mathbb{Q}(\alpha_2,\ldots,\alpha_n)$ , then

$$\mathfrak{N}(X_1 + \alpha_2X_2 + \cdots + \alpha_nX_n) = (\mathfrak{N}_L(X_1 + \alpha_2X_2 + \cdots + \alpha_nX_n))^{[K:L]},$$

where $\mathfrak{N}_L$ is the norm from $L$ to $\mathbb{Q}$ and $[K:L]$ is the degree of $K$ over $L$ . Hence $K = L$ is necessary for the form to be irreducible. It is also sufficient, for if $K = L$ , then by the Theorem on the Primitive Element (Van der Waerden (1955), §43) we have $K = \mathbb{Q}(\beta)$ where $\beta = c_2\alpha_2 + \cdots + c_n\alpha_n$ for certain rational $c_2,\ldots,c_n$ . Since $\beta$ is of degree $k$ , the form $\mathfrak{N}(X + \beta Y)$ is irreducible, and hence, since $\mathfrak{N}(X + \beta Y) = \mathfrak{N}(X + \alpha_2c_2Y + \cdots + \alpha_nc_nY)$ , the given form is a fortiori irreducible.

An irreducible form $F(X,Y)$ of degree $k$ may be written as

$$F(X,Y) = c(X - \alpha^{(1)}Y)\cdots(X - \alpha^{(k)}Y) = c\,\mathfrak{N}(X - \alpha Y)$$

where $\alpha^{(1)} = \alpha$ , $\alpha^{(2)},\ldots,\alpha^{(k)}$ are the roots of $F(X,1)$ and where $\mathfrak{N}$ is the norm in $K = \mathbb{Q}(\alpha)$ . Hence irreducible forms and irreducible norm forms in two variables are essentially the same. It follows that a "Thue equation" $F(x,y) = a$ can be written as a norm form equation $\mathfrak{N}(\alpha_1x + \alpha_2y) = a'$ , and vice versa. By Thue's Theorem (Theorem 3A of Ch. V) <u>an irreducible norm form equation in two variables and of degree</u> $k \geq 3$ <u>has only a finite number of solutions.</u>

When $n \geq 3$ , an irreducible form is in general not a norm form.

A module $\mathfrak{M}$ is called <u>full</u> if it has rank $n = k$ . A full module which contains the number 1 and which is a ring is called an <u>order</u> of the field $K$ . In particular the algebraic integers in $K$ form an order. If $\mathfrak{O}$ is an order and $\mu \in \mathfrak{O}$ , then $\mu^h \in \mathfrak{O}$ for $h = 1,2,\ldots$ . Hence with $z$ the rational integer mentioned above, $z\mu^h$ is an algebraic integer for $h = 1,2,\ldots$ , so that $\mu$ itself must be an algebraic integer[†)]. Hence every order $\mathfrak{O}$ is contained in the order of algebraic integers, which is therefore called the maximal order of $K$ .

The units in an order $\mathfrak{O}$ are the divisors of $1$ , and it is easily seen that $\varepsilon \in \mathfrak{O}$ is a unit precisely if $\mathfrak{N}(\varepsilon) = \pm 1$ . The units form a group.

LEMMA 1C. *Let $\mathfrak{O}$ be an order in the field $K$ . Then its group of units is infinite except when $K = \mathbb{Q}$ or $K$ is imaginary quadratic.*

This is an immediate consequence of Dirichlet's Unit Theorem when extended to orders. (See Borevich and Shafarevich, Theorem 5 on p. 112).

Given a full module $\mathfrak{M}$ , write $\mathfrak{O}_{\mathfrak{M}}$ for the set of $\lambda \in K$ such that $\lambda\mathfrak{M} \subseteq \mathfrak{M}$ , i.e. that $\lambda\mu \in \mathfrak{M}$ for every $\mu \in \mathfrak{M}$ . Then $\mathfrak{O}_{\mathfrak{M}}$ is easily seen to be an order in $K$ ; it is called the coefficient ring of $\mathfrak{M}$ . Write $\mathfrak{E}_{\mathfrak{M}}$ for the group of the units in $\mathfrak{O}_{\mathfrak{M}}$ which have norm $1$ . It is a subgroup of index $1$ or $2$ in the group of all the units of $\mathfrak{O}_{\mathfrak{M}}$ . Hence it is infinite except when $K$ is rational or imaginary quadratic. Now if $\mu \in \mathfrak{M}$ is a solution of (1.3) , then also $\varepsilon\mu$ is a solution for $\varepsilon \in \mathfrak{E}_{\mathfrak{M}}$ . Thus if $\mathfrak{M}$ is a full module, and if $K$ is not exceptional as indicated above, then given a solution of (1.3) , we have infinitely many solutions.

But full modules are not the only ones for which (1.3) may have infinitely many solutions. Suppose $L$ is a subfield of $K$ amd $\mathfrak{M}_0$ is a submodule of $\mathfrak{M}$ which is proportional to a full module in $L$ , i.e. $\mathfrak{M}_0 = \sigma\mathfrak{L}$ where $\sigma \neq 0$ is fixed and $\mathfrak{L}$ is a full $L$-module. Unless $L$ is exceptional, there will be infinitely many $\lambda \in \mathfrak{L}$ which satisfy

†) This follows from ideal theory. A more direct argument is that the ring $\mathbb{Z}[\mu]$ is a finite $\mathbb{Z}$-module (Zariski-Samuel (1958), §V.1).

$$\mathfrak{N}_L(\lambda) = b$$

for a certain b . But then $\sigma\lambda$ lies in $\mathfrak{M}$ and satisfies

$$\mathfrak{N}(\sigma\lambda) = \mathfrak{N}(\sigma)(\mathfrak{N}_L(\lambda))^{[K:L]} = \mathfrak{N}(\sigma)b^{[K:L]} = a \quad ,$$

say. We therefore define a module $\mathfrak{M}$ in K to be <u>degenerate</u> if it contains a submodule which is proportional to a full module in some subfield L of K , where L is neither rational nor imaginary quadratic. If $\mathfrak{M}$ is degenerate, then for certain values of a we have infinitely many solutions of (1.3) with $\mu \in \mathfrak{M}$ .

We have proved (the easy) half of

THEOREM 1D. *Let $\mathfrak{M}$ be a module in K . There exists an a such that (1.3) , i.e.*

$$\mathfrak{N}(\mu) = a \quad ,$$

*has infinitely many solutions $\mu \in \mathfrak{M}$ , if and only if $\mathfrak{M}$ is degenerate.*

In the next sections we will formulate and prove a result which is stronger than Theorem 1D . In sections 10-14 we will discuss equations of the type

$$\mathfrak{N}(L(\underline{\underline{x}})) = P(\underline{\underline{x}}) \quad .$$

## §2. Full Modules.

Much stronger than Lemma 1C is

LEMMA 2A. *Let $\mathfrak{O}$ be an order in K and $\mathfrak{U}$ the group of units of $\mathfrak{O}$ which have norm 1 . Then*

*(i) $\mathfrak{U}$ is a subgroup of finite index in the group $\mathfrak{U}^K$ of units of K, i.e. the units of the ring of integers in K .*

(ii) <u>There is a constant</u> $\gamma_1 = \gamma_1(\mathfrak{O})$ <u>such that for</u> $\mu \in K$ <u>there is an</u> $\varepsilon \in \mathfrak{E}$ <u>with</u>

$$|(\varepsilon\mu)^{(i)}| \leqq \gamma_1 |\mathfrak{N}(\mu)|^{1/k} \qquad (i = 1,\ldots,k) .$$

<u>Proof</u>. Both assertions are easy consequences of Dirichlet's Unit Theorem for orders.

Now let $\mathfrak{M}$ be a full module, $\mathfrak{O}_{\mathfrak{M}}$ its coefficient ring, $\mathfrak{E}_{\mathfrak{M}}$ the group of units of $\mathfrak{O}_{\mathfrak{M}}$ with norm 1 . Write

$$\mu\mathfrak{E}_{\mathfrak{M}} \tag{2.1}$$

for the set of elements $\mu\varepsilon$ with $\varepsilon \in \mathfrak{E}_{\mathfrak{M}}$ . If $\mu$ is a solution of (1.3), then so is every element of (2.1) . Hence we call (2.1) a <u>family of solutions</u> of (1.3) . By the second assertion of the lemma, every family of solutions contains a solution $\mu$ having

$$|\mu^{(i)}| \leqq \gamma_1 |a|^{1/k} \qquad (i = 1,\ldots,k) .$$

Since there are only finitely many $\mu$ in $\mathfrak{M}$ satisfying these inequalities, <u>there are only finitely many families of solutions</u>. Moreover, a representative of each family may be found in a finite number of steps. Thus the equation (1.3) can be completely solved if $\mathfrak{M}$ is a full module.

## <u>§3. An Example</u>.

It will be illuminating to study an example of a degenerate module. Let $K$ be the field $K = \mathbb{Q}(\sqrt{2},\sqrt{3})$ and $\mathfrak{M}$ the module with basis $\sqrt{2}$, $\sqrt{3}$, $\sqrt{6}$ . We shall express this by writing $\mathfrak{M} = \{x\sqrt{2} + y\sqrt{3} + z\sqrt{6}\}$ . This module is not full in $K$ . Let $L_1, L_2, L_3$ be the following subfields of $K$ :

$$L_1 = \mathbb{Q}(\sqrt{2}) \quad , \quad L_2 = \mathbb{Q}(\sqrt{3}) \quad , \quad L_3 = \mathbb{Q}(\sqrt{6}) \quad ,$$

and let $\mathfrak{M}_1, \mathfrak{M}_2, \mathfrak{M}_3$ be the submodules of $\mathfrak{M}$ given by

$$\mathfrak{M}_1 = \{y\sqrt{3} + z\sqrt{6})\} \ , \ \mathfrak{M}_2 = \{x\sqrt{2} + z\sqrt{6}\} \quad , \ \mathfrak{M}_3 = \{x\sqrt{2} + y\sqrt{3}\} \quad .$$

The module $\mathfrak{M}_i$ is proportional to a full module in $L_i$ : We have $\mathfrak{M}_i = \sigma_i \mathfrak{M}_i'$ $(i = 1,2,3)$ where $\sigma_1 = \sqrt{3}$, $\sigma_2 = \sqrt{2}$, $\sigma_3 = 1/\sqrt{2}$ and where $\mathfrak{M}_1' = \{y + z\sqrt{2}\}$, $\mathfrak{M}_2' = \{x + z\sqrt{3}\}$, $\mathfrak{M}_3' = \{2x + y\sqrt{6}\}$ , so that $\mathfrak{M}_i'$ is full in $L_i$ $(i = 1,2,3)$ .

We now keep $i$ fixed for the moment and study solutions of

$$\text{(3.1i)} \qquad \mathfrak{N}(\mu_i) = a \quad \text{with} \quad \mu_i \in \mathfrak{M}_i \quad .$$

Clearly $\mu_i$ is a solution precisely if $\mu_i = \sigma_i \mu_i'$ where $\mu_i'$ is a solution of

$$\mathfrak{N}(\mu_i') = a\mathfrak{N}(\sigma_i)^{-1} \ .\text{with} \quad \mu_i' \in \mathfrak{M}_i' \quad .$$

We have $\mathfrak{N}(\mu_i') = (\mathfrak{N}_i(\mu_i'))^2$ where $\mathfrak{N}_i$ is the norm from $L_i$ to $\mathbb{Q}$ , and our equation becomes

$$\text{(3.2i)} \qquad \mathfrak{N}_i(\mu_i') = \pm\sqrt{a}\sqrt{\mathfrak{N}(\sigma_i)^{-1}} \quad \text{with} \quad \mu_i' \in \mathfrak{M}_i' \quad .$$

Since $\mathfrak{N}(\sigma_i)$ is easily seen to be a rational square, a necessary condition for the existence of a solution $\mu_i'$ is that $a$ is a rational square. Since $\mathfrak{M}_i'$ is full in $L_i$ , all the solutions of (3.2i) lie in finitely many families $\nu_{i1}\mathfrak{E}_i, \dots, \nu_{ir_i}\mathfrak{E}_i$ , where $\mathfrak{E}_i$ is the full[†)] group of units in $\mathfrak{O}_{\mathfrak{M}_i'} = \mathfrak{O}^{L_i}_{\mathfrak{M}_i'}$,[*)] and is infinite. Hence all the solutions $\mu_i$ of (3.1i) lie in the families

$$\text{(3.3i)} \qquad \mu_{i1}\mathfrak{E}_i, \dots, \mu_{ir_i}\mathfrak{E}_i$$

†) We here use the full group of units because of the $\pm$ sign in (3.2i).
*) I.e. the coefficient ring of $\mathfrak{M}_i'$ in $L_i$ .

with $\mu_{ij} = \sigma_i \nu_{ij}$ $(j = 1,\dots,r_i)$ .

It will follow from Theorem 4B below that all but finitely many solutions of (1.3) lie in one of the submodules $\mathfrak{M}_1, \mathfrak{M}_2, \mathfrak{M}_3$ . Hence all but finitely many solutions lie in one of the families (3.3i) with $i = 1,2,$ or $3$ . In particular, (1.3) has only finitely many solutions unless $a$ is a rational square. All but finitely many solutions of the diophantine equation

$$\mathfrak{N}(x\sqrt{2} + y\sqrt{3} + z\sqrt{6}) = a$$

have $xyz = 0$ .

In the special case when $a = 36$ , the solutions of (3.2i) are easily seen to consist of the single family $\nu_i \mathfrak{U}_i$ where $\nu_1 = \sqrt{2}$ , $\nu_2 = \sqrt{3}$ , $\nu_3 = 6 + 2\sqrt{6}$ . Furthermore, $\mathfrak{U}_i$ consists of $\pm\varepsilon_i^n$ $(n = 0,\pm 1,\cdots)$ where $\varepsilon_1 = 1 + \sqrt{2}$ , $\varepsilon_2 = 2 + \sqrt{3}$ , $\varepsilon_3 = 5 + 2\sqrt{6}$. Hence all but finitely many solutions of

$$(3.4) \qquad \mathfrak{N}(\mu) = 36 \quad \text{with} \quad \mu \in \mathfrak{M}$$

lie in one of the families

$$\pm\sqrt{6}(1 + \sqrt{2})^n , \quad \pm\sqrt{6}(2 + \sqrt{3})^n , \quad \pm(3\sqrt{2} + 2\sqrt{3})(5 + 2\sqrt{6})^n$$

where $n = 0,\pm 1,\cdots$ .

The verification of these assertions concerning (3.4) is left as an Exercise.

## §4. The General Case.

Let $\mathfrak{M}$ be a module in $K$ . Write $\mathbb{Q}\mathfrak{M}$ for the set of products $q\mu$ with $q \in \mathbb{Q}$ , $\mu \in \mathfrak{M}$ ; it is a vector space over $\mathbb{Q}$ . Given a subfield $L$ of $K$ , let

$\mathfrak{M}^L$

consist of the elements $\mu$ of $\mathfrak{M}$ with $L\mu \subseteq \mathbb{Q}\mathfrak{M}$, i.e. with $\lambda\mu \in \mathbb{Q}\mathfrak{M}$ for every $\lambda \in L$. It is easily seen that $\mathfrak{M}^L$ is a submodule of $\mathfrak{M}$. We always have $\mathfrak{M}^{\mathbb{Q}} = \mathfrak{M}$, and $\mathfrak{M}^L \subseteq \mathfrak{M}^{L'}$ if $L \supseteq L'$. We have $\mathfrak{M}^K = \mathfrak{M}$ if $\mathfrak{M}$ is full and $\mathfrak{M}^K = \{0\}$ otherwise.

Remarks. (i) If $\mathfrak{M}$ has the basis $\mu_1,\dots,\mu_n$, if $\mu_1,\dots,\mu_n,\dots,\mu_k$ is a basis of $K$, if $\lambda_1,\dots,\lambda_\ell$ is a basis of $L$, and if

$$\lambda_i\mu_j = \sum_{t=1}^{k} c_{ijt}\mu_t \qquad (1 \leqq i \leqq \ell\,, 1 \leqq j \leqq k)\,,$$

then an element $z_1\mu_1 + \cdots + z_n\mu_n$ of $\mathfrak{M}$ is in $\mathfrak{M}^L$ precisely if

$$\sum_{j=1}^{n} z_j c_{ijt} = 0 \qquad (1 \leqq i \leqq \ell\,, n < t \leqq k)\,.$$

Hence a basis for $\mathfrak{M}^L$ can easily be constructed.

(ii) If $\lambda \in L$, $\mu \in \mathfrak{M}^L$, then $\lambda\mu \in \mathbb{Q}\mathfrak{M}$, so $z\lambda\mu \in \mathfrak{M}$ for a rational $z \neq 0$. If $\lambda' \in L$ then $\lambda'(z\lambda\mu) = (z\lambda'\lambda)\mu \in L\mathfrak{M}^L \subseteq \mathbb{Q}\mathfrak{M}$, and therefore $z\lambda\mu \in \mathfrak{M}^L$ and $\lambda\mu \in \mathbb{Q}\mathfrak{M}^L$. Thus

(4.1) $$L\mathfrak{M}^L = \mathbb{Q}\mathfrak{M}^L\,,$$

and $\mathbb{Q}\mathfrak{M}^L$ is a vector space <u>over L</u>.

(iii) We have

(4.2) $$L\mathfrak{M}^L \cap \mathfrak{M} = \mathfrak{M}^L\,,$$

because $\mu \in L\mathfrak{M}^L \cap \mathfrak{M}$ has $L\mu \subseteq L\mathfrak{M}^L = \mathbb{Q}\mathfrak{M}^L \subseteq \mathbb{Q}\mathfrak{M}$.

(iv) In the example of §3, $K$ has the subfields $\mathbb{Q}, L_1, L_2, L_3, K$. We have $\mathfrak{M}^{\mathbb{Q}} = \mathfrak{M}$, $\mathfrak{M}^{L_1} = \mathfrak{M}_1$, $\mathfrak{M}^{L_2} = \mathfrak{M}_2$, $\mathfrak{M}^{L_3} = \mathfrak{M}_3$, and $\mathfrak{M}^K = \{0\}$.

Now let $\mathfrak{O}^L_{\mathfrak{M}}$ be the ring of coefficients of $\mathfrak{M}^L$, i.e. the set of numbers $\lambda$ in $L$ such that $\lambda\mu \in \mathfrak{M}^L$ for every $\mu \in \mathfrak{M}^L$.

<u>LEMMA 4A</u>. <u>Suppose</u> $\mathfrak{M}^L \neq \{0\}$ . <u>Then</u> $\mathfrak{O}^L_{\mathfrak{M}}$ <u>is an order in</u> L .

<u>Proof</u>. $\mathfrak{O}^L_{\mathfrak{M}}$ clearly is a subring of L containing 1 . Now let $\mu_0$ be an element of $\mathfrak{M}^L$ distinct from 0 . Then $\mu_0 \mathfrak{O}^L_{\mathfrak{M}} \subseteq \mathfrak{M}^L$ , whence $\mathfrak{O}^L_{\mathfrak{M}} \subseteq \mu_0^{-1}\mathfrak{M}^L$ . Now $\mathfrak{M}^L$ and therefore also $\mu_0^{-1}\mathfrak{M}^L$ is a module in K , and hence there is a rational integer $z \neq 0$ such that $z\mu_0^{-1}\mathfrak{M}^L$ and hence a fortiori $z\mathfrak{O}^L_{\mathfrak{M}}$ contains only algebraic integers. This shows that $\mathfrak{O}^L_{\mathfrak{M}}$ is a module in L .

It remains to be shown that $\mathfrak{O}^L_{\mathfrak{M}}$ is full in L . For every $\mu \in \mathfrak{M}^L$ and every $\lambda \in L$ there is by Remark (ii) above a rational integer $z \neq 0$ with $z\lambda\mu \in \mathfrak{M}^L$ . Let $\mu_1,\ldots,\mu_r$ be a basis of $\mathfrak{M}^L$ . For every $\lambda \in L$ there are non-zero rational integers $z_1,\ldots,z_r$ with $z_i\lambda\mu_i \in \mathfrak{M}^L$ $(i = 1,\ldots,r)$ . Thus $z = z_1\cdots z_r$ has $z\lambda\mu_i \in \mathfrak{M}^L$ $(i = 1,\ldots,r)$ , and this implies that $z\lambda\mu \in \mathfrak{M}^L$ for every $\mu$ of $\mathfrak{M}^L$ . Therefore $z\lambda \in \mathfrak{O}^L_{\mathfrak{M}}$ , and since $\lambda$ was arbitrary in L and since $z = z(\lambda) \neq 0$ , the module $\mathfrak{O}^L_{\mathfrak{M}}$ is full in L .

Let $\mathfrak{E}^L$ be the group of units of the ring of integers in L . Let $\mathfrak{E}^L_{\mathfrak{M}}$ be the group of elements $\varepsilon$ in $\mathfrak{O}^L_{\mathfrak{M}}$ with $\mathfrak{N}(\varepsilon) = 1$ . (Here as always $\mathfrak{N}$ denotes the norm from K to $\mathbb{Q}$ , and hence $\mathfrak{E}^L_{\mathfrak{M}} = {}^K\mathfrak{E}^L_{\mathfrak{M}}$ depends on K as well as on L and $\mathfrak{M}$ .) By Lemma 2A the group $\mathfrak{E}^L_{\mathfrak{M}}$ is a subgroup of finite index in $\mathfrak{E}^L$ .

If the equation $\mathfrak{N}(\mu) = a$ has a solution $\mu$ in $\mathfrak{M}^L$ , then every member of $\mu\mathfrak{E}^L_{\mathfrak{M}}$ is a solution. We shall call $\mu\mathfrak{E}^L_{\mathfrak{M}}$ a <u>family</u> of solutions, or more precisely an $(\mathfrak{M},L)$-family of solutions.

<u>THEOREM 4B</u>. <u>The solutions of</u> (1.3) <u>are contained in finitely many families of solutions</u>.

Thus for every subfield L of K there are finitely many $(\mathfrak{M},L)$-

families of solutions, such that $\mu$ is a solution of (1.3) precisely if it lies in one of these families for some $L$ .

Let us again look at the example of §3. The subfields of $K = \mathbb{Q}(\sqrt{2}, \sqrt{3})$ are $K, L_1, L_2, L_3, \mathbb{Q}$ . Since $\mathfrak{M}^K = \{0\}$ , the solutions of (1.3) are contained in a finite number of $(\mathfrak{M}, L_1)$- , $(\mathfrak{M}, L_2)$- , $(\mathfrak{M}, L_3)$- , and $(\mathfrak{M}, \mathbb{Q})$-families. Since each $(\mathfrak{M}, \mathbb{Q})$-family consists of only two elements $\mu$, $-\mu$ , all but finitely many solutions belong to an $(\mathfrak{M}, L_i)$-family, and in particular they lie in one of the submodules $\mathfrak{M}_1, \mathfrak{M}_2$, or $\mathfrak{M}_3$ .

If (1.3) has infinitely many solutions, then by Theorem 4B we must have $\mathfrak{M}^L \neq \{0\}$ for some field $L$ which is not rational or imaginary quadratic. But then if $\sigma \neq 0$ is in $\mathfrak{M}^L$ and if $\lambda_1, \ldots, \lambda_\ell$ is a basis of the field $L$ , we have $z_1\lambda_1\sigma, \ldots, z_\ell\lambda_\ell\sigma$ in $\mathfrak{M}^L$ for suitable nonzero rationals $z_1, \ldots, z_\ell$ , so that $\mathfrak{M}^L$ contains the module $\sigma\mathfrak{L}$ , where $\mathfrak{L}$ is the full module in $L$ with basis $z_1\lambda_1, \ldots, z_\ell\lambda_\ell$ . Thus $\mathfrak{M}$ contains a submodule (namely $\sigma\mathfrak{L}$ ) which is proportional to a full module in $L$ , and hence $\mathfrak{M}$ is degenerate. Therefore Theorem 4B implies Theorem 1D .

It is possible to construct in a finite number of steps a basis for $\mathfrak{M}^L$ and a set of generators of $\mathfrak{E}^L_{\mathfrak{M}}$ for every subfield $L$ of $K$ . But Theorem 4B is ineffective, since in general we have no method to construct a finite number of families of solutions which contain all the solutions. This is a consequence of the fact that our method of proof depends on the preceding chapter, which in turn is a development of the non-effective method of Thue, Siegel, and Roth. In particular, if $\mathfrak{M}$ is degenerate, we have no effective method to decide whether a particular equation (1.3) (with particular $a$ ) has infinitely many solutions. However, by combining the present method with Baker's method one can show that _if the rank of_ $\mathfrak{M}$ _is at most_ 5 , _then one can determine in a finite_ number of steps whether there are infinitely many solutions. For this, as well as the concept of maximal families of solutions (which are shown to be finite in number), see Schmidt (1972).

## §5. Induction on the rank of $\mathfrak{M}$.

We shall prove Theorem 4B by induction on the rank of $\mathfrak{M}$. If $\mathfrak{M}$ has rank 1, then its elements are $x\mu_1$, where $\mu_1 \neq 0$ is fixed and $x$ runs through the rational integers. The equation $\mathfrak{N}(\mu) = a$ with $\mu \in \mathfrak{M}$ becomes $\mathfrak{N}(x\mu_1) = x^k\mathfrak{N}(\mu_1) = a$ where $k$ is the degree of $K$. If $k$ is odd and if there is a solution $\mu = x\mu_1$ at all, then it is the only solution and forms an $(\mathfrak{M},\mathbb{Q})$-family by itself. If $k$ is even and if there is a solution $\mu = x\mu_1$, then there are altogether two solutions $\mu$ and $-\mu$, which form a single $(\mathfrak{M},\mathbb{Q})$-family, since now $\mathfrak{E}^{\mathbb{Q}}_{\mathfrak{M}}$ consists of $+1, -1$.

In what follows we shall assume that rank $\mathfrak{M} = n > 1$, and that Theorem 4B is true for modules of rank less than $n$. The following lemma will be useful.

LEMMA 5A. *Suppose $L$ is a subfield of $K$ and $\mathfrak{E}^L$ is the group of units of the ring of integers of $L$. Let $\mathfrak{E}^L_1$, $\mathfrak{E}^L_2$ be subgroups of finite index in $\mathfrak{E}^L$. Then for every $\kappa \in K$ there are finitely many elements $\kappa_1,\cdots,\kappa_r$ in $\kappa\mathfrak{E}^L_1$ such that $\kappa\mathfrak{E}^L_1$ is contained in the union of the sets*

$$\kappa_1\mathfrak{E}^L_2,\cdots,\kappa_r\mathfrak{E}^L_2 . \tag{5.1}$$

*Proof*. The intersection $\mathfrak{E}^L_0 = \mathfrak{E}^L_1 \cap \mathfrak{E}^L_2$ is a subgroup of finite index in $\mathfrak{E}^L_1$. Hence $\mathfrak{E}^L_1$ is the union of finitely many cosets $\varepsilon_1\mathfrak{E}^L_0,\cdots,\varepsilon_r\mathfrak{E}^L_0$ with $\varepsilon_i \in \mathfrak{E}^L_1$ $(i = 1,\cdots,r)$, and hence it is contained in the union of the sets $\varepsilon_1\mathfrak{E}^L_2,\cdots,\varepsilon_r\mathfrak{E}^L_2$. This implies that $\kappa\mathfrak{E}^L_1$ is contained in the union of the sets (5.1) with $\kappa_i = \kappa\varepsilon_i \in \kappa\mathfrak{E}_1^L$ $(i = 1,\cdots,r)$ .

<u>COROLLARY 5B</u>. <u>Suppose</u> $\mathfrak{M}$ <u>is a module in</u> K <u>and</u> $\mathfrak{M}'$ <u>is a sub-module of</u> $\mathfrak{M}$ . <u>Let</u> L <u>be a subfield of</u> K . <u>Then every</u> $(\mathfrak{M}', L)$-<u>family of solutions is contained in the union of finitely many</u> $(\mathfrak{M}, L)$-<u>families of solutions</u>.

<u>Proof</u>. Let $\mu'\mathfrak{E}^L_{\mathfrak{M}'}$ be an $(\mathfrak{M}', L)$-family of solutions. It is contained in a union $\mu_1\mathfrak{E}^L_{\mathfrak{M}} \cup \cdots \cup \mu_r\mathfrak{E}^L_{\mathfrak{M}}$ where $\mu_i \in \mu'\mathfrak{E}^L_{\mathfrak{M}'} \subseteq \mathfrak{M}'^L \subseteq \mathfrak{M}^L$ .

Given subfields A,B of K , write AB for their compositum, i.e. the smallest subfield of K containing both A and B .

<u>LEMMA 5C</u>. <u>For any two subfields</u> A,B <u>of</u> K , <u>we have</u>

$$(\mathfrak{M}^A)^B = \mathfrak{M}^{AB} .$$

<u>Moreover, if</u> $A \subseteq B$ , <u>then</u>

$$\mathfrak{D}^A_{\mathfrak{M}} \subseteq \mathfrak{D}^B_{\mathfrak{M}} \text{ and } \mathfrak{E}^A_{\mathfrak{M}} \subseteq \mathfrak{E}^B_{\mathfrak{M}} . \tag{5.2}$$

<u>Proof</u>. If $\mu \in \mathfrak{M}^{AB}$ , then $\mu \in \mathfrak{M}^A$ and

$$B\mu \subseteq AB\mathfrak{M}^{AB} = \mathfrak{M}^{AB} \subseteq \mathfrak{M}^A ,$$

so that $\mu \in (\mathfrak{M}^A)^B$ . Conversely, if $\mu \in (\mathfrak{M}^A)^B$ and if $\alpha \in A$ , $\beta \in B$ , then

$$\alpha\beta\mu \in \alpha B\mu \subseteq \alpha\mathfrak{M}^A \subseteq A\mathfrak{M}^A = \mathfrak{M}^A \subseteq \mathfrak{M} .$$

Since every element of AB is of the type $\alpha_1\beta_1 + \cdots + \alpha_t\beta_t$ with $\alpha_i \in A$ , $\beta_i \in B$ , we get $AB\mu \subseteq \mathfrak{M}$ , and in conjunction with $\mu \in \mathfrak{M}$ this yields $\mu \in \mathfrak{M}^{AB}$ .

As for the second assertion of the lemma, suppose that $A \subseteq B$ and

$\alpha \in \mathfrak{L}_{\mathfrak{M}}^{A}$ . Now if $\mu \in \mathfrak{M}^B$ , then $\mu \in \mathfrak{M}^A$ , and $\alpha\mu \in \mathfrak{M}^A \subseteq \mathfrak{M}$ and $\alpha\mu \in B\mathfrak{M}^B$ , so that $\alpha\mu \in B\mathfrak{M}^B \cap \mathfrak{M} = \mathfrak{M}^B$ . This proves that $\alpha \in \mathfrak{L}_{\mathfrak{M}}^{B}$ , so that we have the first relation in (5.2) . The second relation is an immediate consequence of the first.

## §6. Linear Inequalities in a Simplex.

LEMMA 6A. *Suppose* $a_1,\cdots,a_n$ *are positive reals and* $a \geq 0$ , $b > 0$ . *For* $r = 1,\cdots,n$ , *put*

$$f(r) = (a_1 + \cdots + a_r)/r$$

*and*

$$H(r) = (a_1 + \cdots + a_n)b - (a + nb)f(r) \quad .$$

*Denote the maximum of* $f(r)$ *by* $f_1$, *and set* $H_1 = (a_1 + \cdots + a_n)b - (a + nb)f_1$ . *Then the minimum of*

$$H(y_1,\cdots,y_n) = a_1y_1 + \cdots + a_ny_n$$

*in the set of points* $(y_1,\cdots,y_n)$ *with*

(6.1) $$y_1 + \cdots + y_n + a \geqq 0 \quad \text{and} \quad y_1 \leqq \cdots \leqq y_n \leqq b$$

*is* $H_1$ .

*Moreover, if* $r_1 < \cdots < r_g$ *are the values of* $r$ *with* $f(r) = f_1$ , *and if* $g < n$ , *let* $f_2$ *be the maximum of* $f(r)$ *for* $r$ *distinct from* $r_1,\cdots,r_g$ . *Put* $H_2 = (a_1 + \cdots + a_n)b - (a + nb)f_2$ . *Then every* $(y_1,\cdots,y_n)$ *with* (6.1) *has*

(6.2) $$\begin{aligned} &H(y_1,\cdots,y_n) - H_1 \\ &\geqq (a + bn)^{-1}(H_2 - H_1)((y_{r_1} - y_1) + (y_{r_2} - y_{r_1+1}) + \cdots \\ &+ (y_{r_g} - y_{r_{g-1}+1}) + (b - y_{r_g+1})) \quad , \end{aligned}$$

*where we put* $y_{n+1} = b$ *if* $r_g = n$ .

*Proof.* The lemma may be considered as an exercise in Linear Programming. Since $H(y_1,\cdots,y_n)$ decreases and the right hand side of (6.2) does not decrease if we replace $(y_1,\cdots,y_n)$ by $(y_1-\delta,\cdots,y_n-\delta)$ for some $\delta > 0$ , we may restrict ourselves to $(y_1,\cdots,y_n)$ with

$$y_1 + \cdots + y_n + a = 0 \quad \text{and} \quad y_1 \leq \cdots \leq y_n \leq b \quad . \tag{6.3}$$

This set of n-tuples constitutes a simplex in the hyperplane $y_1 + \cdots + y_n + a = 0$ :

The points $\underline{p}_1,\ldots,\underline{p}_n$ given by

$$\underline{p}_r = (b,\ldots,b) - \frac{a+nb}{r}(\underbrace{1,1,\ldots,1}_{r},0,\ldots,0)$$

lie on the hyperplane $y_1 + \cdots + y_n + a = 0$ . Given a point $\underline{y} = (y_1,\ldots,y_n)$ on this hyperplane put $y_{n+1} = b$ and observe that

$$\sum_{r=1}^{n} (y_{r+1} - y_r)r = -y_1 - \cdots - y_n + nb = a + nb$$

and that

$$\sum_{r=1}^{n} (y_{r+1} - y_r)(\underbrace{1,\ldots,1}_{r},0,\ldots,0) = (b-y_1,\ldots,b-y_n) \quad .$$

Thus

$$\begin{aligned}\sum_{r=1}^{n} (y_{r+1} - y_r)r\underline{p}_r &= (a+nb)(b,\ldots,b) + (a+nb)(y_1-b,\ldots,y_n-b) \\ &= (a+nb)\underline{y} \quad ,\end{aligned}$$

so that

$$\underline{y} = \sum_{r=1}^{n} r\,\frac{y_{r+1}-y_r}{a+nb}\,\underline{p}_r \quad .$$

The sum of the coefficients here is 1 , and the coefficients are nonnegative precisely if $\underline{\underline{y}}$ satisfies (6.3). Hence (6.3) defines a simplex on the hyperplane $y_1 + \cdots + y_n + a = 0$ with vertices $\underline{\underline{p}}_1, \ldots, \underline{\underline{p}}_n$. Since

$$H(\underline{\underline{p}}_r) = (a_1 + \cdots + a_n)b - (a + nb)f(r) = H(r) \qquad (1 \leq r \leq n),$$

it is clear that $H_1 = \min(H(1), \ldots, H(n))$ is the minimum of $H(\underline{\underline{y}})$ with $\underline{\underline{y}}$ in (6.3), hence is the minimum with $\underline{\underline{y}}$ satisfying (6.1).

Now suppose that $f(r) = f_1$ for $r = r_1, \ldots, r_g$, and that $f(r) \leq f_2 < f_1$ otherwise. Then $H(\underline{\underline{p}}_r) \geq H_2$ for $r$ distinct from $r_1, \ldots, r_g$, and

$$H(\underline{\underline{y}}) \geq \sum_{r=1}^{n} r \frac{y_{r+1} - y_r}{a + nb} H_1 + \sum_{\substack{r=1 \\ r \neq r_1, \ldots, r_g}}^{n} r \frac{y_{r+1} - y_r}{a + nb}(H_2 - H_1)$$

$$\geq H_1 + \frac{H_2 - H_1}{a + nb}((y_{r_1} - y_1) + (y_{r_2} - y_{r_1+1}) + \cdots + (y_{r_g} - y_{r_{g-1}+1}) + (b - y_{r_g+1})).$$

## §7. Construction of a field L.

Let $\mathfrak{M}$ be a module of rank $n$ in $K$. The elements of $\mathfrak{M}$ are the values of a linear form $M(\underline{\underline{X}}) = \mu X_1 + \cdots + \mu_n X_n$ as the variables run through the integers. The coefficients $\mu_1, \ldots, \mu_n$ lie in $K$ and are linearly independent over $\mathbb{Q}$.

There are $k$ isomorphisms of $K$ into the complex numbers. As in §1 the images of an element $\alpha$ of $K$ under these isomorphisms are denoted by $\alpha^{(1)}, \ldots, \alpha^{(k)}$, and as in §1 we suppose that $\alpha^{(1)} = \alpha$. We shall need the conjugate linear forms

$$M^{(i)}(\underline{\underline{X}}) = \mu_1^{(i)} X_1 + \cdots + \mu_n^{(i)} X_n \qquad (1 \leq i \leq k).$$

<u>LEMMA 7A</u>. <u>The forms</u> $M^{(1)}, \ldots, M^{(k)}$ <u>have rank</u> $n$, <u>i.e. there are</u> $n$ <u>but not more than</u> $n$ <u>linearly independent forms among them.</u>

Proof. Complete $\mu_1,\ldots,\mu_n$ to a basis of $K$ , i.e. choose $\mu_{n+1},\ldots,\mu_k$ such that $\mu_1,\ldots,\mu_k$ is a basis of $K$ . It is then well known that the $(k\times k)$-matrix $\mu_j^{(i)}$ $(1\leq i,j\leq k)$ has rank $k$ , and hence the $(k\times n)$-submatrix $\mu_j^{(i)}$ $(1\leq i\leq k, 1\leq j\leq n)$ has rank $n$ . Hence $M^{(1)},\ldots,M^{(k)}$ have rank $n$ .

Given a subset $S$ of $\{1,\ldots,k\}$ , we shall write $|S|$ for its cardinality, $r(s)$ for the rank of the set of linear forms $M^{(i)}$ with $i\in S$ , and we shall put

$$q(S)=\begin{cases} r(S)/|S| & \text{if } S \text{ is non-empty,}\\ n/k & \text{if } S=\emptyset \text{ , i.e. if } S \text{ is empty.}\end{cases}$$

With this definition we have

$$q(\{1,2,\ldots,k\}) = n/k \quad . \tag{7.1}$$

Now let $\omega$ be a primitive element of $K$ over $\mathbb{Q}$ and put $K^* = \mathbb{Q}(\omega^{(1)},\cdots,\omega^{(k)})$ , so that $K^*$ is the least normal extension of $\mathbb{Q}$ containing $K$ . Let $G$ be the Galois group of $K^*$ over $\mathbb{Q}$ . An element $\varphi$ of $G$ permutes $\omega^{(1)},\cdots,\omega^{(k)}$ , and hence induces a permutation $\hat{\varphi}$ of $\{1,2,\cdots,k\}$ . Given a subset $S$ of $\{1,2,\ldots,k\}$ , the image $\hat{\varphi}(S)$ of $S$ under $\hat{\varphi}$ is again such a subset, and we have $|\hat{\varphi}(S)| = |S|$ , $r(\hat{\varphi}(S)) = r(S)$ and $q(\hat{\varphi}(S)) = q(S)$ .

LEMMA 7B. Let $q_0$ be the minimum value of $q(S)$ for all subsets $S$ of $\{1,2,\cdots,k\}$ . Then

$$q_0 = n/k \quad .$$

Moreover, there are integers $\ell,m$ with $\ell m = k$ , a subset $T$ with $|T| = m$ and elements $\varphi_1,\ldots,\varphi_\ell$ of $G$ such that $\{1,2,\cdots,k\}$ is the

<u>disjoint union of the sets</u>

(7.2) $$\hat{\varphi}_1(T), \cdots, \hat{\varphi}_\ell(T) \quad ,$$

<u>and a set</u> $S$ <u>has</u> $q(S) = q_0$ <u>if and only if it is the union of some of the sets in</u> (7.2) .

<u>Proof</u>. We have $q_0 \leqq n/k$, and since $q(\emptyset) = q(\{1,2,\cdots,k\}) = n/k$ , there is a non-empty set $T$ with $q(T) = q_0$ . Let $T$ be a <u>minimal</u> non-empty set with this property, i.e. suppose that $q(T) = q_0$ and that $q(T') > q_0$ if $0 < |T'| < |T|$ . Now let $S$ be an arbitrary non-empty set with $q(S) = q_0$ , and suppose $\varphi \in G$ . We have

$$|S \cap \hat{\varphi}(T)| + |S \cup \hat{\varphi}(T)| = |S| + |\hat{\varphi}(T)| = |S| + |T| \quad ,$$

$$r(S \cap \hat{\varphi}(T)) + r(S \cup \hat{\varphi}(T)) \leqq r(S) + r(\hat{\varphi}(T)) = r(S) + r(T) \quad .$$

Since $r(S) = q_0|S|$ , $r(T) = q_0|T|$ , and $r(S \cup \hat{\varphi}(T)) \geqq q_0|S \cup \hat{\varphi}(T)|$ , we have $r(S \cap \hat{\varphi}(T)) \leqq q_0|S \cap \hat{\varphi}(T)|$ . Hence either $S \cap \hat{\varphi}(T) = \emptyset$ or $S \cap \hat{\varphi}(T) \neq \emptyset$ and $q(S \cap \hat{\varphi}(T)) = q_0$ . Since $|S \cap \hat{\varphi}(T)| \leqq |T|$ and since $T$ was minimal with $q(T) = q_0$ , we must have $|S \cap \hat{\varphi}(T)| = |T|$ in this case. Hence either $S$ and $\hat{\varphi}(T)$ are disjoint or $S$ contains $\hat{\varphi}(T)$ .

In particular, if $\varphi, \psi$ are elements of $G$ and if we set $S = \hat{\psi}(T)$ , we see that $\hat{\varphi}(T)$ , $\hat{\psi}(T)$ are either disjoint or equal. It follows that $|T|$ is a divisor of $k$ , and if we put

$$m = |T| \quad , \quad \ell = k/m \quad ,$$

then there are elements $\varphi_1, \cdots, \varphi_\ell$ of $G$ such that $\{1,2,\cdots,k\}$ is the disjoint union of the sets (7.2) . Every non-empty set $S$ with $q(S) = q_0$ is a union of sets from (7.2) . Conversely, if $S$ is a union of $s \geqq 1$ of the sets (7.2) , then $|S| = sm$ , $r(S) \leqq sr(T)$ , whence

$q(S) \leq r(T)/m = q(T) = q_0$ , so that $q(S) = q_0$ . In particular, since $\{1,2,\cdots,k\}$ is the union of the sets (7.2) , we have $q_0 = q(\{1,2,\cdots,k\}) = n/k$ by (7.1) . Finally, since $q_0 = n/k$ , all the statements made above remain true if $S$ is the empty set, and the proof of Lemma 7B is complete.

It is clear that the set $T$ in the lemma may be replaced by any of the sets in (7.2) . One of these sets contains 1 , so that we may assume that $1 \in T$ . One of the sets in (7.2) is $T$ itself, and we may suppose that $\hat{\varphi}_1(T) = T$ . We may now replace $\varphi_1$ by the identity mapping. Then we have

(7.3) $\qquad 1 \in T$ , and $\varphi_1$ is the identity mapping .

Let $G_L$ be the subgroup of $G$ consisting of elements $\varphi$ with $\hat{\varphi}(T) = T$ . The left cosets of $G_L$ in $G$ are $\varphi_1 G_L = G_L$ , $\varphi_2 G_L, \cdots, \varphi_\ell G_L$ , so that $G_L$ has index $\ell$ in $G$ . Let $L$ be the fixed field of $G_L$ , i.e. the subfield of $K^*$ consisting of elements $\lambda$ having $\varphi(\lambda) = \lambda$ for every $\varphi \in G_L$ . Then $L$ is a field of degree $\ell$ . Now let $G_K$ be the subgroup of $G$ consisting of elements $\varphi$ with $\hat{\varphi}(1) = 1$ , i.e. with $\varphi(\omega^{(1)}) = \omega^{(1)}$ . It is clear that $K$ is the fixed field of $G_K$ . Now if $\varphi \in G_K$ , then $\hat{\varphi}(1) = 1$ , and since $1 \in T$ , and $\hat{\varphi}$ permutes the sets (7.2) , this implies that $\hat{\varphi}(T) = T$ , whence that $\varphi \in G_L$ . Hence $G_K \subseteq G_L$ and $L \subseteq K$ .

The conjugates of an element $\lambda \in L$ over $\mathbb{Q}$ are $\lambda = \varphi_1(\lambda), \cdots, \varphi_\ell(\lambda)$ . The conjugates of an element $\kappa \in K$ over $\mathbb{Q}$ are $\varphi_i(\kappa^{(j)})$ with $1 \leq i \leq \ell$ and $j \in T$ . The conjugates of an element $\kappa \in K$ over $L$ are $\kappa^{(j)}$ with $j \in T$ .

Suppose $T = \{t_1, \cdots, t_m\}$ . It will sometimes be convenient to write

$$(7.4)\qquad \begin{aligned} \lambda^{[i]} &= \varphi_i(\lambda) && (i = 1,\cdots,\ell) \text{ for } \lambda \in L \;, \\ \kappa^{[i,j]} &= \varphi_i(\kappa^{(t_j)}) && (i = 1,\cdots,\ell;\ j = 1,\cdots,m) \text{ for } \kappa \in K \;. \end{aligned}$$

Then every $\lambda \in L$ has

$$\lambda^{[i,j]} = \lambda^{[i]} \qquad (i = 1,\cdots,\ell;\ j = 1,\cdots,m).$$

Now let $P$ be a subfield of $K$. Let $p$ be the degree of $P$ (over $\mathbb{Q}$) and $s$ the degree of $K$ over $P$, so that $ps = k$. Let $\omega = \omega^{(1)}, \omega^{(i_2)}, \cdots, \omega^{(i_s)}$ be the conjugates of $\omega$ over $P$, and put $S = S(P) = \{1, i_2, \cdots, i_s\}$.

<u>LEMMA 7C</u>. $q(S(P)) = q_0$ *if and only if* $P \subseteq L$.

<u>Proof</u>. We have $q(S(P)) = q_0$ precisely if $S(P)$ is a union of sets (7.2). Now if $S(P)$ is such a union, then we have $T \subseteq S(P)$ since $1 \in T$ and $1 \in S(P)$. For every $\pi \in P$ we have $\pi^{(i)} = \pi$ for $i \in S(P)$, whence $\pi^{(i)} = \pi$ for $i \in T$, whence $\varphi(\pi) = \pi^{(\hat{\varphi}(1))} = \pi$ if $\hat{\varphi}(1) \in T$, whence $\varphi(\pi) = \pi$ if $\varphi \in G_L$, and $\pi$ must lie in $L$. Thus $P \subseteq L$. Conversely, suppose that $P \subseteq L$. Let $\lambda$ be a primitive element of $L$ (over $\mathbb{Q}$) and let $\lambda = \varphi_1(\lambda) = \varphi_{u_1}(\lambda)$, $\varphi_{u_2}(\lambda), \cdots, \varphi_{u_v}(\lambda)$ be the conjugates of $\lambda$ over $P$. Then the conjugates of $\omega$ over $P$ are $\varphi_{u_i}(\omega^{(j)})$ with $1 \leq i \leq v$ and $j \in T$, and $S(P) = \hat{\varphi}_{u_1}(T) \cup \cdots \cup \hat{\varphi}_{u_v}(T)$, so that $q(S(P)) = q_0$ by Lemma 7B.

<u>LEMMA 7D</u>. *A subfield* $P$ *of* $K$ *has* $\mathfrak{M}^P = \mathfrak{M}$ *if and only if* $P \subseteq L$.

<u>Proof</u>. Suppose $\mathfrak{M}$ contains $r$ but not more than $r$ linearly independent elements over $P$. Let $\delta_1, \cdots, \delta_r$ be such elements. Then

$$M(\underline{X}) = \delta_1 M_1(\underline{X}) + \cdots + \delta_r M_r(\underline{X}) \quad ,$$

where $M_1(\underline{\underline{X}}),\cdots,M_r(\underline{\underline{X}})$ are linearly independent linear forms with coefficients in $P$. The elements $\delta_1,\cdots,\delta_r$ can be extended to a basis $\delta_1,\cdots,\delta_r,\cdots,\delta_s$ of $K$ over $P$, and since the matrix $(\delta_j^{(i)})$ $(i \in S(P)$, $1 \leq j \leq s)$ is nonsingular, the matrix $(\delta_j^{(i)})$ $(i \in S(P)$, $1 \leq j \leq r)$ has rank $r$. Hence $r(S(P)) = r$ and $q(S(P)) = r/s$.

Now if $\mathfrak{M} = \mathfrak{M}^P$, then since $\delta_i \in \mathfrak{M}$ $(i = 1,\ldots,r)$ it follows that $P\delta_i \in \mathbb{Q}\mathfrak{M}$, and hence there is a basis $\pi_1,\ldots,\pi_p$ of $P$ over $\mathbb{Q}$ such that $\mathfrak{M}$ contains the $rp$ elements $\delta_i\pi_j$ $(1 \leq i \leq r,\ 1 \leq j \leq p)$, and $n = rp$. Conversely, if $n = rp$, then each of the forms $M_i(\underline{\underline{X}})$ above represents a basis†) of $P$ over $\mathbb{Q}$, and hence there is a basis $\pi_1,\ldots,\pi_p$ of $P$ such that $\mathfrak{M}$ contains $\delta_i\pi_j$ $(1 \leq i \leq r,\ 1 \leq j \leq p)$, that in fact $\mathbb{Q}\mathfrak{M}$ is spanned by these $rp$ elements over $\mathbb{Q}$, so that $\mathbb{Q}\mathfrak{M} = P\mathfrak{M}$ and $\mathfrak{M} = \mathfrak{M}^P$. Thus $\mathfrak{M} = \mathfrak{M}^P$ precisely if $n = rp = r(k/s) = kq(S(P))$, i.e. $q(S(P)) = n/k = q_0$. By Lemma 7C this holds if and only if $P \subseteq L$.

We note that the existence of a maximal subfield $L$ of $K$ with $\mathfrak{M}^L = \mathfrak{M}$ follows from Lemma 5C. But we now have more precise information on $L$.

## §8. The Main Lemma.

For an element $\kappa$ of our field $K$ put

$$(8.1)\qquad \lceil\kappa\rceil = \max(|\kappa^{(1)}|,\cdots,|\kappa^{(k)}|) .$$

Suppose we are given the equation (1.3), i.e.

$$(8.2)\qquad \mathfrak{N}(\mu) = \mu^{(1)}\cdots\mu^{(k)} = a \quad \text{with } \mu \in \mathfrak{M} .$$

LEMMA 8A. *Suppose* $\varepsilon > 0$. *There are finitely many submodules* $\mathfrak{M}_1,\cdots,\mathfrak{M}_w$ *of* $\mathfrak{M}$, *each of rank less than* $n = \operatorname{rank}\mathfrak{M}$, *such that every*

---

†) i.e. the value set of $M_1(\underline{\underline{x}})$ with $\underline{\underline{x}} \in \mathbb{Z}^n$ contains a basis.

solution $\mu$ of (8.2) either lies in one of these submodules, or satisfies

$$|\mu^{[i,h]}| < \overline{|\mu|}^{\varepsilon} |\mu^{[i,j]}| \qquad (1 \leq i \leq \ell ; 1 \leq h,j \leq m). \tag{8.3}$$

Proof. We shall at first deal only with solutions $\mu$ having

$$|\mu^{(1)}| \leq \cdots \leq |\mu^{(k)}| \quad . \tag{8.4}$$

We may suppose that $\varepsilon < 1$. Pick $\delta > 0$ so small that

$$\delta < \varepsilon/(6kn^2) \quad . \tag{8.5}$$

The elements of $\mathfrak{M}$ are the values of a linear form $M(\underline{\underline{x}})$ as $\underline{\underline{x}}$ runs through the integer points.

There is a form of this type whose coefficients are linearly independent over the rationals. Hence the forms $M^{(1)},\cdots,M^{(k)}$ have rank $n$ by Lemma 7A, and we have $|\underline{\underline{x}}| \leq \gamma_1 \overline{|M(\underline{\underline{x}})|}$ with a constant $\gamma_1$ independent of $\underline{\underline{x}}$. We are concerned with solutions $\underline{\underline{x}}$ of

$$M^{(1)}(\underline{\underline{x}}) \cdots M^{(k)}(\underline{\underline{x}}) = a \tag{8.6}$$

with

$$|M^{(1)}(\underline{\underline{x}})| \leq \cdots \leq |M^{(k)}(\underline{\underline{x}})| \quad . \tag{8.7}$$

Put $i_0 = 0$. Let $i_1$ be the largest integer such that $M^{(1)}$, $M^{(2)}, \cdots, M^{(i_1)}$ have rank 1. Let $i_2$ be the largest integer such that $M^{(1)},M^{(2)},\cdots,M^{(i_2)}$ have rank 2 ; and so on. We obtain numbers $0 = i_0 < i_1 < \cdots < i_n = k$. In view of (8.7) and since $M^{(i_j)}$ is a linear combination of $M^{(1)},\cdots,M^{(i_{j-1}+1)}$, we have

$$|M^{(i_j)}(\underline{\underline{x}})| \leq \gamma_2 |M^{(i_{j-1}+1)}(\underline{\underline{x}})| \qquad (j = 1,\cdots,n)$$

with constant $\gamma_2$. Hence all but finitely many integer points $\underline{\underline{x}}$ with (8.7) have

$$(8.8)\quad |M^{(i_j)}(\underline{\underline{x}})||\underline{\underline{x}}|^{-\delta} \leq |M^{(i_{j-1}+1)}(\underline{\underline{x}})| \leq |M^{(i_{j-1}+2)}(\underline{\underline{x}})| \leq \cdots \leq |M^{(i_j)}(\underline{\underline{x}})| \quad (j = 1,\cdots,n)\ .$$

Also, all but finitely many integer points have

$$(8.9)\quad |M^{(k)}(\underline{\underline{x}})| = |M^{(i_n)}(\underline{\underline{x}})| \leq \overline{|\underline{\underline{x}}|}^{1+\delta} \text{ and } 1 < \overline{|\underline{\underline{x}}|} \leq \overline{|M(\underline{\underline{x}})|}^{2}\ ,$$

and all but finitely many solutions of (8.6) have

$$(8.10)\quad |M^{(1)}(\underline{\underline{x}}) \cdots M^{(k)}(\underline{\underline{x}})| \leq \overline{|\underline{\underline{x}}|}^{\delta}\ .$$

Clearly $M^{(1)}, M^{(i_1+1)},\cdots, M^{(i_{n-1}+1)}$ are $n$ linearly independent linear forms, so that by the Subspace Theorem (Th. 1F of Ch. VI),

$$(8.11)\quad |M^{(1)}(\underline{\underline{x}})M^{(i_1+1)}(\underline{\underline{x}}) \cdots M^{(i_{n-1}+1)}(\underline{\underline{x}})| \geq \overline{|\underline{\underline{x}}|}^{-\delta}\ ,$$

unless $\underline{\underline{x}}$ lies in one of a finite number of proper rational subspaces determined by the linear form $M$. As $\underline{\underline{x}}$ runs through the integer points of a proper rational subspace, $M(\underline{\underline{x}})$ runs through a submodule of $\mathfrak{M}$ of rank less than $n$. Hence if we disregard solutions $\mu = M(\underline{\underline{x}})$ of (8.2), (8.4) which lie in a finite number of submodules $\mathfrak{M}_1,\cdots,\mathfrak{M}_z$ of rank less than $n$, then (8.8), (8.9), (8.10) and (8.11) hold.

Assume now, until the last paragraph of this section, that there is a solution $\mu = M(\underline{\underline{x}})$ of (8.2), (8.4) (and hence with $\underline{\underline{x}}$ satisfying (8.6), (8.7)), which is not in $\mathfrak{M}_1,\cdots,\mathfrak{M}_z$. In what follows, $\underline{\underline{x}}$ will be a solution of (8.6), (8.7) with $M(\underline{\underline{x}})$ outside of $\mathfrak{M}_1,\cdots,\mathfrak{M}_z$, so that (8.8), (8.9), (8.10) and (8.11) will hold. Define $c_1 = c_1(\underline{\underline{x}}),\cdots,c_n = c_n(\underline{\underline{x}})$ by

$$|M^{(i_j)}(\underline{\underline{x}})| = |\underline{\underline{x}}|^{c_j} \qquad (j = 1,\cdots,n) .$$

Then

(8.12) $$c_1 \leqq \cdots \leqq c_n \leqq 1 + \delta$$

by (8.7) and (8.9), and

(8.13) $$c_1 + \cdots + c_n + \delta \geqq 0$$

by (8.7) and (8.11). By (8.8) and (8.10) we have

$$i_1 c_1 + (i_2 - i_1)c_2 + \cdots + (i_n - i_{n-1})c_n$$

(8.14) $$\leqq \delta + (i_1 - 1)\delta + (i_2 - i_1 - 1)\delta + \cdots + (i_n - i_{n-1} - 1)\delta$$

$$= \delta + (k - n)\delta \leqq k\delta \quad .$$

We are going to apply Lemma 6A with $a = \delta$, $b = 1 + \delta$ and with $a_1 = i_1 = i_1 - i_0$, $a_2 = i_2 - i_1, \cdots, a_n = i_n - i_{n-1} = k - i_{n-1}$. Then $f(r) = i_r/r$ $(r = 1,\cdots,n)$. Let $S_r$ be the subset of $\{1,2,\cdots,k\}$ given by

(8.15) $$S_r = \{1,2,\cdots,i_r\} \qquad (r = 1,\cdots,n).$$

Then $|S_r| = i_r$, $r(S_r) = r$, $q(S_r) = 1/f(r)$, whence $f(r) \leqq 1/q_0 = k/n$ by Lemma 7B, and in particular we have $f_1 \leqq k/n$. By Lemma 6A, the minimum of $H(c_1,\cdots,c_n)$ with $(c_1,\cdots,c_n)$ subject to (8.12) and (8.13) is $H_1 = k(1 + \delta) - (\delta + n(1 + \delta))f_1$. On the other hand, the minimum is $\leqq k\delta$ by (8.14), so that $H_1 \leqq k\delta$ and $f_1 n(1+2\delta) \geqq f_1(\delta + n(1+\delta)) \geqq k(1+\delta) - k\delta = k$, whence

$$f_1 \geqq (k/n)(1+2\delta)^{-1} > (k/n)(1 - 2\delta) > (k/n)(1 - (1/kn)) = (k/n) - (1/n^2)$$

by (8.5). Now $f_1$ is a rational number with denominator at most $n$, and the rational numbers with denominator at most $n$ have mutual distances greater than $1/n^2$. Hence $f_1 \geqq k/n$, and by what we said above we have

$$(8.16) \qquad f_1 = k/n \quad \text{and} \quad H_1 = k(1+\delta) - (\delta + n(1+\delta))f_1 = -\delta k/n \quad .$$

Suppose that $f(r) = f_1 = k/n$ for $r = r_1, \cdots, r_g$ with $1 \leqq r_1 < r_2 < \cdots < r_g \leqq n$, and for no other values of $r$. Since $f(n) = k/n$, we have $r_g = n$.

Now suppose that $g < n$. Since $f_2$ (as defined in Lemma 6A) is another rational number with denominator at most $n$, we have $f_2 < f_1 - (1/n^2)$. By (8.14) and (8.16) we have

$$H(c_1, \cdots, c_n) - H_1 \leqq k\delta + (\delta k/n) < 2k\delta \quad ,$$

and hence (6.2) yields (on noting that $r_g = n$),

$$(8.17) \qquad \begin{aligned} &(c_{r_1} - c_1) + (c_{r_2} - c_{r_1+1}) + \cdots + (c_{r_g} - c_{r_{g-1}+1}) \\ &\leqq 2k\delta\,(a+nb)(H_2 - H_1)^{-1} = 2k\delta\,(f_1 - f_2)^{-1} < 2k\delta n^2 < \varepsilon/3 \end{aligned}$$

by (8.5). This inequality is trivially true if $g = n$, i.e. if $r_1 = 1, \cdots, r_g = n$, and hence is generally true.

Putting $r_0 = 0$, we have $0 \leqq c_{r_e} - c_{r_{e-1}+1} < \varepsilon/3 \quad (e = 1, \cdots, g)$ and hence $|c_h - c_j| < \varepsilon/3$ if $r_{e-1} + 1 \leqq h,j \leqq r_e$. Put differently,

$$|M^{(i_h)}(\underline{\underline{x}})| < \overline{|\underline{\underline{x}}|}^{\varepsilon/3} |M^{(i_j)}(\underline{\underline{x}})| \quad \text{if} \quad r_{e-1} + 1 \leqq h,j \leqq r_e \quad .$$

Combining this with (8.8) and observing that $\delta + (\varepsilon/3) < \varepsilon/2$, we obtain

$$|M^{(u)}(\underline{\underline{x}})| < \overline{|\underline{\underline{x}}|}^{\varepsilon/2} |M^{(v)}(\underline{\underline{x}})| \quad \text{if} \quad i_{r_{e-1}} + 1 \leqq u,v \leqq i_{r_e}$$

for some $e$ in $1 \leqq e \leqq g$ , which in turn by (8.9) yields

$$|M^{(u)}(\underline{\underline{x}})| < \overline{|M(\underline{\underline{x}})|}^{\varepsilon}|M^{(v)}(\underline{\underline{x}})| \quad \text{if} \quad i_{r_{e-1}} + 1 \leqq u,v \leqq i_{r_e} \quad .$$

With $S_r$ defined by (8.15) , put

$$V_1 = S_{r_1} , V_2 = S_{r_2} \sim S_{r_1}, \cdots , V_g = S_{r_g} \sim S_{r_{g-1}} = \{1,\cdots,k\} \sim S_{r_{g-1}} \quad ,$$

where $\sim$ denotes the set theoretic difference. We have

$$|M^{(u)}(\underline{\underline{x}})| < \overline{|M(\underline{\underline{x}})|}^{\varepsilon}|M^{(v)}(\underline{\underline{x}})| \quad \text{if} \quad u,v \in V_e$$

for some $e$ in $1 \leqq e \leqq g$ . Thus $\mu = M(\underline{\underline{x}})$ satisfies

$$|\mu^{(u)}| < \overline{|\mu|}^{\varepsilon}|\mu^{(v)}| \quad \text{if} \quad u,v \in V_e \quad .$$

Now $q(S_{r_e}) = 1/f(r_e) = 1/f_1 = q_0$ $(e = 1,\cdots,g)$ , and hence each set $S_{r_e}$ is the disjoint union of some of the sets (7.2) . Hence also each set $V_e$ $(e = 1,\cdots,g)$ is the union of some sets (7.2) . Thus each set $\hat{\varphi}_i(T)$ of (7.2) $(i = 1,\cdots,\ell)$ is contained in one of the sets $V_e$ , and we have

$$|\mu^{(u)}| < \overline{|\mu|}^{\varepsilon}|\mu^{(v)}| \quad \text{if} \quad u,v \in \hat{\varphi}_i(T)$$

for some $i$ . Now if again $T = \{t_1,\cdots,t_m\}$ as in §7 , we have

$$|\mu^{(\hat{\varphi}_i(t_h))}| < \overline{|\mu|}^{\varepsilon}|\mu^{(\hat{\varphi}_i(t_j))}| \quad \text{if} \quad 1 \leqq i \leqq \ell \quad \text{and} \quad 1 \leqq h,j \leqq m \quad .$$

In the notation of (7.4) this becomes (8.3) .

All this holds if $\mu$ is a solution of (8.2) with (8.4) and if $\mu$ does not lie in certain submodules $\mathfrak{M}_1,\cdots,\mathfrak{M}_z$ of rank less than $n$ . The same is true if (8.4) is replaced by any other ordering. Hence there is a finite number of submodules $\mathfrak{M}_1,\cdots,\mathfrak{M}_w$ , each of rank less than $n$ , such that (8.3) holds for every solution $\mu$ of (8.2) which is outside of these submodules.

## §9. Proof of the Main Theorem.

We shall apply Lemma 8A with some fixed $\varepsilon$ in $0 < \varepsilon < 1$ ; say with $\varepsilon = 1/2$ . By our inductive assumption (see §5) , and since each of the modules $\mathfrak{M}_1, \cdots, \mathfrak{M}_w$ of Lemma 8A has rank less than $n$ , the solutions $\mu$ of $\mathfrak{N}(\mu) = a$ with $\mu \in \mathfrak{M}_i$ $(i = 1, \cdots, w)$ lie in finitely many families of solutions. By Corollary 5B , every $(\mathfrak{M}_i, P)$-family of solutions where $P$ is a subfield of $K$ , is contained in the union of finitely many $(\mathfrak{M}, P)$-families of solutions. Hence there are finitely many families of solutions

$$\mu_1 \mathfrak{E}_{\mathfrak{M}}^{P_1}, \cdots, \mu_z \mathfrak{E}_{\mathfrak{M}}^{P_z} \tag{9.1}$$

with subfields $P_1, \cdots, P_z$ of $K$ and with $\mu_i \in \mathfrak{M}^{P_i}$ $(i = 1, \cdots, z)$ , such that every solution $\mu$ of (1.3) satisfies (8.3) unless it lies in one of these families.

Since $\mathfrak{M} = \mathfrak{M}^L$ by Lemma 7D , we have $\mathfrak{M}^{P_i L} = (\mathfrak{M}^L)^{P_i} = \mathfrak{M}^{P_i}$ by Lemma 5C . Thus $\mu_i \in \mathfrak{M}^{P_i L}$ $(i = 1, \cdots, z)$ , and since $\mathfrak{E}_{\mathfrak{M}}^{P_i} \subseteq \mathfrak{E}_{\mathfrak{M}}^{P_i L}$ $(i = 1, \cdots, z)$, by Lemma 5C again, the family $\mu_i \mathfrak{E}_{\mathfrak{M}}^{P_i}$ is contained in the family $\mu_i \mathfrak{E}_{\mathfrak{M}}^{P_i L}$ . Hence if we replace $\mu_i \mathfrak{E}_{\mathfrak{M}}^{P_i}$ in (9.1) by $\mu_i \mathfrak{E}_{\mathfrak{M}}^{P_i L}$ and change the notation back from $P_i L$ to $P_i$ , we obtain families (9.1) where the fields $P_i$ satisfy

$$L \subseteq P_i \subseteq K \qquad (i = 1, \cdots, z) . \tag{9.2}$$

Now suppose that an $(\mathfrak{M}, L)$-family and an $(\mathfrak{M}, P_i)$-family have a common element $\mu$ . Then the $(\mathfrak{M}, L)$-family is $\mu \mathfrak{E}_{\mathfrak{M}}^L$ and the $(\mathfrak{M}, P_i)$-family is $\mu \mathfrak{E}_{\mathfrak{M}}^{P_i}$ . Now in view of (9.2) and by Lemma 5C , we have $\mathfrak{E}_{\mathfrak{M}}^L \subseteq \mathfrak{E}_{\mathfrak{M}}^{P_i}$ , so that the $(\mathfrak{M}, L)$-family is contained in the $(\mathfrak{M}, P_i)$-family.

Since $\mathfrak{M} = \mathfrak{M}^L$ , every solution of (1.3) is contained in an $(\mathfrak{M}, L)$-family of solutions. If such an $(\mathfrak{M}, L)$-family intersects one of the families (9.1) , then it is contained in it. Therefore in order to prove Theorem 4B , it will suffice to show the following

<u>LEMMA 9A</u>. <u>Every</u> $(\mathfrak{M},L)$-<u>family of solutions of</u> (1.3) <u>which is disjoint from the families</u> (9.1) <u>has an element</u> $\mu$ <u>with</u>

$$\overline{|\mu|} \leq (\gamma_1|a|^{1/k})^{1/(1-\varepsilon)} , \tag{9.3}$$

<u>where</u> $\gamma_1 = \gamma_1(K,\mathfrak{M})$ .

Theorem 4B follows, since there are only finitely many elements $\mu$ in $\mathfrak{M}$ with $\overline{|\mu|}$ below a given bound.

<u>Proof of Lemma 9A</u> . Suppose we are given an $(\mathfrak{M},L)$-family of solutions which is disjoint from the families (9.1) , and let $\mu$ be a member of the family for which $\overline{|\mu|}$ is least possible. Since it lies outside the families (9.1) , the element $\mu$ satisfies (8.3) .

We saw in §7 that $\varphi_1,\cdots,\varphi_\ell$ are the isomorphisms of L into the complex numbers, and for $\lambda \in L$ we put $\lambda^{[i]} = \varphi_i(\lambda)$ $(i = 1,\cdots,\ell)$ , so that $\lambda^{[1]},\cdots,\lambda^{[\ell]}$ are the conjugates of $\lambda$ . We also introduced the notation $\mu^{[i,j]}$ for the conjugates of elements $\mu \in K$ . Now given the $\mu$ of the last paragraph, write $\lambda = \mu^{[1,1]}\mu^{[1,2]}\cdots\mu^{[1,m]}$ . Then $\lambda$ lies in L and

$$\lambda^{[i]} = \mu^{[i,1]}\mu^{[i,2]}\cdots\mu^{[i,m]} \qquad (i = 1,\cdots,\ell) .$$

We have

$$\mathfrak{N}(\mu) = \mathfrak{N}_L(\lambda) = \lambda^{[1]}\cdots\lambda^{[\ell]}$$

where $\mathfrak{N}_L$ is the norm from L to $\mathbb{Q}$ .

By Dirichlet's unit theorem (see Lemma 2A), there is for every

$\nu \in L$ an element $\varepsilon$ in the group $\mathfrak{E}^L$ of units of the ring of integers in $L$ such that

$$|(\varepsilon\nu)^{[i]}| \leqq \gamma_2 |\mathfrak{N}_L(\nu)|^{1/\ell} \qquad (i = 1,\cdots,\ell) ,$$

with a constant $\gamma_2$ depending on $L$ only. Since the group $(\mathfrak{E}^L_{\mathfrak{M}})^m$ of elements $\eta = \varepsilon^m$ with $\varepsilon \in \mathfrak{E}^L_{\mathfrak{M}}$ is of finite index in $\mathfrak{E}^L$, there is also an element $\eta \in (\mathfrak{E}^L_{\mathfrak{M}})^m$ with $|(\eta\nu)^{[i]}| \leqq \gamma_3 |\mathfrak{N}_L(\nu)|^{1/\ell}$ $(i = 1,\cdots,\ell)$. Here $\gamma_3$ depends only on $K, L, \mathfrak{M}$ and hence (since $K$ has only a finite number of subfields) only on $K$ and $\mathfrak{M}$. In particular, we may apply this to $\nu = \lambda$. There is an $\varepsilon \in \mathfrak{E}^L_{\mathfrak{M}}$ with

$$(9.4) \qquad |(\varepsilon^m\lambda)^{[i]}| \leqq \gamma_3 |\mathfrak{N}_L(\lambda)|^{1/\ell} = \gamma_3 |\mathfrak{N}(\mu)|^{1/\ell} = \gamma_3 |a|^{1/\ell} \qquad (i = 1,\cdots,\ell) .$$

On the other hand, (8.3) remains valid if the right hand side is replaced by $\overline{|\mu|}^{\,\varepsilon}\, |\mu^{[i,1]}\cdots\mu^{[i,m]}|^{1/m}$, and substituting $\varepsilon\mu$ for $\mu$ we obtain

$$|(\varepsilon\mu)^{[i,j]}| \leqq \overline{|\varepsilon\mu|}^{\,\varepsilon} |(\varepsilon^m\lambda)^{[i]}|^{1/m} \qquad (i = 1,\cdots,\ell;\ j = 1,\cdots,m)^{\dagger)} .$$

By this and by (9.4), the element $\mu' = \varepsilon\mu$ satisfies

$$\overline{|\mu'|} = \max(|\mu'^{[1,1]}|,\cdots,|\mu'^{[\ell,m]}|) \leqq \overline{|\mu'|}^{\,\varepsilon} \gamma_3^{1/m} |a|^{1/(m\ell)} .$$

Since $\mu$ had the minimum value of $\overline{|\mu|}$ in our family, we get

$$\overline{|\mu|} \leqq \overline{|\mu|}^{\,\varepsilon} \gamma_1 |a|^{1/k}$$

with $\gamma_1 = \gamma_3^{1/m}$, and (9.3) follows.

This completes the proof of Lemma 9A and hence of Theorem 4B.

## §10. Equations $\mathfrak{N}(M(\underline{\underline{x}})) = P(\underline{\underline{x}})$.

We saw in Theorem 3B of Ch. V that if $F(X,Y)$ is a form of

---

†) There should be no confusion of the unit $\varepsilon$ and the exponent $\varepsilon$.

degree $k \geq 3$ without multiple factors and if $\nu < k-2$, then there are only finitely many integer points $\underline{\underline{x}} = (x,y)$ with

$$(10.1) \qquad 0 < |F(x,y)| < |\underline{\underline{x}}|^{\nu} \quad .$$

Our aim here is to generalize this as far as possible to equations in $n$ variables.

Let $K$ be a number field of degree $k$. Suppose $K = \mathbb{Q}(\omega)$, and set $K^* = \mathbb{Q}(\omega^{(1)},\cdots,\omega^{(k)})$, so that $K^*$ is a normal extension of the rationals. Further let $G$ be the Galois group of $K^*$. We shall say that $K$ is $h$ <u>times transitive</u> if $G$ is $h$ times transitive on the set $\{\omega^{(1)},\cdots,\omega^{(k)}\}$, i.e. if $1 \leq h \leq k$ and if given distinct integers $i_1,\cdots,i_h$ between $1$ and $k$, there is a $\varphi \in G$ with $\varphi(\omega^{(1)}) = \omega^{(i_1)},\cdots,$ $\varphi(\omega^{(h)}) = \omega^{(i_h)}$. It is easily seen that this definition is independent of the choice of the primitive element $\omega$ of $K$.

<u>THEOREM 10A</u>. <u>Suppose</u> $n < k$ <u>and</u> $M(\underline{\underline{X}}) = \alpha_1 X_1 + \cdots + \alpha_n X_n$ <u>has coefficients in</u> $K$ <u>where</u> $K$ <u>is</u> $n-1$ <u>times transitive</u>. <u>Further assume that any</u> $n$ <u>conjugates</u> $M^{(i_1)},\cdots,M^{(i_n)}$ <u>of</u> $M$ <u>are linearly independent</u>. <u>Then for</u>

$$(10.2) \qquad \nu < k-n$$

<u>there are only finitely many integer points</u> $\underline{\underline{x}}$ <u>with</u>

$$|\mathfrak{N}(M(\underline{\underline{x}}))| \leq |\underline{\underline{x}}|^{\nu} \quad .$$

<u>COROLLARY 10B</u>. <u>Suppose</u> $M(\underline{\underline{X}})$ <u>is as above, and</u> $P(\underline{\underline{X}})$ <u>is a polynomial of total degree less than</u> $k-n$. <u>Then the diophantine equation</u>

$$\mathfrak{N}(M(\underline{\underline{x}})) = P(\underline{\underline{x}})$$

<u>has only finitely many solutions</u>.

<u>Remarks</u>. (i) The theorem contains the assertion concerning (10.1)

in the case when $F(X,Y)$ is irreducible: For in that case, $F(X,Y)$ is, except for a constant factor, a norm form $\mathfrak{N}(M(\underline{X}))$ in 2 variables, as was seen in §1 . In fact the form $M(\underline{X})$ may be taken to be of the type $M(\underline{X}) = X + \alpha Y$ where $K = Q(\alpha)$ . It is always true that $K$ is $1 = 2 - 1$ times transitive, and any two conjugates $M^{(i_1)}, M^{(i_2)}$ of $M$ are linearly independent.

(ii) The hypotheses of Theorem 10A hold if $K$ is $n$ times transitive and if the coefficients $\alpha_1,\cdots,\alpha_n$ of $M(\underline{X})$ are linearly independent over $Q$ . For if $M^{(i_1)},\cdots,M^{(i_n)}$ were linearly dependent for some $n$-tuple $i_1,\cdots,i_n$ , then by the transitivity this would be true for any $n$-tuple, and $M^{(1)},\cdots,M^{(k)}$ would have rank less than $n$ , in contradiction to Lemma 7A .

(iii) The condition (10.2) is best possible unless $n = 1$ , or $n = 2$ , and $M^{(1)},\cdots,M^{(k)}$ are complex conjugate in pairs.†)

For either $n \geq 2$ and one of the forms, say $M^{(1)}$, is real. Then by Minkowski's Theorem on linear forms, there is for $Q > 0$ an integer point $\underline{x} \neq \underline{0}$ with

$$|M^{(1)}(\underline{x})| \leq \gamma Q^{1-n} \quad , \quad |x_i| \leq Q \qquad (i = 2,\ldots,n) \; ,$$

where the coordinates were arranged so that $M^{(1)}, X_2,\cdots,X_n$ are linearly independent. It follows that $\overline{|\underline{x}|} \ll Q$ , whence $|M^{(j)}(\underline{x})| \ll Q$ $(j = 2,\ldots,k)$ , and $|\mathfrak{N}(M(\underline{x}))| \ll Q^{1-n}\overline{|\underline{x}|}^{k-1}$ , and

$$|\mathfrak{N}(M(\underline{x}))| \ll \overline{|\underline{x}|}^{k-n} \quad . \tag{10.3}$$

Since $Q$ may be taken arbitrarily large, one gets infinitely many solutions of (10.3) .

Or $n \geq 3$ and two of the forms, say $M^{(1)}, M^{(2)}$ , are complex conjugate. Let $R, I$ be the real part and the imaginary part of $M^{(1)}$ .

†) In the exceptional case $|\mathfrak{N}(M(\underline{x}))| \gg \overline{|\underline{x}|}^{k}$, as is easily seen. Compare with the end of §12.

Again by Minkowski's Theorem, there is for $Q > 0$ an integer point $\underline{x} \neq \underline{0}$ with

$$|R(\underline{x})| \ll Q^{1-(n/2)} \quad , \quad |I(\underline{x})| \ll Q^{1-(n/2)} \quad , \quad |x_i| \leq Q \quad (i=3,\ldots,n),$$

where the coordinates were arranged so that $R, I, X_3, \cdots, X_n$ are independent. It follows that $|M^{(1)}(\underline{x})| = |M^{(2)}(\underline{x})| \ll Q^{1-(n/2)}$ and $|\underline{x}| \ll Q$, whence $|M^{(j)}(\underline{x})| \ll Q$ $(j = 3,\ldots,k)$, and hence again (10.3) holds.

A form $F(\underline{X}) = F(X_1,\ldots,X_n)$ is called decomposable if it is a product of linear forms with algebraic coefficients. In particular, a norm form or a product of norm forms is decomposable. Conversely it is easily seen that a decomposable form with rational coefficients is a product of a constant and one or several norm forms. Of course, a form in two variables is always decomposable, while forms in $n \geq 3$ variables are in general not.

THEOREM 10C. *Suppose* $1 < n \leq \ell$ *and* $2(\ell - 1) < t$. *Suppose* $F(\underline{X}) = F(X_1,\ldots,X_n)$ *is a decomposable form of degree* $t$ *with rational coefficients. Suppose that* $F(\underline{X})$ *is not divisible by a rational form of degree less than* $\ell$, *and that any* $\ell$ *linear factors in the product* $F(\underline{X}) = M_1(\underline{X}) \ldots M_t(\underline{X})$ *have rank* $n$. *Then for any* $\nu$ *with*

$$\nu < t - 2(\ell - 1) \quad , \tag{10.4}$$

*there are only finitely many integer points* $\underline{x}$ *having*

$$|F(\underline{x})| < |\underline{x}|^{\nu} \quad . \tag{10.5}$$

One obtains an obvious corollary concerning equations $F(\underline{x}) = P(\underline{x})$, where $P$ is a polynomial of degree less than $t - 2(\ell - 1)$. When $n = \ell = 2$, our result is essentially the same as Theorem 3B of Ch. V.

THEOREM 10D*. Suppose the field K of degree k is primitive, i.e. it has no non-trivial subfields. Suppose that

$$N(\underline{X}) = \mathfrak{N}(\alpha_1 X_1 + \cdots + \alpha_n X_n)$$

is an irreducible norm form with respect to K . Here $\alpha_1, \cdots, \alpha_n$ lie in K , and the norm refers to K ). Now if $\alpha_1, \cdots, \alpha_n$ are linearly independent over Q , and if

$$\nu < k - \frac{n}{n-1} 2^{(n-2)/2} k^{(n-2)/(n-1)} , \tag{10.6}$$

then there are only finitely many integer points $\underline{x}$ with

$$|N(\underline{x})| < |\underline{x}|^{\nu} .$$

A proof of this theorem will not be given here. See Schmidt (1973). When $n = 2$ , then (10.6) reduces to $\nu < k - 2$ again. When $n = 3$ , the condition (10.6) becomes $\nu < k - (3/\sqrt{2})\sqrt{k}$ .

In order to prove Theorems 10A and 10C we shall need

## §11. Another Theorem on Linear Forms.

Consider linear forms $M(\underline{X})$ in $\underline{X} = (X_1, \cdots, X_n)$ with real or complex coefficients. Call the form real if all its coefficients are real, otherwise call it complex. A system of linear forms $M_1, \cdots, M_t$ will be called symmetric if every complex form among them occurs as often as its complex conjugate.

THEOREM 11A. Suppose $M_1, \cdots, M_t$ is a symmetric system of linear forms with algebraic coefficients. Given $\eta > 0$ , the following two conditions are equivalent.

(a) There are infinitely many integer points $\underline{x}$ with

$$|M_1(\underline{x}) \dots M_t(\underline{x})| \leq \gamma_1 |\underline{x}|^{t-\eta} \tag{11.1}$$

where $\gamma_1 = \gamma_1(M_1,\cdots,M_t,\eta)$ .

(b) There is a $d$-dimensional rational subspace $S^d \neq \{\underline{0}\}$ of $\mathbb{R}^n$ and there is a symmetric (sub-)system of forms $M_{i_1},\dots,M_{i_m}$ with $1 \leq m \leq t$ and $i_1 < \cdots < i_m$ whose restrictions to $S^d$ have rank $r$ with

$$r \leq dm/\eta \quad \text{and} \quad r < d \quad . \tag{11.2}$$

This Theorem could equally well have been included in Chapter VI . The preceding sections of the present chapter, with the exception of §6, are not needed for the proof.

The part (b) ⇒ (a) of the proof is†) relatively easy: We may suppose that $M_1,\cdots,M_m$ are a symmetric system whose rank $r$ on $S^d$ satisfies (11.2) . The real parts and imaginary parts of these forms will then also have rank $r$ on $S^d$ , hence on $S^d$ will be linear combinations of certain real forms $R_1,\cdots,R_r$ . Suppose, say, that the linear forms $R_1,\dots,R_r,X_{r+1},\dots,X_d$ are linearly independent on $S^d$ . By applying Minkowski's linear forms theorem to the lattice of integer points in $S^d$ , we obtain for given $Q > 0$ an integer point $\underline{x} \neq \underline{0}$ in $S^d$ having

$$|R_i(\underline{x})| \ll Q^{r-d} \quad (1 \leq i \leq r) \quad \text{and} \quad |x_j| \ll Q^r \quad (r+1 \leq j \leq d) \quad . \tag{11.3}$$

It follows that $|\underline{x}| \ll Q^r$ and $|M_i(\underline{x})| \ll Q^{r-d}$ for $1 \leq i \leq m$ . Thus $|M_i(\underline{x})| \ll |\underline{x}|^{(r-d)/r}$ $(1 \leq i \leq m)$ and

$$|M_1(\underline{x}) \dots M_t(\underline{x})| \ll |\underline{x}|^{(m/r)(r-d)+t-m} = |\underline{x}|^{t-(md/r)} \leq |\underline{x}|^{t-\eta} \quad .$$

Since $Q$ in (11.3) may be chosen arbitrarily large, we get infinitely many solutions of (11.1) , i.e. (a) holds.

†) The trivial case $r = 0$ is excluded in what follows.

If there is an integer point $\underline{\underline{x}} \neq \underline{\underline{0}}$ having $M_j(\underline{\underline{x}}) = 0$ for one of our forms $M_j$, then (b) holds with $S^d = S^1$ the subspace spanned by $\underline{\underline{x}}$, with $r = 0$, and with the symmetric subsystem consisting of $M_j$ alone if $M_j$ is real, and of $M_j$ and its complex conjugate if $M_j$ is complex. Hence we may suppose that

$$M_1(\underline{\underline{x}}) \dots M_t(\underline{\underline{x}}) \neq 0 \tag{11.4}$$

for nonzero integer points $\underline{\underline{x}}$.

## §12. Proof of the Theorem on Linear Forms.

It remains to do the implication (a) ⇒ (b). We start with the case when our linear forms are real. There is a smallest $d$ in $1 \leq d \leq n$ and a rational subspace $S^d$ such that for a certain constant $\gamma_2$ the relation

$$|M_1(\underline{\underline{x}}) \dots M_t(\underline{\underline{x}})| \leq \gamma_2 |\underline{\underline{x}}|^{t-\eta} \tag{12.1}$$

has infinitely many solutions in integer points in $S^d$. Now $d = 1$ is possible only if $M_1(\underline{\underline{x}}) \dots M_t(\underline{\underline{x}}) = 0$ for some $\underline{\underline{x}} \neq \underline{\underline{0}}$, which is ruled out by (11.4). So $2 \leq d \leq n$. After rearrangement of the linear forms, we have infinitely many integer points $\underline{\underline{x}}$ in $S^d$ with both (12.1) and

$$0 < |M_1(\underline{\underline{x}})| \leq \dots \leq |M_t(\underline{\underline{x}})| . \tag{12.2}$$

Suppose $M_1, \dots, M_t$ have rank $u$ on $S^d$. Let $i_1$ be the smallest number so that $M_{i_1} \neq 0$ on $S^d$; in view of (11.4) we have $i_1 = 1$. Let $i_2$ be the smallest number so that $M_{i_1}, M_{i_2}$ have rank 2 on $S^d$, and so forth. We obtain†) numbers $1 = i_1 < i_2 < \dots < i_u$.

†) The choice of the $i_j$ is similar to but not the same as the one in §8.

The points $\underline{\underline{x}}$ of $S^d$ with (12.2) have

(12.3)
$$|M_1(\underline{\underline{x}}) \ldots M_t(\underline{\underline{x}})| \gg\ll |M_1(\underline{\underline{x}})|^{i_2-1} |M_{i_2}(\underline{\underline{x}})|^{i_3-i_2} \ldots |M_{i_{u-1}}(\underline{\underline{x}})|^{i_u-i_{u-1}} |M_{i_u}(\underline{\underline{x}})|^{t+1-i_u} .$$

If $u = d$, then the Subspace Theorem (Th. 1F of Ch. VI) may be applied to the $d$ linear forms $M_{i_1},\ldots,M_{i_u}$ on $S^d$. Since $S^d$ was chosen minimal, it follows that given $\varepsilon > 0$ we have $|M_{i_1}\ldots M_{i_u}| \geq \overline{|\underline{\underline{x}}|}^{-\varepsilon}$ for all but finitely many of our points $\underline{\underline{x}}$. If $u < d$, we may augment $M_{i_1},\ldots,M_{i_u}$ by $d-u$ of the variables to get $d$ linearly independent forms on $S^d$, and we may conclude that

(12.4)
$$|M_{i_1}(\underline{\underline{x}}) \ldots M_{i_u}(\underline{\underline{x}})| \overline{|\underline{\underline{x}}|}^{d-u} \geq \overline{|\underline{\underline{x}}|}^{-\varepsilon} .$$

Thus (12.4) holds for every $u \leq d$. Put

$$v = \begin{cases} u & \text{if} \quad u < d , \\ u-1 = d-1 & \text{if} \quad u = d . \end{cases}$$

It is clear that $|M_{i_u}(\underline{\underline{x}})| \leq \overline{|\underline{\underline{x}}|}^{1+\varepsilon}$ if $\overline{|\underline{\underline{x}}|}$ is large. Thus no matter whether $u < d$ or $u = d$, we have

(12.5)
$$|M_{i_1}(\underline{\underline{x}}) \ldots M_{i_v}(\underline{\underline{x}})| \overline{|\underline{\underline{x}}|}^{d-v} \geq \overline{|\underline{\underline{x}}|}^{-2\varepsilon} .$$

Determine $c_1,\ldots,c_v$ by

$$|M_{i_j}(\underline{\underline{x}})| = \overline{|\underline{\underline{x}}|}^{c_j} \qquad (1 \leq j \leq v) ,$$

so that $c_1,\ldots,c_v$ are functions of $\underline{\underline{x}}$. If $\overline{|\underline{\underline{x}}|}$ is large, then $c_j \leq 1+\varepsilon$, and from (12.2), (12.5) we obtain

(12.6)
$$c_1 + \cdots + c_v + d - v + 2\varepsilon \geq 0 \quad \text{and} \quad c_1 \leq \cdots \leq c_v \leq 1 + \varepsilon .$$

On the other hand, (12.1) and (12.3) yield

$$(i_2 - 1)c_1 + (i_3 - i_2)c_2 + \cdots + (i_v - i_{v-1})c_{v-1} + (t+1-i_v)c_v \leqq t - \eta + \varepsilon$$

if $u < d$ , and

$$(i_2 - 1)c_1 + (i_3 - i_2)c_2 + \cdots + (i_{v+1} - i_v)c_v + (t+1-i_{v+1})(1-\varepsilon) \leqq t - \eta + \varepsilon$$

if $u = d$ . Setting

$$i_{v+1} = t + 1 \quad \text{if} \quad u < d \quad ,$$

we get in both cases

$$H \leqq i_{v+1} - 1 - \eta + 2t\varepsilon \quad , \tag{12.7}$$

where $H = H(c_1,\ldots,c_v) = (i_2 - 1)c_1 + \cdots + (i_{v+1} - i_v)c_v$ .

By Lemma 6A , applied with $a = d - v + 2\varepsilon$ , $b = 1 + \varepsilon$ , we see that the minimum of $H$ with $c_1,\ldots,c_v$ subject to (12.6) is

$$H_1 = (a_1 + \cdots + a_v)b - (a+vb)f_1 = (i_{v+1} - 1)(1+\varepsilon) - (d + v\varepsilon + 2\varepsilon)f_1 \quad ,$$

where $f_1 = f(r)$ for some $r$ in $1 \leqq r \leqq v$ and where

$$f(r) = (a_1 + \cdots + a_r)/r = (i_{r+1} - 1)/r \quad .$$

Combining this with (12.7) we get

$$(i_{v+1} - 1)(1+\varepsilon) - (d + v\varepsilon + 2\varepsilon)(i_{r+1} - 1)/r \leqq i_{v+1} - 1 - \eta + 2t\varepsilon \quad ,$$

and therefore

$$(d/r)(i_{r+1} - 1) \geqq \eta - 3dt\varepsilon - 2t\varepsilon \quad .$$

We have $r \leqq v < d$ . Put $m = i_{r+1} - 1$ and consider the $m$ linear forms $M_1,\ldots,M_{i_{r+1}-1}$. These forms have rank $r$ on $S^d$, with $md/r \geqq \eta - 5dt\varepsilon$. So for every $\varepsilon > 0$ there is a number $m$ and there are $m$ linear forms among $M_1,\ldots,M_t$ which have a rank $r$ satisfying $r < d$ and $md/r \geqq \eta - 5dt\varepsilon$. The number of possibilities for $m,r$ and the $m$ forms among $M_1,\ldots,M_t$ is finite, and hence there is a number $m$ and there are $m$ forms among $M_1,\ldots,M_t$ whose rank $r$ on $S^d$ satisfies (11.2). Theorem 1B is proved for real linear forms.

In the general case, for complex forms, note that a complex form may be written as $M = M^R + iM^I$ where $M^R, M^I$ are real forms.

The proof in the complex case up to (12.2) is just as the real case. In particular, $S^d$ is a minimal subspace of $\mathbb{R}^n$ such that an inequality like (12.1) has infinitely many solutions in integer points in $S^d$. We note that the forms can be ordered in such a way that complex conjugate forms stand next to each other in (12.2).

We are going to choose real forms $R_1,\ldots,R_t$ as follows. If $M_j$ is real, set $R_j = M_j$. If $M_j$ and $M_{j+1}$ are complex conjugate, then infinitely many of our points $\underline{x} \in S^d$ with (12.1), (12.2) have $|M_j^R(\underline{x})| \geqq |M_j^I(\underline{x})|$, or there are infinitely many points with the reversed inequality. Without loss of generality suppose, say, that we have $|M_j^R(\underline{x})| \geqq |M_j^I(\underline{x})|$. We further may suppose that $M_j^R(\underline{x})M_j^I(\underline{x}) \geqq 0$. Set

$$R_j = M_j^R \quad \text{and} \quad R_{j+1} = M_j^R + M_j^I \quad ,$$

<u>except</u> if the restriction of $M_j^R$ to $S^d$ is a linear combination of $M_1,\ldots,M_{j-1}$, but the restriction of $M_j^I$ is not such a linear combination. In the exceptional case set

$$R_j = M_j^R + M_j^I \quad \text{and} \quad R_{j+1} = M_j^R \quad .$$

This choice of $R_j$ , $R_{j+1}$ guarantees that if the restriction of $R_{j+1}$ to $S^d$ is independent of $R_1,\ldots,R_j$ , then the restriction of $R_j$ is independent of $R_1,\ldots,R_{j-1}$ . In either case we have

$$|M_j(\underline{x})| = |M_{j+1}(\underline{x})| \leq 2|R_j(\underline{x})| \leq 4|R_{j+1}(\underline{x})| \leq 8|M_j(\underline{x})| ,$$

so that

$$|R_j(\underline{x})| \leq 2|R_{j+1}(\underline{x})| \qquad (1 \leq j < t) . \tag{12.8}$$

If (12.1) has infinitely many solutions, then so does

$$|R_1(\underline{x}) \ldots R_t(\underline{x})| \leq \gamma_3 \overline{|\underline{x}|}^{\,t-\eta} .$$

Since $R_1,\ldots,R_t$ are real, and since the theorem has been proved for real forms, there will be a subspace $S^d$ of dimension $d$ and $m$ forms among $R_1,\ldots,R_t$ , whose rank satisfies (11.2) . In fact in the real case we saw that if (12.2) holds, then we can take the first $m$ forms. Now (12.8) is of course just as good as (12.2) , and we get forms $R_1,\ldots,R_m$ whose rank $r$ on $S^d$ satisfies (11.2) . We distinguish three cases:

(i) The system $M_1,\ldots,M_m$ is symmetric. Its rank $r$ on $S^d$ satisfies (11.2) and we are finished.

(ii) The system $M_1,\ldots,M_m$ is not symmetric and $R_{m+1}$ on $S^d$ is a linear combination or $R_1,\ldots,R_m$ . Then $R_1,\ldots,R_{m+1}$ have rank $r$ on $S^d$ , hence so does the symmetric system $M_1,\ldots,M_{m+1}$ , and we have $r < d$ and $r \leq md/\eta < (m+1)d/\eta$ .

(iii) The system is not symmetric and $R_{m+1}$ on $S^d$ is not a linear combination of $R_1,\ldots,R_m$ . Then $R_m$ is not a linear combination of $R_1,\ldots,R_{m-1}$ , so that $R_1,\ldots,R_{m-1}$ have rank $r-1$ on $S^d$ , and in particular $r \geq 1$ . Now if $\eta > n$ , then $r \leq md/\eta < md/n \leq m$ and

$m \geqq 2$ . The $m-1$ forms $M_1,\ldots,M_{m-1}$ have rank $r-1 < r\frac{m-1}{m} \leqq (m-1)d/\eta$ (since $r < m$) , and we are finished.

There remains the case when $\eta \leqq n$ . This case is completely elementary. We distinguish two subcases.

(I) We say that <u>we have sufficiently many variables</u> if either $n \geqq 3$ , or $n = 2$ and at least one of the forms is proportional to a real form. In this case there is a symmetric system $M_{i_1},\ldots,M_{i_m}$ where either $m < n$ , or $m = n = 2$ and both $M_{i_1}, M_{i_2}$ are proportional to the same real form. The system $M_{i_1},\ldots,M_{i_m}$ has rank $r$ with $r < n$ and $r \leqq m = nm/n \leqq nm/\eta$ . So (11.2) is true with $S^d = \mathbb{R}^n$ and $d = n$ .

(II) If we don't have sufficiently many variables, then either $n = 1$ and $|M_1(\underline{\underline{x}}) \ldots M_t(\underline{\underline{x}})| \gg |\overline{\underline{\underline{x}}}|^t$ (by (11.4)) . Or $n = 2$ and none of the given linear forms is proportional to a real form. Then $M_1,\ldots,M_t$ occur in pairs of complex conjugate forms, with each pair of rank 2 . Therefore again $|M_1(\underline{\underline{x}}) \ldots M_t(\underline{\underline{x}})| \gg |\overline{\underline{\underline{x}}}|^t$ , and (a) can in fact not hold for any $\eta > 0$ .

## §13. Proof of Theorem 10A.

Now that we have Theorem 11A , the proof is quite easy.

By the hypothesis of our theorem, linear forms $M^{(i_1)},\cdots,M^{(i_m)}$ with $i_1 < \cdots < i_m$ and with $m \geqq n$ have rank $n$ , and hence forms $M^{(i_1)},\cdots,M^{(i_m)}$ with $i_1 < \cdots < i_m$ and with $m \leqq n$ have rank $m$ . Thus the rank $r$ equals $\min(n,m)$ . Let $r(M^{(i_1)},\cdots,M^{(i_m)};S^d)$ be the rank of the restrictions of $M^{(i_1)},\cdots,M^{(i_m)}$ to a rational subspace $S^d$ of dimension $d$ . We claim that

$$r(M^{(i_1)},\cdots,M^{(i_m)};S^d) = \min(d,m) \quad . \tag{13.1}$$

It is clear that $r(M^{(i_1)},\cdots,M^{(i_m)};S^d) \leqq \min(d,m)$ . Equality certainly holds if $m \geqq n$ , for then $r(M^{(i_1)},\cdots,M^{(i_m)};\mathbb{R}^n) = n$ , whence $r(M^{(i_1)},\cdots,M^{(i_m)};S^d) = d$ . On the other hand, when $m < n$ , the field $K$ is $m$ times transitive, so that $r(M^{(i_1)},\cdots,M^{(i_m)};S^d) = r(M^{(e_1)},\cdots,M^{(e_m)};S^d)$ for any $e_1 < \cdots < e_m$ . If this rank is less than $m$ , then it follows that $r(M^{(i_1)},\cdots,M^{(i_m)};S^d) = r(M^{(1)},\cdots, M^{(k)};S^d) = d = \min(d,m)$ .

Suppose the inequality of Theorem 10A had infinitely many solutions. We may apply Theorem 11A to the forms $M^{(1)},\cdots,M^{(k)}$ with $\eta = k - \nu$ . By this theorem there is a rational subspace $S^d$ of $\mathbb{R}^n$ with dimension $d$ in $1 \leqq d \leqq n$ , and there are linear forms $M^{(i_1)},\cdots,M^{(i_m)}$ such that $r = r(M^{(i_1)},\cdots,M^{(i_m)};S^d)$ satisfies

$$r \leqq dm/(k-\nu) \quad \text{and} \quad r < d \ .$$

In particular, since $\nu < k - n \leqq k - d$ , we have $r < m$ and $r < d$ , which contradicts (13.1) .

## §14. Proof of Theorem 10C .

We have to show that under the hypotheses of Theorem 10C , (11.1) cannot hold with $\eta > 2(\ell - 1)$ . In view of Theorem 11A , it will be enough to verify that we cannot have $r < dm/(2(\ell-1))$ and $r < d$ , i.e. to show that

$$(14.1) \qquad r < d \quad \text{implies} \quad dm \leqq 2(\ell - 1)r \ .$$

If $M(\underline{\underline{X}})$ is a linear form with algebraic coefficients and with its first non-zero coefficient equal to 1 , write $\mathfrak{Z}(M)$ for the number field generated by the coefficients of $M$ . We shall say that two such forms $M(\underline{\underline{X}}), M'(\underline{\underline{X}})$ are <u>conjugate</u>, if there is an isomorphism from $\mathfrak{Z}(M)$ onto

$\mathfrak{J}(M')$ which maps the coefficients of $M(\underline{\underline{X}})$ into the respective coefficients of $M'(\underline{\underline{X}})$. A form $M(\underline{\underline{X}})$ has precisely $f$ conjugates where $f$ is the degree of $\mathfrak{J}(M)$, and no two conjugates are proportional.

In proving Theorem 10C, we may assume without loss of generality that all the linear forms $M_i$ in $F(\underline{\underline{X}}) = M_1(\underline{\underline{X}}) \dots M_t(\underline{\underline{X}})$ have their first non-zero coefficient equal to 1. With a form $M_i(\underline{\underline{X}})$, also all its conjugates divide $F(\underline{\underline{X}})$. If we block together conjugate linear forms, we may write

$$F(\underline{\underline{X}}) = F_1(\underline{\underline{X}}) \dots F_p(\underline{\underline{X}}) ,$$

where each $F_j$ is the product of conjugate linear forms. Each form $F_j$, being a product of conjugate linear forms, has rational coefficients, and hence has degree $\geqq \ell$ by the hypothesis of Theorem 10C. Hence every linear form $M_i(\underline{\underline{x}})$ has at least $\ell$ conjugates.

Now if $M_i(\underline{\underline{x}}) = 0$ for some integer point $\underline{\underline{x}}$, then also $M_{i_j}(\underline{\underline{x}}) = 0$ for the conjugates $M_{i_1} = M_i, M_{i_2}, \dots, M_{i_f}$ of $M_i$. We have $f \geqq \ell$, so that by our hypothesis the forms $M_{i_1}, \dots, M_{i_f}$ have rank $n$, and $\underline{\underline{x}}$ must be $\underline{\underline{0}}$. Therefore $M_i(\underline{\underline{x}}) \neq 0$ $(1 \leqq i \leqq d)$ for every integer point $\underline{\underline{x}} \neq \underline{\underline{0}}$. Hence we have rank $r > 0$ on every rational subspace $S^d$ of dimension $d \geqq 1$. We therefore have to show (14.1) only for $r > 0$, i.e. we have to show that

(14.2) $$0 < r < d \text{ implies } dm \leqq 2(\ell - 1)r .$$

Let $S^d$ be an arbitrary fixed rational subspace of $\mathbb{R}^n$ of dimension $d \geqq 2$; we shall prove (14.2) for $S^d$. Let $m(r)$ be the maximum number of linear forms $M_{i_1}, \dots, M_{i_m}$ whose restrictions to $S^d$ have rank $r$, and put $\pi(r) = m(r)/r$. We have to show that

$$\pi(r) \leq 2(\ell - 1)/d \qquad (0 < r < d) \quad . \tag{14.3}$$

Let $\pi_0$ be the maximum value of $\pi(r)$ in $0 < r < d$ . Let $r_0$ be the largest number in $0 < r < d$ with $\pi(r) = \pi_0$ . Put $m = m(r_0)$ ; there exists an $m$-tuple $M_{i_1},\ldots,M_{i_m}$ of linear forms whose restrictions to $S^d$ have rank $r_0$ , hence rank $< d$ . It follows that the forms $M_{i_1},\ldots,M_{i_m}$ have rank $< n$ on the full $n$-dimensional space $\mathbb{R}^n$ . By the hypothesis of Theorem 10C , we may conclude that

$$m < \ell \quad . \tag{14.4}$$

Now if $M_i$ occurs in an $m$-tuple of linear forms whose restrictions to $S^d$ have rank $r_0$ , then so does every conjugate of $M_i$ . Since $M_i$ has at least $\ell$ conjugates and since $m < \ell$ , there must be two distinct $m$-tuples $M_{i_1},\ldots,M_{i_m}$ and $M_{j_1},\ldots,M_{j_m}$ whose restrictions to $S^d$ have rank $r_0$ . Let the intersection of these two $m$-tuples consist of $m^-$ linear forms with rank $r^-$ on $S^d$ , and let the union of these two $m$-tuples consist of $m^+$ linear forms with rank $r^+$ on $S^d$ . Then $m^+ > m$ ,

$$m^- + m^+ = 2m \text{ and } r^- + r^+ \leq 2r_0 \quad . \tag{14.5}$$

Now if $m^- = 0$ , then we have $r^- = 0$ and $\pi(r^+) \geq m^+/r^+ \geq m/r_0 = \pi_0$ . If $m^- > 0$ , then $0 < r^- \leq r_0$ and $m^-/r^- \leq \pi(r^-) \leq \pi(r_0) = \pi_0$ , whence again $\pi(r^+) \geq m^+/r^+ \geq \pi_0$ . Now clearly $r^+ \geq r_0$ , and if we had $r^+ = r_0$ , then $\pi(r^+) = \pi_0 = m/r_0$ and $m^+ = m$ , which is impossible. Hence $r^+ > r_0$ , and since $r_0$ was maximal in $0 < r < d$ with $\pi(r) = \pi_0$ , we must have $r^+ = d$ . Now (14.4) , (14.5) yield $d \leq 2r_0$ and

$$\pi_0 = m/r_0 \leq (\ell - 1)/r_0 \leq 2(\ell - 1)/d \quad ,$$

whence (14.3) .

## VIII. Approximation By Algebraic Numbers.

References: Mahler (1932), Koksma (1939), Wirsing (1961), Davenport and Schmidt (1967).

§1. The Setting. In the first chapters we studied approximation to real numbers by rationals. We now take up approximation to real numbers by algebraic numbers. This is quite different from the questions e.g. considered in Chapter V on approximation to algebraic numbers by rationals.

There are two basic questions here. Firstly, we may suppose a real algebraic number field $K$ is given, and we study approximations to a real number $\alpha$ by elements of $K$. (A variant of this is when a complex (non-real) algebraic number field is given, and one studies approximations to a complex number $\alpha$ by elements of $K$. But this variant will not be taken up in these Notes.) Secondly, we may suppose a natural number $k$ is given, and we are interested in approximations to a real number $\alpha$ by real algebraic numbers of degree $\leq k$. (Again there is a complex variant, which will not be developed here.)

With each of these two basic questions, one may strive for Dirichlet type theorems, i.e. theorems which assert that every real $\alpha$ may be approximated to a certain extent. Or one may be looking, e.g., for Roth type theorems, which assert that a given algebraic $\xi$ cannot be approximated too well. Finally, there are questions which have no counterparts in the theory developed so far, and which I want to call questions of comparisons. Here one compares the approximation properties of a number $\alpha$, either with regard to approximations in different fields, or with regard to approximations of various degrees.

The presentation of these questions will be as follows.

| | Dirichlet type results | Roth type results | Questions of comparison |
|---|---|---|---|
| Given a field K | §2 | §7,8 | §10 |
| Given a bound on the degree | §3,4,5,6. | §9 | |

In §11 a different viewpoint will be pursued.

Questions about simultaneous approximation, etc., will not be taken up in the context of this chapter.

§2. Field Height and Approximation by Elements of a Given Number Field. As usual, if $P = P(X) = a_0X^n + \cdots + a_n$ is a polynomial, we set $\lceil P \rceil = \max(|a_0|, \ldots, |a_n|)$ . If P is a nonzero polynomial with rational coefficients, then there is a polynomial $P_0$ proportional to P whose coefficients are coprime rational integers, and $P_0$ is unique up to a factor ±1 . Define the height of P by

$$H(P) = \lceil P_0 \rceil \quad .$$

Then if P has rational integer coefficients, $H(P) \leqq \lceil P \rceil$ .

Now let K be a real algebraic number field of degree k . There are k isomorphisms $\sigma_1$ (= identity map), $\sigma_2, \ldots, \sigma_k$ of K into the complex numbers. Write $\sigma_i(\beta) = \beta^{(i)}$ $(i = 1, \ldots, k)$ for $\beta \in K$ ; then $\beta^{(1)} = \beta, \beta^{(2)}, \ldots, \beta^{(k)}$ are the conjugates of $\beta$ . The field polynomial of $\beta$ is $P_0(X)$ where

$$P(X) = (X - \beta^{(1)}) \cdots (X - \beta^{(k)}) \quad ,$$

and its field height is

$$H_K(\beta) = \lceil P_0 \rceil = H(P) \quad .$$

Given a bound $C$ , there are only finitely many $\beta$ in $K$ with $H_K(\beta) \leq C$ .

THEOREM 2A. Let $K$ be a real algebraic number field. Then there exists a constant $c_K$ such that for every real $\alpha$ not in $K$ there are infinitely many $\beta \in K$ having

$$|\alpha - \beta| < c_K \max(1, |\alpha|^2) H_K(\beta)^{-2} \quad .$$

It will follow from Theorem 8A that this result is essentially best possible. For a generalization, see Schmidt (1975) .

The connection between the cases $|\alpha| < 1$ and $|\alpha| > 1$ is as follows. Suppose the theorem is true for $|\alpha| < 1$ with some constant $c_K'$ , i.e. there are infinitely many $\beta \in K$ with $|\alpha - \beta| < c_K' H_K(\beta)^{-2}$ . Then if $|\alpha| > 1$ , there are infinitely many $\gamma \in K$ with $|(1/\alpha) - \gamma| < c_K' H_K(\gamma)^{-2}$ . Clearly $|\gamma| > 1/(|\alpha|(1+\varepsilon))$ if $\varepsilon > 0$ and if $H_K(\gamma)$ is large. Setting $\beta = 1/\gamma$ we have $H_K(\beta) = H_K(\gamma)$ and

$$|\alpha - \beta| = |\alpha\beta| |(1/\alpha) - \gamma| < c_k' |\alpha|^2 (1+\varepsilon) H_K(\beta)^{-2} \quad .$$

Hence Theorem 2A is true for any given constant $c_K > c_K'$ .

In the case when $K = \mathbb{Q}$ , the rationals, the height $H_{\mathbb{Q}}(p/q) = \max(|p|, |q|)$ if $p/q$ is in its lowest terms. If $\alpha \notin \mathbb{Q}$ and $|\alpha| < 1$ , Dirichlet's Theorem yields infinitely many $p/q$ with $|\alpha - (p/q)| < 1/q^2$ , and since here $|p| < |\alpha q| + 1/q < q$ for large $q$ , we then have $H_{\mathbb{Q}}(p/q) = q$ and $|\alpha - (p/q)| < H_{\mathbb{Q}}(p/q)^{-2}$ . Hence we may take $c_{\mathbb{Q}}' = 1$ , and $c_{\mathbb{Q}}$ any constant $> 1$ . If we use Hurwitz' Theorem instead of Dirichlet's, then we may take any constant $> 1/\sqrt{5}$ .

Proof of Theorem 2A. We may suppose that $|\alpha| < 1$ . Let $\theta_1,\dots,\theta_k$ be an integer basis of $K$ . Given large $Q$ we apply Minkowski's Linear Forms Theorem (Theorem 2C of Ch. II) to find rational integers $q_1,\dots,q_k,p_1,\dots,p_k$ , not all zero, with

$$(2.1)\qquad |\alpha\theta_1 q_1 + \cdots + \alpha\theta_k q_k - \theta p_1 - \cdots - \theta_k p_k|\,|\theta_k|^{-1} < Q^{-2k+1} ,$$

$$(2.2)\qquad |q_1|,\dots,|q_k|,|p_1|,\dots,|p_{k-1}| \leqq Q .$$

Then $|p_k| \ll Q$ ; here and below, the constant in $\ll$ depends only on $\theta_1,\dots,\theta_k$ , hence ultimately only on $K$ . If we had $q_1 = \cdots = q_k = 0$ , then by (2.1) , (2.2) , the algebraic integer $\theta_1 p_1 + \cdots + \theta_k p_k$ would have norm $\ll Q^{-2k+1}Q^{k-1} = Q^{-k}$ , whence we would get the contradiction $p_1 = \cdots = p_k = 0$ . Thus not all of $q_1,\dots,q_k$ are zero, and $\theta_1 q_1 + \cdots + \theta_k q_k \neq 0$ . Putting

$$\beta = \frac{\theta_1 p_1 + \cdots + \theta_k p_k}{\theta_1 q_1 + \cdots + \theta_k q_k} ,$$

we have

$$(2.3)\qquad |\alpha - \beta| < Q^{-2k+1}|\theta_1 q_1 + \cdots + \theta_k q_k|^{-1} .$$

The field polynomial of $\beta$ is proportional to the polynomial

$$P(X) = \prod_{i=1}^{k}((\theta_1^{(i)}q_1 + \cdots + \theta_k^{(i)}q_k)X - (\theta_1^{(i)}p_1 + \cdots + \theta_k^{(i)}p_k)) .$$

Since $P$ has rational integer coefficients,

$$(2.4)\qquad H_K(\beta) \leqq \boxed{P} .$$

To estimate $\boxed{P}$ we observe that $|\theta_1 p_1 + \cdots + \theta_k p_k| \ll |\theta_1 q_1 + \cdots + \theta_k q_k|$ by (2.1)†), and that both $|\theta_1^{(i)}p_1 + \cdots + \theta_k^{(i)}p_k|$ and $|\theta_1^{(i)}q_1 + \cdots + \theta_k^{(i)}q_k|$ are $\ll Q$ for $i = 2,\dots,k$ by (2.2) . It follows that

†) Note that $|\theta_1 p_1 + \cdots + \theta_k p_k| \gg Q^{-k+1}$ by Lemma 1A of Ch. VI .

$$H_K(\beta) \leq \lceil P \rceil \ll |\theta_1 q_1 + \cdots + \theta_k q_k| Q^{k-1} \ll Q^k \quad .$$

Combining this with (2.3) , (2.4) we get

$$|\alpha - \beta| \ll Q^{-k} H_K(\beta)^{-1} \ll H_K(\beta)^{-2} \quad .$$

Since $\alpha \notin K$ , the left hand side is not zero, and hence as $Q \to \infty$ , we obtain infinitely many distinct $\beta$ with $|\alpha - \beta| \ll H_K(\beta)^{-2}$ .

<u>Exercise</u>. Show that if $K$ is a complex algebraic number field and if $\alpha$ is complex but not in $K$ , then there are infinitely many $\beta \in K$ with

$$|\alpha - \beta| < \frac{c_K \max(1, |\alpha|^2)}{H(\beta)} \quad .$$

## §3. Absolute Height and Approximation by Algebraic Numbers of Bounded Degree.

Let

$$P(X) = a_0 X^n + \cdots + a_n = a_0 (X - \alpha^{(1)}) \cdots (X - \alpha^{(n)})$$

be a polynomial. There are many quantities besides $\lceil P \rceil$ to measure the "size" of $P$ . Among them are

$$L(P) = |a_0| + \cdots + |a_n|$$

and Mahler's

$$M(P) = |a_0| \prod_{i=1}^{n} \max(1, |\alpha^{(i)}|) \quad .$$

We have $\lceil P \rceil \leq L(P) \leq (n+1) \lceil P \rceil$ and

(3.1) $$L(PQ) \leq L(P)L(Q) \quad , \quad \lceil PQ \rceil \leq (n+1) \lceil P \rceil \lceil Q \rceil \quad ,$$

while $M$ satisfies the very satisfying relation

$$M(PQ) = M(P)M(Q) \quad . \tag{3.2}$$

Also,

$$\boxed{P} \leq L(P) \leq |a_0| \prod_{i=1}^{n} (1 + |\alpha_i|) \leq 2^n M(P) \quad . \tag{3.3}$$

In the opposite direction we wish to show that

$$M(P) \ll_n \boxed{P} \quad . \tag{3.4}$$

Suppose that

$$|\alpha_1| \leq \cdots \leq |\alpha_n| \quad .$$

If

$$|\alpha_1| \leq 4^n , |\alpha_2| \leq 4^n |\alpha_1| , \ldots , |\alpha_n| \leq 4^n |\alpha_{n-1}| \quad , \tag{3.5}$$

then $M(P) \ll |a_0| \leq \boxed{P}$ . Otherwise, suppose the i-th inequality (3.5) is the first one which fails to hold. Then

$$M(P) \ll |a_0| |\alpha_i| \cdots |\alpha_n| \quad . \tag{3.6}$$

On the other hand

$$\begin{aligned} a_{n-i+1}/a_0 &= \alpha_i \alpha_{i+1} \cdots \alpha_n + \alpha_{i-1} \alpha_{i+1} \cdots \alpha_n + \cdots + \alpha_1 \cdots \alpha_{n-i+1} \\ &= \pi_1 + \cdots + \pi_t \quad , \end{aligned}$$

say, where $t = 1$ if $i = 1$ , and where $t = \binom{n}{i-1} \leq 2^n$ and $|\pi_j|/|\pi_1| \leq 4^{-n}$ $(j = 2,\ldots,t)$ if $i > 1$ . Thus

$$\begin{aligned} \boxed{P} \geq |a_{n-i+1}| &\geq |a_0| |\pi_1| (1 - (2^n/4^n)) \\ &\geq \tfrac{1}{2} |a_0| |\alpha_i| |\alpha_{i+1}| \cdots |\alpha_n| \quad , \end{aligned}$$

which in conjunction with (3.6) yields (3.4) . From (3.1) , (3.2) , (3.3) , (3.4) we see that

$$(3.7)\qquad L(PQ) \leqq L(P)L(Q) \ll L(PQ)$$

and

$$(3.8)\qquad \overline{|PQ|} \ll \overline{|P|}\,\overline{|Q|} \ll \overline{|PQ|} \quad ,$$

with constants in $\ll$ which depend only on the degrees of $P$ and of $Q$ .

More explicit estimates than these were given by Mahler (1960) and others. $L(P)$ and $\overline{|P|}$ may be defined for polynomials in any number of variables, and (3.7) , (3.8) remain true with constants which depend on the degrees and on the number of variables. It was discovered by Enflo and reproved more simply by Montgomery (1975/76) that (3.7) is true with constants which depend only on the total degrees of $P$ and of $Q$ , but not on the number of variables.

Now if $\alpha$ is algebraic of precise degree $n$ , it satisfies an equation $P_0(\alpha) = 0$ where $P_0(X)$ is a nonzero polynomial of degree $n$ with coprime integer coefficients. The <u>absolute height</u>, or briefly <u>height</u>, of $\alpha$ is given by

$$H(\alpha) = H(P_0) = \overline{|P_0|} \quad .$$

Also define

$$M(\alpha) = M(P_0) \quad .$$

Further if $\alpha$ lies in a number field $K$ , then the degree $[K : \mathbb{Q}] = k$ , say, is a multiple of $n$ . If $F_0$ is the field polynomial of $\alpha$ , then it is seen that

$$F_0 = P_0^{k/n} ,$$

so that by (3.8) ,

$$H_K(\beta) \gg_k \ll (H(\beta))^{k/n} . \tag{3.9}$$

CONJECTURE. *Suppose* $\alpha$ *is real, but is not algebraic of degree* $\leq k$ . *Suppose* $\varepsilon > 0$ . *Then there are infinitely many real algebraic* $\beta$ *of degree* $\leq k$ *with*

$$|\alpha - \beta| < \frac{c_1}{H(\beta)^{k+1-\varepsilon}} , \tag{3.10}$$

*where* $c_1 = c_1(k,\alpha,\varepsilon)$ .

Perhaps even a stronger form of the conjecture is true, without the $\varepsilon$ , and with (3.10) replaced by

$$|\alpha - \beta| < \frac{c_2}{H(\beta)^{k+1}} , \tag{3.11}$$

where $c_2 = c_2(k,\alpha)$ . Such a strong form is true for $k = 1$ , by Dirichlet's Theorem. The strong form is also true for $k = 2$ :

THEOREM 3A. (Davenport and Schmidt (1967)). *For any real* $\alpha$ *which is not rational or quadratic over* $\mathbb{Q}$ , *there exists infinitely many rational or real quadratic* $\beta$ *such that*

$$|\alpha - \beta| < c_3 H(\beta)^{-3} ,$$

*where* $c_3 = c_3(\alpha)$ .

In fact Davenport and Schmidt proved a more explicit result with $c_3(\alpha) = c_4 \max(1, |\alpha|^2)$ , where $c_4$ is any fixed constant greater than $160/9$ . For $k > 2$ the conjectured exponent $k + 1$ or $k + 1 - \varepsilon$

has not yet been obtained, but we have

THEOREM 3B. (Wirsing (1961)). For any real $\alpha$ which is not algebraic of degree $\leq k$, then there are infinitely many real algebraic $\beta$ of degree $\leq k$ having

$$|\alpha - \beta| < c_5(k,\alpha)H(\beta)^{-(k+3)/2} .$$

Wirsing in fact obtained the slightly better exponent

$$\frac{k+6}{4} + \frac{1}{4}\sqrt{k^2 + 4k - 4} .$$

One could also try to approximate by algebraic integers. A natural conjecture is that if $\alpha$ is not algebraic of degree $\leq k$, then there are infinitely many real algebraic integers $\beta$ of degree $\leq k$ with

$$|\alpha - \beta| < \frac{c_6(k,\alpha,\varepsilon)}{H(\beta)^{k+\varepsilon}} .$$

The following will not be proved in these Notes.

THEOREM 3C*. (Davenport and Schmidt (1969)). Suppose $k \geq 3$ and suppose $\alpha$ is real, but not algebraic of degree $\leq (k-1)/2$. Then there are infinitely many algebraic integers $\beta$ of degree $\leq k$ with

$$|\alpha - \beta| < c_7(k,\alpha)H(\beta)^{-[(k+1)/2]} .$$

Incidentally, (3.11), if true, would be best possible. For if $\alpha$ is algebraic of precise degree $k+1$, the resultant $R(\alpha,\beta)$ is a nonvanishing integer, and if

$$a_0(X - \alpha^{(1)}) \cdots (X - \alpha^{(k+1)}) \quad \text{and} \quad b_0(X - \beta^{(1)}) \cdots (X - \beta^{(d)})$$

are the minimal polynomials of $\alpha,\beta$, then (Van der Waerden (1955), §30)

$$1 \leq |R(\alpha,\beta)| = |a_0^d b_0^{k+1} \prod_{i=1}^{k+1} \prod_{j=1}^{d} | \alpha^{(i)} - \beta^{(j)}|$$
$$= |\alpha - \beta| |b_0^{k+1} Q(\beta^{(1)},\dots,\beta^{(d)})| ,$$

where $Q$ is a polynomial with coefficients depending on $\alpha$ , and of degree at most $k+1$ in each $\beta^{(i)}$ . Thus with Mahler's $M(\beta)$ ,

$$1 \ll |\alpha - \beta| M(\beta)^{k+1} \ll |\alpha - \beta| H(\beta)^{k+1} ,$$

or

$$|\alpha - \beta| > c_8(\alpha) H(\beta)^{-k-1} .$$

For algebraic $\alpha$ of degree $> k+1$ , see §9 . It can be shown that for almost all $\alpha$ and for $\varepsilon > 0$ ,

$$|\alpha - \beta| > c_9(k,\alpha,\varepsilon) H(\beta)^{-k-1-\varepsilon} , \tag{3.12}$$

for $\beta$ of degree $\leq k$ , and

$$|\alpha - \beta| > c_{10}(k,\alpha,\varepsilon) H(\beta)^{-k-\varepsilon}$$

for algebraic integers $\beta$ of degree $\leq k$ . For more information (involving Hausdorff dimension) on (3.12) see Baker and Schmidt (1970).

## §4. Approximation by Quadratic Irrationals. A Theorem on Two Linear Forms.

In this section write $\underline{\underline{X}} = (X,Y,Z)$ ,

$$\underline{\underline{x}} = (x,y,z) \text{ and } \overline{|\underline{\underline{x}}|} = \max(|x|,|y|,|z|) .$$

Let $L(\underline{\underline{X}}) = \alpha X + \beta Y + \gamma Z$ be a linear form. It follows, e.g. from Minkowski's Linear Forms Theorem, that there are infinitely many nonzero integer points $\underline{\underline{x}}$ having

$$|L(\underline{x})| \ll \overline{|\underline{x}|}^{-2} ,$$

with the constant in $\ll$ depending on L . We will prove the stronger

THEOREM 4A. *Suppose* L,P *are independent linear forms. Then there are infinitely many integer points* $\underline{x} \neq \underline{0}$ *with*

(4.1) $$|L(\underline{x})| \ll \overline{|\underline{x}|}^{-4} |P(\underline{x})|^2 ,$$

*and with a constant in* $\ll$ *depending on* L *and on* P .

A more general result was given by Davenport and Schmidt (1970).

To deduce Theorem 3A , put

$$L(\underline{X}) = \alpha^2 X + \alpha Y + Z \quad \text{and} \quad P(\underline{X}) = 2\alpha X + Y .$$

Given $\underline{x}$ with (4.1) , the polynomial

$$B(T) = xT^2 + yT + z$$

has

(4.2) $$|B(\alpha)| \ll \overline{|\underline{x}|}^{-4} |P(\underline{x})|^2 \ll \overline{|\underline{x}|}^{-3} |P(\underline{x})| \ll \overline{|\underline{x}|}^{-3} |B'(\alpha)| .$$

Now either deg B = 1 . Then B(T) has the rational root $\beta$ with

$$0 = B(\beta) = B(\alpha) + (\beta - \alpha)B'(\alpha) ,$$

so that

$$|\alpha - \beta| = |B(\alpha)/B'(\alpha)| \ll \overline{|\underline{x}|}^{-3} \ll \overline{|B|}^{-3} \ll H(\beta)^{-3} .$$

Or deg B = 2 . Then the roots $\beta$ of B satisfy

$$0 = B(\beta) = B(\alpha) + (\beta - \alpha)B'(\alpha) + \frac{1}{2}(\beta - \alpha)^2 B''(\alpha) .$$

Solving this by the quadratic formula for $\beta - \alpha$ , we see from (4.2) that the roots $\beta$ are real, and that one of the roots $\beta$ has

$$\begin{aligned} |\beta - \alpha| &= |-B'(\alpha) + (B'^2(\alpha) - 2B(\alpha)B''(\alpha))^{1/2}|\,|B''(\alpha)|^{-1} \\ &\ll |B(\alpha)/B'(\alpha)| \\ &\ll \overline{|\underline{\underline{x}}|}^{-3} \ll \overline{|B|}^{-3} \ll H(\beta)^{-3} \quad . \end{aligned}$$

To prove Theorem 4A we may suppose that $L(\underline{\underline{x}}) \neq 0$ for integer points $\underline{\underline{x}} \neq \underline{\underline{0}}$ . Choose a linear form $F$ such that $F, L, P$ are independent. Putting

$$(4.3) \qquad \langle \underline{\underline{x}} \rangle = \max(|F(\underline{\underline{x}})|, |L(\underline{\underline{x}})|, |P(\underline{\underline{x}})|) \quad ,$$

we have

$$(4.4) \qquad \langle \underline{\underline{x}} \rangle \gg \ll \overline{|\underline{\underline{x}}|} \quad .$$

For each real $X > 0$ we consider the finite set of integer points $\underline{\underline{x}} \neq \underline{\underline{0}}$ satisfying

$$\langle \underline{\underline{x}} \rangle \leq X \quad .$$

For large $X$ this set is not empty, and by our hypothesis on $L$ , the values of $L$ at the points of this set are distinct. We choose the unique point $\underline{\underline{x}}$ in this set for which $|L(\underline{\underline{x}})|$ is miminal and the first nonzero element of the triple $F(\underline{\underline{x}}), L(\underline{\underline{x}}), P(\underline{\underline{x}})$ is positive, and we call this the minimal point corresponding to $X$ .

It is obvious that if $\underline{\underline{x}}$ is the minimal point corresponding to $X^*$ , then it is the minimal point corresponding to any $X$ in some interval $X^* \leq X < X^{**}$ . There is a sequence of numbers

(4.5) $$X_1 < X_2 < \cdots$$

which tend to infinity, and a sequence of points

(4.6) $$\underline{\underline{x}}_1, \underline{\underline{x}}_2, \ldots$$

such that $\underline{\underline{x}}_i$ is the minimal point corresponding to all $X$ in the range $X_i \leqq X < X_{i+1}$, but to no other $X$. It is clear that

(4.7) $$\langle \underline{\underline{x}}_i \rangle = X_i \quad .$$

We introduce the notations

(4.8) $$F_i = F(\underline{\underline{x}}_i) \quad , \quad L_i = L(\underline{\underline{x}}_i) \quad , \quad P_i = P(\underline{\underline{x}}_i) \quad .$$

Plainly

(4.9) $$|L_1| > |L_2| > \cdots \quad .$$

By our construction of the sequences (4.5), (4.6), there is no integer point $\underline{\underline{x}} \neq \underline{\underline{0}}$ with

(4.10) $$\langle \underline{\underline{x}} \rangle < X_{i+1} \quad , \quad |L(\underline{\underline{x}})| < |L_i| .$$

The inequalities (4.10) define a symmetric convex set of volume $\gg X_{i+1}^2 |L_i|$, so that by Minkowski's Theorem (Th. 2B of Ch. II) we have $X_{i+1}^2 |L_i| \ll 1$, or

(4.11) $$|L_i| \ll X_{i+1}^{-2} \quad .$$

Our assertion concerning (4.1) will be proved if we can show that $|L_i| \ll X_i^{-4} P_i^2$ for infinitely many $i$. We shall assume indirectly that

(4.12) $$P_i^2 = o(|L_i| X_i^4) \quad ,$$

and eventually we shall reach a contradiction.

§5. Approximation by Quadratic Irrationals, Continued. As a consequence of (4.11) and (4.12) we obtain

(5.1) $$|P_i| = o(X_i^2 X_{i+1}^{-1}) = o(X_i) \quad .$$

Since $L_i \to 0$ and since $\langle \underline{x}_i \rangle = \max(|F_i|,|L_i|,|P_i|)$ , it follows from (4.7) that

(5.2) $$F_i = X_i \quad .$$

LEMMA 5A. The signs of the $L_i$ alternate when $i$ is large.

Proof. The point

$$\underline{y} = \underline{x}_{i+1} - \underline{x}_i$$

has $0 < F(\underline{y}) < X_{i+1}$ by (4.5) and (5.2) . Since $|L(\underline{y})| \ll 1$ and since $|P(\underline{y})| \leq |P_{i+1}| + |P_i| = o(X_{i+1})$ by (5.1) , we have

$$\langle \underline{y} \rangle < X_{i+1}$$

if $i$ is large. Since (4.10) has no nonzero solution, we must have

$$|L_{i+1} - L_i| = |L(\underline{y})| \geq |L_i| \quad .$$

In conjunction with (4.9) this yields $L_i L_{i+1} < 0$ .

LEMMA 5B. Suppose $i$ is large and

(5.3) $$|P_{i+1}| \leq \frac{1}{2} X_i \quad .$$

Then

$$\underline{\underline{x}}_{i+1} = t\underline{\underline{x}}_i + \underline{\underline{x}}_{i-1} \quad ,$$

*where* $t$ *is a positive integer.*

*Proof.* Define integers $t,u$ by

$$t = [X_{i+1}/X_i] \quad , \quad u = [|L_{i-1}/|L_i|] \quad ,$$

and integer points $\underline{\underline{y}},\underline{\underline{z}}$ by

$$\underline{\underline{y}} = \underline{\underline{x}}_{i+1} - t\underline{\underline{x}}_i \quad , \quad \underline{\underline{z}} = \underline{\underline{x}}_{i-1} + u\underline{\underline{x}}_i \quad .$$

Clearly $\underline{\underline{x}}_i$ and $\underline{\underline{x}}_{i+1}$ are independent, and therefore $\underline{\underline{y}} \neq \underline{\underline{0}}$ . Similarly $\underline{\underline{z}} \neq \underline{\underline{0}}$ .

We have $0 \leq F(\underline{\underline{y}}) < X_i$ from (4.5) and (5.2) . It follows from (4.11) that

$$|L(\underline{\underline{y}})| < \frac{3}{4}X_i$$

if $i$ is large. Finally from (5.1) , (5.3) ,

$$|P(\underline{\underline{y}})| \leq |P_{i+1}| + X_{i+1}X_i^{-1}|P_i| \leq \frac{1}{2}X_i + o(X_{i+1}X_i^{-1}X_i^2X_{i+1}^{-1}) < \frac{3}{4}X_i$$

if $i$ is large. Thus $\langle\underline{\underline{y}}\rangle < X_i$ , but since (4.10) has no nonzero solution, we must have

$$|L(\underline{\underline{y}})| \geq |L_{i-1}| \quad .$$

Thus $|L_{i-1}| \leq |L_{i+1}| + t|L_i|$ , whence

$$u \leq t + |L_{i+1}|/|L_i| < t + 1 \quad ,$$

whence $u \leq t$ .

Turning to $\underline{\underline{z}}$ , we have

$$(5.4) \qquad |L(\underline{z})| = |L_{i-1}| - u|L_i| < |L_i|$$

since $L_{i-1}, L_i$ are of opposite sign. Since (4.10) has no nonzero solution, $\langle \underline{z} \rangle \geq X_{i+1}$. Since $|L(\underline{z})| = o(X_{i+1})$ and since

$$|P(\underline{z})| \leq |P_{i-1}| + u|P_i| \leq |P_{i-1}| + t|P_i| = o(X_{i+1})$$

by (5.1), we must have $|F(\underline{z})| \geq X_{i+1}$. Thus

$$X_{i-1} + uX_i \geq X_{i+1} ,$$

whence $t < u + 1$, whence $t \leq u$. Combining this with the opposite inequality shown above, we get

$$u = t .$$

Finally consider the point

$$(5.5) \qquad \underline{w} = \underline{x}_{i+1} - t\underline{x}_i - \underline{x}_{i-1} = \underline{y} - \underline{x}_{i-1} = \underline{x}_{i+1} - \underline{z} .$$

From the form $\underline{y} - \underline{x}_{i-1}$, from what we already know about $\underline{y}$, and from (5.1), we get $|L(\underline{w})|, |P(\underline{w})| < X_i$ if $i$ is large. From $0 \leq F(\underline{y}) < X_i$ we get $|F(\underline{w})| < X_i$, so that

$$\langle \underline{w} \rangle < X_i < X_{i+1} .$$

Also, from the form $\underline{x}_{i+1} - \underline{z}$, on noting that $L(\underline{z})$ has the same sign as $L_{i-1}$, and so the same sign as $L_{i+1}$, we obtain

$$|L(\underline{w})| \leq \max(|L_{i+1}|, |L(\underline{z})|) < |L_i| .$$

Thus $\underline{w}$ satisfies (4.10), which has the consequence that $\underline{w} = \underline{0}$.

The proof of Theorem 4A is now completed as follows. Suppose (5.3) is satisfied for some large $i$. Then by (5.2) and by Lemma 5B,

$$|X_i L_{i+1} - X_{i+1} L_i| = |X_{i-1} L_i - X_i L_{i-1}|$$

and

$$|P_i L_{i+1} - P_{i+1} L_i| = |P_{i-1} L_i - P_i L_{i-1}| \quad .$$

Now suppose that (5.3) holds for every $i$ in $h < i < k$ . Then

$$|X_h L_{h+1} - X_{h+1} L_h| = |X_{k-1} L_k - X_k L_{k-1}| \tag{5.6}$$

and

$$|P_h L_{h+1} - P_{h+1} L_h| = |P_{k-1} L_k - P_k L_{k-1}| \quad .$$

The latter gives

$$|P_{h+1} L_h| \leqq |P_h L_{h+1}| + |P_{k-1} L_k| + |P_k L_{k-1}| \quad . \tag{5.7}$$

These relations remain trivially true if $k = h + 1$ .

The left hand side of (5.6) is $|L(X_h \underline{x}_{h+1} - X_{h+1} \underline{x}_h)|$ , and this is not $0$ . But by (4.11) , the right hand side of (5.6) tends to $0$ as $k \to \infty$ . Hence (5.3) cannot be satisfied for all large $i$ , and so there are infinitely many $i$ for which (5.3) fails to hold.

Suppose that (5.3) is satisfied for every $i$ in $h < i < k$ , but not for $i = h$ or $i = k$ . Then $|P_{h+1}| > \frac{1}{2} X_h$ , and by (5.7) and by (4.12) , (5.1) we get

$$\begin{aligned} \frac{1}{2} X_h |L_h| &\leqq |P_h L_{h+1}| + |P_{k-1} L_k| + |P_k L_{k-1}| \\ &= o(X_h |L_{h+1}| + |L_{k-1}|^{1/2} X_{k-1}^2 |L_k| + |L_k|^{1/2} X_k^2 |L_{k-1}|) \; . \end{aligned}$$

Since $|L_{h+1}| < |L_h|$ and $|L_k| < |L_{k-1}|$ , this yields

$$X_h |L_h| = o(X_k^2 |L_k|^{1/2} |L_{k-1}|) \quad ,$$

which by $|L_{k-1}| \leqq |L_h|$ further implies

$$X_h |L_h|^{1/2} = o(X_k^2 |L_{k-1}|^{1/2} |L_k|^{1/2}) = o(X_k |L_k|^{1/2})$$

by (4.11) . In particular, if $h$ , and hence $k$ , is large, then

$$X_k |L_k|^{1/2} > 2X_h |L_h|^{1/2} .$$

But this is impossible, since it leads to an infinite sequence of values of $j$ for which $X_j |L_j|^{1/2}$ increases to infinity, while this expression is bounded by (4.11) .

§6. Proof of Wirsing's Theorem. Throughout, $\alpha$ will be a fixed real which is not algebraic of degree $\leqq k$ , and the constants in $\ll$ will depend on $\alpha$ and on $k$ .

Given a polynomial $P(X) = b_0(X - \beta_1) \dots (X - \beta_d)$ with $b_0 > 0$ and of degree $d \leqq k$ , put

$$p_i = |\alpha - \beta_i| \qquad (i = 1, \dots, d)$$

and

$$M_\alpha(P) = b_0 \prod_{i=1}^{d} \max(1, p_i) .$$

Then $M_\alpha(P)$ is equal to Mahler's $M(P_\alpha)$ where $P_\alpha(X) = P(X + \alpha)$ , so that

$$(6.1) \qquad M_\alpha(P) = M(P_\alpha) \gg \ll \lceil P_\alpha \rceil \gg \ll \lceil P \rceil$$

by (3.3) , (3.4) . Since

$$M_\alpha(P) \prod_{i=1}^{d} \min(1, p_i) = b_0 \prod_{i=1}^{d} p_i = |P(\alpha)| ,$$

we get

$$|P(\alpha)| \gg\ll \lceil P \rceil \prod_{i=1}^{d} \min(1,p_i) \quad . \tag{6.2}$$

LEMMA 6A. *Suppose*

$$P(X) = b_0(X-\beta_1)\ldots(X-\beta_d) \quad , \quad Q(X) = c_0(X-\gamma_1)\ldots(X-\gamma_e)$$

*are polynomials with integral coefficients and without common factor, of respective degrees* $d,e$ *where* $1 \leqq d, e \leqq k$ . *Put*

$$p_i = |\alpha-\beta_i| \quad (i=1,\ldots,d) \quad , \quad q_j = |\alpha-\gamma_j| \quad (j=1,\ldots,e) \quad ,$$

*and suppose that*

$$p_1 \leqq \ldots \leqq p_d \quad \text{and} \quad q_1 \leqq \ldots \leqq q_e \quad .$$

*Further suppose that*

$$p_1 \leqq q_1 < 1 \quad . \tag{6.3}$$

*Then one of the following relations holds.*

(i) $p_1 \ll |P(\alpha)| \lceil P \rceil^{-1}$ , *and* $\beta_1$ *is real,*

(ii) $q_1 \ll |Q(\alpha)| \lceil Q \rceil^{-1}$ , *and* $\gamma_1$ *is real,*

(iii) $1 \ll Q(\alpha)^2 \lceil P \rceil^{k} \lceil Q \rceil^{k-2}$ ,

(iv) $p_1^2 \ll |P(\alpha)| Q(\alpha)^2 \lceil P \rceil^{k-1} \lceil Q \rceil^{k-2}$ , *and* $\beta_1$ *is real,*

(v) $p_1^2 \ll P(\alpha)^2 |Q(\alpha)| \lceil P \rceil^{k-2} \lceil Q \rceil^{k-1}$ , *and* $\beta_1$ *is real.*

*Proof.* If $d = 1$ or if $d > 1$ and $p_2 \geqq 1$ , then (6.2) yields (i) . In particular $\beta_1$ is real, since $p_1 < p_2$ in the case $d > 1$ . Similarly, if $e = 1$ or if $e > 1$ and $q_2 \geqq 1$ , then (ii) follows. We may thus suppose that

(6.4) $$2 \leqq d,e \leqq k \quad , \text{ and that } \quad p_2 < 1, q_2 < 1 \quad .$$

Writing $R$ for the resultant of $P,Q$, we have

(6.5) $$1 \leqq |R| = b_0^e c_0^d \prod_{i,j} |\beta_i - \gamma_j| \ll b_0^k c_0^k \prod_{i,j} \max(p_i, q_j) = AB \quad ,$$

say, where

$$A = \prod_{\substack{i \\ p_i<1}} \prod_{\substack{j \\ q_j<1}} \max(p_i, q_j)$$

and where

$$B \leqq b_0^k c_0^k \prod_{i,j} \max(1,p_i)\max(1,q_j) \leqq M_\alpha(P)^k M_\alpha(Q)^k \ll \overline{|P|}^k \overline{|Q|}^k$$

by (6.1) .

Now if $p_2 \leqq q_1$ , then $p_1 \leqq p_2 \leqq q_1 < 1$ and

$$A \ll \prod_{\substack{j \\ q_j<1}} q_j^2 \ll Q(\alpha)^2 \overline{|Q|}^{-2}$$

by (6.2) , and (iii) follows.

We may thus suppose that $q_1 < p_2$ . We have the two subcases

(a) $p_1 \leqq q_1 < p_2 \leqq q_2$ ,

(b) $p_1 \leqq q_1 \leqq q_2 < p_2$ .

In the case (a) we have

$$A \ll \left( \prod_{\substack{i \geqq 2 \\ p_i<1}} p_i \right) q_1 \left( \prod_{\substack{j \geqq 2 \\ q_j<1}} q_j \right)^2 \quad .$$

Multiplying (6.5) by $p_1 q_1$ we get

$$p_1^2 \leqq p_1 q_1 \ll \left( \prod_{p_i<1} p_i \right) \left( \prod_{q_j<1} q_j \right)^2 \overline{|P|}^k \overline{|Q|}^k$$
$$\ll |P(\alpha)| Q(\alpha)^2 \overline{|P|}^{k-1} \overline{|Q|}^{k-2} \quad ,$$

from (6.2) . Thus (iv) holds in this case. (Note that $\beta_1$ is real since $p_1 < p_2$ .) Finally in the case (b) we multiply (6.5) by $p_1^2$ to get (v) .

The proof of Wirsing's Theorem 3B is now completed as follows. It is an immediate consequence of Minkowski's Linear Forms Theorem that there are infinitely many nonzero polynomials $P(X)$ of degree $\leq k$ and with rational integer coefficients having

$$|P(\alpha)| \ll \lceil P \rceil^{-k} .$$

If $P = P_1 \dots P_t$ is the factorization of $P$ into irreducible factors, then

$$|P_1(\alpha) \dots P_t(\alpha)| \ll (\lceil P_1 \rceil \dots \lceil P_t \rceil)^{-k}$$

by (3.8) . Since $P(\alpha) \neq 0$ by hypothesis, it is not very difficult to deduce that

there are infinitely many distinct polynomials $P$ of degree $\leq k$ with coprime rational integer coefficients, which are irreducible over the rationals and which have

$$|P(\alpha)| \ll \lceil P \rceil^{-k} . \tag{6.6}$$

If $P$ is such a polynomial, and if $Q$ is a polynomial of degree $\leq k$ with integral coefficients which is a multiple of $P$ , then $\lceil Q \rceil \gg \lceil P \rceil$ by (3.8)†). Let's say $\lceil Q \rceil \geq c_1 \lceil P \rceil$ . The inequalities

$$|q_0|, \dots, |q_k| \leq \frac{1}{2} c_1 \lceil P \rceil ,$$

$$|q_0 \alpha^k + \dots + q_k| \leq c_2 \lceil P \rceil^{-k}$$

define a symmetric convex set in $(k+1)$-dimensional space of $(k+1)$-

†) but with a different meaning of $Q$ .

tuples $(q_0,\dots,q_k)$ , of volume $\gg c_1^k c_2$ . Hence if $c_2$ is chosen sufficiently large, then our inequalities have a nontrivial integral solution. The polynomial

$$Q(X) = q_0 X^k + \cdots + q_k$$

has

$$(6.7) \qquad \lceil Q \rceil \ll \lceil P \rceil \quad \text{and} \quad |Q(\alpha)| \ll \lceil P \rceil^{-k} .$$

$Q$ is not a multiple of $P$ , and by the irreducibility of $P$ , the two polynomials $P,Q$ have no common factor.

Introducing the quantitites $p_i, q_j$ as in Lemma 6A , we have $p_1 < 1$ , $q_1 < 1$ from (6.6) , (6.7) . Suppose that $p_1 \leqq q_1$ , i.e. (6.3) holds. In the cases (i) and (ii) we have

$$|\alpha - \beta_1| \ll |P(\alpha)| \lceil P \rceil^{-1} \ll \lceil P \rceil^{-k-1} \ll H(\beta_1)^{-k-1}$$

and

$$|\alpha - \gamma_1| \ll |Q(\alpha)| \lceil Q \rceil^{-1} \ll \lceil P \rceil^{-k} \lceil Q \rceil^{-1} \ll \lceil Q \rceil^{-k-1} \ll H(\gamma_1)^{-k-1} ,$$

respectively. The case (iii) leads to

$$1 \ll \lceil P \rceil^{-2k} \lceil P \rceil^{2k-2} ,$$

hence is impossible for large $\lceil P \rceil$ . The case (iv) yields

$$|\alpha - \beta_1| \ll \lceil P \rceil^{-k/2} \lceil P \rceil^{-k} \lceil P \rceil^{(k-1)/2} \lceil P \rceil^{(k-2)/2} = \lceil P \rceil^{-(k+3)/2} \ll H(\beta_1)^{-(k+3)/2} ,$$

and the case (v) is similar. The situation is analogous when $q_1 < p_1$ .

<u>§7. A Subspace Theorem for Number Fields.</u> Before turning to a Roth type theorem for number fields, we need to obtain a generalized

Subspace Theorem.

Let $K$ be a number field of degree $k$ and $I$ the ring of integers of $K$. Denote the images of an element $\beta \in K$ under the isomorphisms of $K$ into $\mathbb{C}$ by $\beta^{(1)} = \beta$, $\beta^{(2)},\ldots,\beta^{(k)}$. Given

$$\underline{\underline{\beta}} = (\beta_1,\ldots,\beta_n)$$

with components in $K$, put $\underline{\underline{\beta}}^{(j)} = (\beta_1^{(j)},\ldots,\beta_n^{(j)})$ and

$$\overline{|\underline{\underline{\beta}}|} = \max(|\beta_1^{(1)}|,\ldots,|\beta_n^{(1)}|,\ldots,|\beta_1^{(k)}|,\ldots,|\beta_n^{(k)}|) .$$

THEOREM 7A. *Suppose that for each $j$ in $1 \leqq j \leqq k$ we are given $n$ linearly independent linear forms $M_{j1},\ldots,M_{jn}$ in $n$ variables, with real or complex algebraic coefficients. Suppose $\varepsilon > 0$. Then there are finitely many proper subspaces $T_1,\ldots,T_w$ of $n$-dimensional space $K^n$ such that every nonzero solution $\underline{\underline{\beta}} \in I^n$ of*

$$\text{(7.1)} \qquad \prod_{j=1}^{k} |M_{j1}(\underline{\underline{\beta}}^{(j)}) \ldots M_{jn}(\underline{\underline{\beta}}^{(j)})| < \overline{|\underline{\underline{\beta}}|}^{-\varepsilon}$$

*lies in one of these subspaces.*

*Proof*. Let $\theta_1,\ldots,\theta_k$ be an integer basis of $K$. We have

$$\beta_i = x_{i1}\theta_1 + \cdots + x_{ik}\theta_k \qquad (i = 1,\ldots,n)$$

with rational integers $x_{11},\ldots,x_{nk}$, and hence $\underline{\underline{x}} = (x_{11},\ldots,x_{nk})$ has

$$\text{(7.2)} \qquad \overline{|\underline{\underline{x}}|} \gg \ll \overline{|\underline{\underline{\beta}}|} .$$

Each

$$M_{ji}(\underline{\underline{\beta}}^{(j)}) = L_{ji}(\underline{\underline{x}}) ,$$

i.e. it is a linear form in $\underline{\underline{x}}$. If there were a nontrivial relation of linear dependence, say $\sum_{i,j} c_{ji} L_{ji} = 0$ , then

$$\sum_{i=1}^{n} c_{ji} M_{ji} = 0 \qquad (j = 1,\ldots,k) ,$$

contradicting our hypothesis on $M_{j1},\ldots,M_{jn}$ . So $L_{11},\ldots,L_{kn}$ are linearly independent.

Now suppose that a vector $\underline{\underline{\beta}} \in I^n$ satisfies (7.1) with $\varepsilon/2$ in place of $\varepsilon$ . Then by (7.2) ,

$$\prod_{j=1}^{k} \prod_{i=1}^{n} |L_{ij}(\underline{\underline{x}})| < \overline{|\underline{\underline{x}}|}^{-\varepsilon/3}$$

if $\overline{|\underline{\underline{x}}|}$ is large. By the rational Subspace Theorem (Ch. VI, Theorem 1F), applied with $kn$ in place of $n$, there are finitely many rational subspaces $S_1,\ldots,S_h$ of $\mathbb{Q}^{kn}$ , each of dimension $kn - 1$ , such that $\underline{\underline{x}}$ lies in one of these subspaces. Each subspace $S_t$ is defined by a linear relation in $x_{11},\ldots,x_{nk}$ . It is also defined by a linear relation in the $\beta_i^{(j)}$ :

$$(7.3) \quad c_{t11}\beta_1^{(1)} + \cdots + c_{t1n}\beta_n^{(1)} + \cdots + c_{tk1}\beta_1^{(k)} + \cdots + c_{tkn}\beta_n^{(k)} = 0 .$$

Pick $u = kh$ elements $\lambda_1,\ldots,\lambda_u$ in $I$ such that any $k$ among them are linearly independent over $\mathbb{Q}$ . Suppose $\underline{\underline{\beta}} \in I^n$ satisfies (7.1). If $\overline{|\underline{\underline{\beta}}|}$ is large, then each $\underline{\underline{\beta}}_s = \lambda_s \underline{\underline{\beta}}$ $(s = 1,\ldots,u)$ satisfies (7.1) with $\varepsilon$ replaced by $\varepsilon/2$ . Hence each $\underline{\underline{\beta}}_s$ satisfies an equation (7.3). That is, we have

$$(7.4) \quad \lambda_s^{(1)}(c_{t11}\beta_1^{(1)} + \cdots + c_{t1n}\beta_n^{(1)}) + \cdots + \lambda_s^{(k)}(c_{tk1}\beta_1^{(k)} + \cdots + c_{tkn}\beta_n^{(k)}) = 0$$

for some $t$ in $1 \leqq t \leqq h$ . Since $u = kh$ , there is a $t$ in $1 \leqq t \leqq h$ for which (7.4) is satisfied for at least $k$ values of $s$ in $1 \leqq s \leqq u$ . Say it is satisfied for $s = 1,\ldots,k$ . The linear independence of $\lambda_1,\dots,\lambda_k$ over $Q$ implies the nonvanishing of the determinant $|\lambda_s^{(i)}|$ $(1 \leqq i,s \leqq k)$, and it follows that

$$c_{tj1}\beta_1^{(j)} + \cdots + c_{tjn}\beta_n^{(j)} = 0 \qquad (j = 1,\dots,k) \tag{7.5}$$

There is a $j_0 = j_0(t)$ for which the equation (7.5) is nontrivial. Since $\beta_1^{(j_0)},\dots,\beta_n^{(j_0)}$ lie in the field $K^{(j_0)}$, the nontrivial relation (7.5) with $j = j_0$ implies a nontrivial relation

$$d_{t1}\beta_1^{(j_0)} + \cdots + d_{tn}\beta_n^{(j_0)} = 0$$

with coefficients $d_{t1},\dots,d_{tn}$ in $K^{(j_0)}$. From this we get a nontrivial relation

$$e_{t1}\beta_1 + \cdots + e_{tn}\beta_n = 0 \tag{7.6}$$

with coefficients $e_{t1},\dots,e_{tn}$ in $K$.

Thus every solution $\underline{\underline{\beta}}$ of (7.1) with large $\overline{|\underline{\underline{\beta}}|}$ satisfies (7.6) for some $t$ in $1 \leqq t \leqq h$, hence lies in one of $h$ proper subspaces $T_1,\dots,T_h$ of $K^n$. By adding a finite number of additional subspaces $T_{h+1},\dots,T_w$, we may take care of $\underline{\underline{\beta}}$ with small $\overline{|\underline{\underline{\beta}}|}$.

## §8. Approximation to Algebraic Numbers by Elements of Number Fields.

THEOREM 8A. *Let $\alpha$ be a real algebraic number and $K$ a real algebraic number field. Given $\delta > 0$, there are at most finitely many elements $\beta \in K$ having*

$$|\alpha - \beta| < H_K(\beta)^{-2-\delta} . \tag{8.1}$$

This was first proved by LeVeque (1955) who carried over Roth's arguments to number fields. Here we will deduce it from the Subspace

Theorem. A generalization to simultaneous approximation was carried out by Schmidt (1975). By applying the theorem to each subfield of K, we see that in fact

$$|\alpha - \beta| < H(\beta)^{-2-\delta}$$

has only finitely many solutions $\beta \in K$, where $H(\beta)$ is the absolute height.

Theorem 8A may be used to generalize the theory of Thue Equations (Theorem 3A of Ch. V) to number fields.

If $\underline{\underline{\beta}} = (\beta_0, \beta_1) \in K^2$, write

$$(8.2) \qquad \overline{|\underline{\underline{\beta}}|} = \max(|\beta_0^{(1)}|, |\beta_1^{(1)}|, \ldots, |\beta_0^{(k)}, \beta_1^{(k)}|) ,$$

where again $k = \deg K$ and the upper indices denote conjugation.

<u>LEMMA 8B</u>. <u>Given</u> $\beta$ <u>in</u> K, <u>there is a vector</u> $\underline{\underline{\beta}} = (\beta_0, \beta_1)$ <u>with components in the ring</u> I <u>of integers of</u> K, <u>with</u> $\beta_0 \neq 0$, <u>with</u>

$$\beta = \beta_1/\beta_0 ,$$

<u>and with</u>

$$(8.3) \qquad H_K(\beta) \gg\ll \overline{|\underline{\underline{\beta}}|}^k .$$

<u>Proof</u>. Let $\mathfrak{B}$ be the (possibly fractional) ideal generated by $1, \beta$. In the ideal class of $\mathfrak{B}$ there is an integral ideal $\mathfrak{C}$ with norm $\mathfrak{N}(\mathfrak{C}) \ll 1$. We have $\mathfrak{C} = \mathfrak{B}(\theta)$ with a principal ideal $(\theta)$, and $\mathfrak{C}$ is generated by $\gamma_0, \gamma_1$ where $\gamma_0 = \theta$, $\gamma_1 = \beta\theta$. By Dirichlet's unit theorem there is a unit $\varepsilon$ (of the ring I) such that the quantities $\beta_0 = \gamma_0 \varepsilon$, $\beta_1 = \gamma_1 \varepsilon$ have

$$\max(|\beta_0^{(i)}|,|\beta_1^{(i)}|) \gg\ll \left(\prod_{j=1}^{k} \max(|\gamma_0^{(j)}|,|\gamma_1^{(j)}|)\right)^{1/k} \qquad (i = 1,\ldots,k) .$$

Thus $\underline{\underline{\beta}} = (\beta_0,\beta_1)$ satisfies

$$(8.4) \qquad \lceil\underline{\underline{\beta}}\rceil \gg\ll \max(|\beta_0^{(i)}|,|\beta_1^{(i)}|) \qquad (i = 1,\ldots,k) .$$

The polynomial

$$B(X) = \prod_{i=1}^{k} (\beta_0^{(i)}X - \beta_1^{(i)})$$

is an integral multiple of the field polynomial $P_0$ of $\beta$ (see §2). In fact from the analogue in $K$ of Gauss' Lemma,

$$B(X) = \mathfrak{N}(\mathfrak{C})P_0(X) \quad ,$$

since $\beta_0,\beta_1$ generate $\mathfrak{C}$ . We obtain

$$H_K(\beta) = \lceil P_0\rceil \gg\ll \lceil B\rceil \gg\ll M(B) = \prod_{i=1}^{k} \max(|\beta_0^{(i)}|,|\beta_1^{(i)}|) \gg\ll \lceil\underline{\underline{\beta}}\rceil^k .$$

<u>Proof of Theorem 8A</u>. Given $\beta$ with (8.1) , construct $\underline{\underline{\beta}}$ as in Lemma 8B . After multiplying (8.1) by

$$|(\beta_0^{(1)})^2\beta_0^{(2)}\beta_1^{(2)} \ldots \beta_0^{(k)}\beta_1^{(k)}| \quad ,$$

we obtain

$$|(\alpha\beta_0^{(1)} - \beta_1^{(1)})\beta_0^{(1)}\beta_0^{(2)}\beta_1^{(2)} \ldots \beta_0^{(k)}\beta_1^{(k)}| \ll \lceil\underline{\underline{\beta}}\rceil^{2k-2k-\delta k} = \lceil\underline{\underline{\beta}}\rceil^{-\delta k} \quad ,$$

from (8.3) .

By Theorem 7A the solutions $\underline{\underline{\beta}} \in I^2$ of this inequality lie in finitely many one dimensional subspaces. But $\beta = \beta_1/\beta_0$ is constant in each such subspace.

It is possible to give results in which both $\alpha$ and $\beta$ are allowed to vary in a given number field $K$ . One can construct a constant

$c_1 = c_1(d,\delta)$ such that in any number field $K$ of degree $d$ there are only finitely many pairs $\alpha,\beta$ with

$$H(\beta) > H(\alpha)^{c_1} \quad \text{and} \quad |\alpha - \beta| < H(\beta)^{-2-\delta} \quad .$$

## §9. Approximation to Algebraic Numbers by Algebraic Numbers of Bounded Degree.

THEOREM 9A. *Suppose* $\alpha$ *is a real algebraic number, and suppose* $k \geq 1$ , $\delta > 0$ . *Then there are only finitely many algebraic numbers* $\beta$ *of degree* $\leq k$ *with*

$$|\alpha - \beta| < H(\beta)^{-k-1-\delta} \quad .$$

It may be shown that the $-k-1$ in the exponent is best possible. Roth (1955a,b) had conjectured and Wirsing (1971) had proved a result with $-2k$ in place of $-k-1$ . For $k = 1$ the theorem becomes Roth's Theorem.

Mahler (1932) introduced for arbitrary real $\alpha$ the quantity $\omega_k = \omega_k(\alpha)$ which is the least upper bound of the numbers $\omega$ such that

$$0 < |P(\alpha)| < \lceil P \rceil^{-\omega}$$

has infinitely many solutions in polynomials $P$ of degree $\leq k$ with integer coefficients. Clearly

$$\omega_1 \leq \omega_2 \leq \dots \quad .$$

If $\alpha$ is not algebraic of degree $\leq k$ , then $\alpha^k,\dots,\alpha,1$ are linearly independent over $\mathbb{Q}$ , and it follows from Corollary 1D of Ch. II that

$$\omega_k \geq k \quad . \tag{9.1}$$

If $\alpha$ is algebraic of degree $d$, then an obvious norm argument (like Lemma 1A of Ch. VI) shows that

$$\omega_k \leqq d - 1 \quad .$$

On the other hand it follows from Corollary 1E in Ch. VI that $0 < |P(\alpha)| < \overline{|P|}^{-k-\delta}$ has only finitely many solutions in polynomials of degree $\leqq k$ if $\alpha$ is algebraic, so that

$$\omega_k \leqq k \quad . \tag{9.2}$$

Hence

$$\omega_k = \min(k, d - 1) \tag{9.3}$$

if $\alpha$ is of degree $d$.

Koksma (1939) introduced the quantity $\omega_k^* = \omega_k^*(\alpha)$ which is the least upper bound of the numbers $\omega^*$ such that

$$|\alpha - \beta| < H(\beta)^{-\omega^*-1} \tag{9.4}$$

has infinitely many solutions in real algebraic numbers $\beta$ of degree $\leqq k$. Again

$$\omega_1^* \leqq \omega_2^* \leqq \dots \quad .$$

Now if $\beta \neq \alpha$ and if $P$ is the defining polynomial of $\beta$, then (9.4) implies

$$0 < |P(\alpha)| \ll |\alpha - \beta| \overline{|P|} \ll \overline{|P|}^{-\omega^*} \quad .$$

Hence always

$$\omega_k \geqq \omega_k^* \quad . \tag{9.5}$$

Thus from (9.2) , $\omega_k^* \leqq k$ is $\alpha$ is algebraic, and Theorem 9A follows.

We make a few additional remarks. The conjecture of §3 is that (in analogy to (9.1) )

$$\omega_k^* \geqq k ,$$

unless $\alpha$ is algebraic of degree $\leqq k$ , but Theorem 3B gives only

(9.6) $$\omega_k^* \geqq (k+1)/2 .$$

Wirsing (1961) proved for $\alpha$ which are not algebraic of degree $\leqq k$ that

(9.7) $$\omega_k^* \geqq (\omega_k + 1)/2 ,$$

which complements (9.5) , and which in conjunction with (9.1) yields (9.6) .

Sprindzuk (1965) (see also (1977)) proved a conjecture of Mahler (1932), to the effect that

(9.8) $$\omega_k(\alpha) = \omega_k^*(\alpha) = k$$

for almost all $\alpha$ , in the sense of Lebesgue measure.

<u>§10. Mahler's Classification of Transcendental Numbers</u>. We give a report without proofs. It follows from the last section that $\alpha$ is algebraic if and only if the sequence $\omega_1, \omega_2, \ldots$ is bounded, i.e. if

$$\omega_k \ll 1 .$$

Mahler (1932) calls these numbers A - <u>numbers</u>. He divides the remaining (i.e. the transcendental) real numbers into three classes as follows. The S - <u>numbers</u> are characterized by

(S) $\omega_k \not\ll 1$ but $\omega_k \ll k$ ,

the T - numbers by

(T) $\omega_k \not\ll k$ , but each $\omega_k < \infty$ ,

the U - numbers by

(U) some $\omega_k = \infty$ .

If (following Koksma (1939)) one defines an analogous classification in terms of the $\omega_k^*$ , then this classification in fact coincides with Mahler's classification by (9.5) , (9.7) . Almost every number is an S - number, which may be seen e.g. from Sprindzuk's result concerning (9.8) . Mahler proved that is $\alpha, \beta$ belong to two distinct classes among S,T,U , then $\alpha, \beta$ are algebraically independent.

The existence of T - numbers had been left open by Mahler, but was proved by Schmidt (1968) (see also Schmidt (1971a) and Baker (1975)). Schmidt (1971a) constructs T - numbers with $\omega_k \ll k^3$ , and Baker asserts the existence of T - numbers with $\omega_k \ll k^2$ , but so far the existence of T - numbers with $\omega_k = o(k^2)$ has not been established.

An U - number is of degree k if $\omega_k = \infty$ but $\omega_j < \infty$ for $j < k$ . The number $\alpha$ constructed in Chapter V below Liouville's Theorem is a U - number of degree 1 . LeVeque (1953) establishes the existence of U - numbers of prescribed degrees.

§11. A Theorem of Mignotte. Suppose $\alpha, \beta$ are algebraic but not conjugates†). Suppose they are of respective degrees d,e , with defining polynomials

†) For differences of conjugate algebraic numbers see Mignotte and Payafar (1979) .

$$A(X) = a_0(X-\alpha^{(1)}) \dots (X-\alpha^{(d)}) \quad \text{and} \quad B(X) = b_0(X-\beta^{(1)}) \dots (X-\beta^{(e)}) \quad .$$

The resultant R of A and B is not zero. It follows that

$$1 \leqq |R| = |\alpha-\beta| a_0^e b_0^d \prod_{\substack{i=j \\ (i,j)\neq(1,1)}}^{d} \prod_{j=1}^{e} |\alpha^{(i)} - \beta^{(j)}|$$

$$\leqq |\alpha-\beta| 2^{de} M(A)^e M(B)^e \quad ,$$

where $M(A)$ , $M(B)$ are Mahler's constants' of §3. Thus if we put $M(\alpha) = M(A)$ and $M(\beta) = M(B)$ , then

$$|\alpha-\beta| \geqq 2^{-de} M(\alpha)^{-e} M(\beta)^{-d} \quad .$$

In particular, if $\alpha = 1$ , then $d = 1$ , $M(\alpha) = 1$ , and

(11.1) $$|1-\beta| \geqq 2^{-e} M(\beta)^{-1} \quad .$$

<u>THEOREM 11A</u>. (Mignotte (to appear)). <u>Suppose</u> $\alpha,\beta$ <u>are as above, and put</u> $\alpha^* = \max(1,|\alpha|)$ . <u>Suppose further that</u>

(11.2) $$4d \log(8\alpha^* e M(\beta)) \leqq e \log 2e \quad .$$

<u>Then</u>

(11.3) $$|\alpha-\beta| \geqq (2e)^{-d} M(\alpha)^{-2e} \exp\left(-3(de \log 2e)^{1/2} (\log(8\alpha^* e M(\beta)))^{1/2}\right) \quad .$$

In the estimates given earlier, we had assumed that $\alpha$ was fixed and that $\beta$ was of bounded degree. The present estimate is different. It is most useful when $M(\beta)$ is fixed and the degree $e$ of $\beta$ varies. For example, when $\alpha = 1$ and when $e$ is large compared to $M(\beta)$ , then

(11.4) $$|1-\beta| \geqq \exp(-4e^{1/2} \log 8eM(\beta)) \quad ,$$

which is (for bounded $M(\beta)$ and large $e$ ) better than (11.1) .

The proof has some points in common with a recent result of Dobrowolski (1979) that for $\varepsilon > 0$ and for an algebraic integer $\beta$ of degree $e > c_1(\varepsilon)$ ,

$$M(\beta) > 1 + (1 - \varepsilon)(\log\log e/\log e)^3 ,$$

unless $\beta$ is a root of unity.

LEMMA 11B*. *Suppose* $A(X) = a_0(X - \alpha^{(1)}) \ldots (X - \alpha^{(d)})$ *is the defining polynomial of* $\alpha$ . *Then if* $\{i_1,\ldots,i_s\}$ *is a subset of* $\{1,\ldots,d\}$ ,

$$a_0\alpha^{(i_1)} \ldots \alpha^{(i_s)}$$

*is an algebraic integer*.

This auxiliary lemma is well known and will not be proved here. See e.g. Schneider (1957, Hilfssatz 17).

LEMMA 11C. *Let* $P_{ij}(X)$ $(1 \leqq i \leqq m , 1 \leqq j \leqq n)$ *be polynomials of degree* $\leqq p$ *with rational integer coefficients*. *Write* $L(P_{ij})$ *for the sum of the absolute values of the coefficients of* $P_{ij}$ , *and put*

$$L_i = L(P_{i1}) + \cdots + L(P_{in}) \qquad (i = 1,\ldots,m) .$$

*Let* $\alpha$ *be algebraic of degree* $d$ . *If* $L_i \neq 0$ $(i = 1,\ldots,m)$ *and if*

$$(11.5) \qquad n > md ,$$

*then there is a nonzero integer point* $\underline{\underline{x}} = (x_1,\ldots,x_n)$ *with*

$$(11.6) \qquad P_{i1}(\alpha)x_1 + \cdots + P_{in}(\alpha)x_n = 0 \qquad (i = 1,\ldots,m)$$

*having*

$$(11.7) \qquad |\underline{\underline{x}}| \leq [\left((L_1 \ldots L_m)^d M(\alpha)^{mp}\right)^{1/(n-md)}] + 1 \quad .$$

This is a variant on Siegel's Lemma (Lemma 5B of Ch. V ). See Mignotte and Waldschmidt (1978) .

Proof. Construct linear forms

$$\mathfrak{L}_{ij}(\underline{\underline{X}}) = P_{i1}(\alpha^{(j)})X_1 + \cdots + P_{in}(\alpha^{(j)})X_n \qquad (1 \leq i \leq m\ ,\ 1 \leq j \leq d)\ .$$

Denote the right hand side of (11.7) by B . The n + md inequalities in $\underline{\underline{x}} = (x_1, \ldots, x_n)$ ,

$$(11.8) \qquad |x_i| \leq B \qquad (i = 1,\ldots,n)\ ,$$

$$(11.9) \qquad |\mathfrak{L}_{ij}(\underline{\underline{x}})| < a_0^{-p/d} \qquad (1 \leq i \leq m\ ,\ 1 \leq j \leq d)\ ,$$

where $a_0$ is the leading coefficient of A[†)], define a symmetric convex set $\mathfrak{K}$ in $\mathbb{R}^n$ . The inequalities (11.8) alone define a cube $\mathfrak{C}$ of volume $(2B)^n$ . Now consider a fixed inequality (11.9) , and suppose that $\alpha^{(j)}$ is real. Then the inequality (11.9) defines a "strip", i.e. a symmetric set bounded by two parallel hyperplanes. For $\underline{\underline{x}} \in \mathfrak{C}$ ,

$$|\mathfrak{L}_{ij}(\underline{\underline{x}})| \leq BL(\mathfrak{L}_{ij}) \quad ,$$

where $L(\mathfrak{L}_{ij})$ is the sum of the absolute values of the coefficients of $\mathfrak{L}_{ij}$ . Hence the intersection of $\mathfrak{C}$ with the set defined by this particular inequality (11.9) will cut down the volume by a factor

$$\geq \min(1, a_0^{-p/d} B^{-1} L(\mathfrak{L}_{ij})^{-1}) \quad .$$

Or suppose that $\alpha^{(i)}, \alpha^{(j+1)}$ are conjugate complex, and consider a pair of inequalities (11.9) involving $\mathfrak{L}_{ij}$ and $\mathfrak{L}_{i,j+1}$ . These inequalities

---

†) A is the defining polynomial of $\alpha$ .

define a circular cylinder. For $\underline{\underline{x}} \in \mathfrak{C}$ ,

$$|\mathfrak{L}_{ij}(\underline{\underline{x}})| = |\mathfrak{L}_{i,j+1}(\underline{\underline{x}})| \leqq BL(\mathfrak{L}_{ij}) = BL(\mathfrak{L}_{i,j+1}) \quad .$$

Hence the intersection with the set defined by this pair of inequalities will cut down the volume by a factor

$$\geqq \min(1, a_0^{-2p/d} B^{-2} L(\mathfrak{L}_{ij})^{-1} L(\mathfrak{L}_{i,j+1})^{-1}) \quad .$$

It follows that $\mathfrak{K}$ has volume

$$V(\mathfrak{K}) \geqq (2B)^n \prod_{i=1}^{m} \prod_{j=1}^{d} \min(1, a_0^{-p/d} B^{-1} L(\mathfrak{L}_{ij})^{-1})$$

$$\geqq 2^n B^{n-md} \prod_{i=1}^{m} \prod_{j=1}^{d} (\max(1, a_0^{p/d} L(\mathfrak{L}_{ij})))^{-1} \quad .$$

Now $L(\mathfrak{L}_{ij}) \leqq L_i(\max(1, |\alpha^{(j)}|))^p$ , whence we get

$$V(\mathfrak{K}) \geqq 2^n B^{n-md} (L_1 \ldots L_m)^{-d} M(\alpha)^{-mp} \quad .$$

By our choice of $B$ , $V(\mathfrak{K}) > 2^n$ , so that according to Minkowski there is a nontrivial solution $\underline{\underline{x}}$ of (11.8) and (11.9) . Now by Lemma 11B ,

$$a_0^p \, \mathfrak{L}_{i1}(\underline{\underline{x}}) \ldots \mathfrak{L}_{id}(\underline{\underline{x}}) \qquad (i = 1, \ldots, m)$$

is an algebraic integer, and in fact it must be a rational integer. On the other hand by (11.9) it has absolute value $< 1$ , so that it must be zero. Thus (11.6) is satisfied.

LEMMA 11D. Suppose $\alpha$ is algebraic of degree $d$ . For $n > md$ , there exists a nonzero polynomial $P(X)$ of degree $< n$ with rational integer coefficients which has a zero at $x = \alpha$ of order $\geqq m$ , and which has

(11.10) $$\lceil P \rceil \le \left[\left(\left(\prod_{i=1}^{m} \binom{n}{i}\right)^{d} M(\alpha)^{mn}\right)^{1/(n-md)}\right] + 1 \quad .$$

Proof. Set $P(X) = p_0X^{n-1} + \cdots + p_{n-2}X + p_{n-1}$ . After dividing a typical condition $P^{(i)}(\alpha) = 0$ by $i!$ , we get the conditions

$$p_0\binom{n-1}{i}\alpha^{n-1-i} + p_1\binom{n-2}{i}\alpha^{n-2-i} + \cdots + a_i = 0 \qquad (i = 0,1,\ldots,m-1).$$

We may apply Lemma 11C with $p = n - 1$ and with

$$L_i = \binom{n-1}{i} + \binom{n-2}{i} + \cdots + 1 = \binom{n}{i+1} \qquad (i = 0,1,\ldots,m-1) \quad .$$

Lemma 11D follows.

Henceforth denote the right hand side of (11.10) by $C$ . It is clear that

(11.11) $$|P(z)| \le nC\max(1,|z|^n)$$

for every complex number $z$ . On the other hand, by the maximum principle, the function $P(z)/(z-\alpha)^m$ takes its maximum modulus in the disk $|z| \le 2\alpha^*$ on the boundary, so that for $|z| \le 2\alpha^*$ ,

(11.12) $$|P(z)| \le |z-\alpha|^m nC(2\alpha^*)^n|2\alpha^*-\alpha|^{-m} \le |z-\alpha|^m nC2^n\alpha^{*n-m}$$

Suppose now that $\alpha,\beta,A,B,d,e$ are as in Theorem 11A , and that

(11.13) $$n = md + e \quad .$$

Then $P$ , since it is of degree $< n$ , and since it is divisible by $A^m$ , cannot be divisible by $B$ , and hence $P(\beta) \neq 0$ . Suppose also that

(11.14) $$|\beta| \le 2\alpha^* \quad .$$

Estimating $P(\beta)$ by (11.12) , and the conjugates $P(\beta^{(j)})$ $(j = 2,\ldots,e)$

by (11.11) , we get

$$|P(\beta)| \leqq |\alpha-\beta|^m nC2^n \alpha^{*n-m} ,$$

$$|P(\beta^{(j)})| \leqq nC\max(1,|\beta^{(j)}|^n) \qquad (j = 2,\ldots,e) .$$

Now if $b_0$ is the leading coefficient of the defining polynomial of $\beta$, then $b_0^n P(\beta^{(1)}) \ldots P(\beta^{(e)})$ is an algebraic integer by Lemma 11B . In fact it is a nonzero rational integer. It follows that

$$\begin{aligned} 1 &\leqq |\alpha-\beta|^m (nC)^e\, 2^n\, \alpha^{*n-m}\, b_0^n \prod_{j=2}^{e} \max(1,|\beta^{(j)}|^n) \\ &\leqq |\alpha-\beta|^m (nC)^e\, 2^n\, \alpha^{*n-m}\, M(\beta)^n . \end{aligned} \qquad (11.15)$$

Now suppose that

$$m \leqq \frac{1}{2}n . \qquad (11.16)$$

Then $\binom{n}{i} \leqq \binom{n}{m} \leqq n^m$ for $i \leqq m$ , and by (11.13) and the definition of $C$ ,

$$C^e \leqq 2^e n^{m^2 d} M(\alpha)^{mn} .$$

Substituting this into (11.15) we get

$$1 \leqq |\alpha-\beta|^m M(\alpha)^{mn} M(\beta)^n n^{m^2 d+e} 2^{n+e} \alpha^{*n} ,$$

whence (since $e < n$ ),

$$1 \leqq |\alpha-\beta|\, M(\alpha)^n (4\alpha^* nM(\beta))^{n/m} n^{md} . \qquad (11.17)$$

We now choose $m$ to make the right hand side small. We put

$$m = [(e\log(8\alpha^* eM(\beta))/(d\log 2e))^{1/2}] + 1 .$$

Then

$$md \leqq \left(de \log(8\alpha^{*}eM(\beta))/\log 2e\right)^{1/2} + d \leqq \frac{1}{2}e + \frac{1}{2}e = e$$

by (11.2) . In conjunction with (11.13) we get (11.16) . We also have $n \leqq 2e$ , and therefore

$$(4\alpha^{*}nM(\beta))^{n/m} \leqq \exp\left(2\left(de(\log 2e)(\log(8\alpha^{*}eM(\beta)))\right)^{1/2}\right) .$$

Finally

$$n^{md} \leqq (2e)^{md} \leqq (2e)^{d}\exp\left(\left(de(\log 2e)(\log(8\alpha^{*}eM(\beta)))\right)^{1/2}\right) .$$

Substituting our estimates into (11.17) we get the desired (11.3) .

If (11.14) fails to hold, then $|\alpha - \beta| \geqq 1$ and (11.3) is trivially true.

## REFERENCES


A. Baker (1967). Simultaneous approximation to certain algebraic numbers. Proc. Camb. Phil. Soc. 63, 693-702.

_______ (1967/68). Contributions to the theory of Diophantine equations. I. On the representation of integers by binary forms. Philos. Trans. Roy. Soc. London Ser A 263, 173-191.

_______ (1975). Transcendental Number Theory. Cambridge University Press.

A. Baker and W.M. Schmidt (1970). Diophantine Approximation and Hausdorff Dimension. Proc. London Math. Soc. (3) 21, 1-11.

A.S. Besicovitch (1934). Sets of fractional dimension (IV): On rational approximations to real numbers. J. London Math. Soc. 9, 126-131.

H.F. Blichfeldt (1914). A new principle in the geometry of numbers with some applications. Trans. A.M.S. 15, 227-235.

E. Borel (1903a). Sur l'approximation des nombres par des nombres rationnels. C. R. Acad. Sci. Paris 136, 1054-1055.

_______ (1903b). Contribution à l'analyse arithmétique du continu. J. Math. pures appl. (5) 9, 329-375.

Z.I. Borevich and I.R. Shafarevich (1966). Number Theory. (Translated from the Russian). Academic Press.

J.W.S. Cassels (1950a). Some metrical theorems in diophantine approximation I. Proc. Camb. Phil. Soc. 46, 209-218.

___________ (1950b). Some metrical theorems in diophantine approximation III. Ibid. 219-225.

___________ (1955a). Simultaneous diophantine approximation.

Journ. London Math. Soc. 30, 119-121.

_____________ (1955b). Simultaneous diophantine approximation II. Proc. Lon. Math. Soc. (3) 5, 435-448.

_____________ (1956). On a result of Marshall Hall. Mathematika 3 109-110.

_____________ (1957). An introduction to diophantine approximation. Cambridge Tracts 45, Cambridge Univ. Press.

_____________ (1959). An introduction to the Geometry of Numbers. Springer Grundlehren 99. Berlin - Göttingen-Heidelberg.

T.W. Cusick (1975). The connection between the Lagrange and Markoff Spectra. Duke Math. J. 42, 507-517.

H. Davenport (1937). Note on a result of Siegel. Acta Arith. 2, 262-265.

___________ (1952). Simultaneous Diophantine approximation. Proc. London Math. Soc. (3) 2, 406-416.

___________ (1954). Simultaneous Diophantine approximation. Mathematika 1, 51-72.

___________ (1962). A note on Diophantine approximation. Studies in Math. Analysis and Related Topics, Stanford Univ. Press, 77-81.

___________ (1964). A note on Diophantine approximation II. Mathematika 11, 50-58.

___________ (1968). A note on Thue's Theorem. Mathematika 15, 76-87.

H. Davenport and K. Mahler (1946). Simultaneous diophantine approximation. Duke Math. J. 43, 105-111.

H. Davenport and K.F. Roth (1955). Rational approximation to algebraic numbers. Mathematika 2, 160-167.

H. Davenport and W.M. Schmidt (1967). Approximation to real numbers by quadratic irrationals. Acta. Arith. 13, 169-176.

______________ (1968). Dirichlet's Theorem on diophantine approximation. Rendic. Convegno Teoria dei numeri, Rome. 113-132.

______________ (1969). Approximation to real numbers by algebraic integers. Acta Arith. 15, 393-416.

______________ (1970). Supplement to a theorem on linear forms. Colloquia Math. Soc. János Bolyai (Number Th., Debrecen, Hung. 1968), 15-25.

L.G.P. Dirichlet (1842). Verallgemeinerung eines Satzes aus der Lehre von den Kettenbrüchen nebst einigen Anwendungen auf die Theorie der Zahlen. S. B. Preuss. Akad. Wiss. 93-95.

E. Dobrowolski (1979). On a question of Lehmer and the number of irreducible factors of a polynomial. Acta. Arith. 34, 391-401.

F.J. Dyson (1947a). The approximation to algebraic numbers by rationals. Acta Math. Acad. Sci. Hung. 79, 225-240.

______ (1947b). On simultaneous Diophantine approximations. Proc. Lond. Math. Soc. (2) 49, 409-420.

P. Erdös (1970). On the Distribution of the Convergents of Almost All Real Numbers. J. of Number Theory 2, 425-441.

L. Euler (1737). De fractionibus continuis. Commentarii Acad. Sci. Imperiali Petropolitanae 9.

_____ (1748). Introductio in analysin infinitorum I.

N.I. Feldman (1951). The approximation of certain transcendental numbers. I. (Russian). Izv. Akad. Nauk SSSR Ser Mat. 15, 53-74.

__________ (1970). Bounds for linear forms of certain algebraic numbers (Russian). Mat. Zametki 7, 569-580. English Transl. Math. Notes 7 (1970), 343-349.

__________ (1971). An effective power sharpening of a theorem of Liouville (Russian). Izv. Akad. Nauk SSSR Ser. Mat. 35, 973-990.

P. Furtwängler (1927-28). Über die Simultane Approximation von Irrationalzahlen I, II. Math. Ann. 96, 169-175 and 99, 71-83.

P.X. Gallagher (1962). Metric simultaneous diophantine approximation. J. Lon. Math. Soc. 37, 387-390.

__________ (1965). Metric simultaneous diophantine approximation. II Mathematika 12, 123-127.

A.O. Gelfond (1952). Transcendental and Algebraic Numbers. (Russian). Moscow. English transl. (1969), Dover Publications, New York.

W.H. Greub (1967). Multilinear Algebra. Springer Grundlehren 136.

M. Hall (1947). On the sum and product of continued fractions. Ann. of Math. (2) 48, 966-993.

G.H. Hardy and E.M. Wright (1954). An introduction to the theory of numbers. (3rd ed.) Oxford, Clarendon Press.

C.J. Hightower (1970). The minima of indefinite binary quadratic forms. J. Number Theory 2, 364-378.

A. Hurwitz (1891). Über die angenäherte Darstellung der Irrationalzahlen durch rationale Brüche. Math. Ann. 39, 279-284.

__________ (1906). Über eine Aufgabe der unbestimmten Analysis. Arch. Math. Phys. 11 (1906), 185-196.

V. Jarnik (1929). Diophantische Approximationen und Hausdorffsches Mass. Recueil math. Moscow 36, 371-382.

________ (1959). Eine Bemerkung zum Übertragungssatz. Bulgar. Akad. Nauk Izv. Mat. Inst. 3, No 2, 169-175.

F. John (1948). Extremum problems with inequalities as subsidiary conditions. Studies and Essays presented to R. Courant. Interscience, New York, 187-204.

A. Khintchine (1924). Einige Sätze über Kettenbrüche, mit Anwendungen auf die Theorie der Diophantischen Approximationen. Math. Ann. 92, 115-125.

____________ (1925). Zwei Bemerkungen zu einer Arbeit des Herrn Perron. Math. Z. 22, 274-284.

____________ (1926a). Zur metrischen Theorie der diophantischen Approximationen. Math. Z. 24, 706-714.

____________ (1926b). Über eine Klasse linearer Diophantischer Approximationen. Rend. Circ. Math. Palermo 50, 170-195.

J.F. Koksma (1939). Über die Mahlersche Klasseneinteilung der transzendenten Zahlen und die Approximation komplexer Zahlen durch algebraische Zahlen. Mh. Math. Physik 48, 176-189.

L. Kuipers and H. Niederreiter (1974). Uniform distribution of sequences. John Wiley & Sons. New York-London-Sydney-Toronto.

J.L. Lagrange (1770a). Additions aux éléments d'algebra d'Euler.

____________ (1770b). Additions au mémoire sur la résolution des équations numériques. Mém. Berl. 24.

S. Lang (1965). Report on diophantine approximations. Bull. de la Soc. Math. de France 93, 117-192.

C.G. Lekkerkerker (1969). Geometry of Numbers. Wolters-Noordhoff Publishing and American Elsevier Publishing.

W.J. LeVeque (1953). On Mahler's U-numbers. J. London Math. Soc. 28, 220-229.

___________ (1955). Topics in Number Theory. Addison-Wesley Publ. Co.: Reading, Mass.

J. Liouville (1844). Sur des classes très-étendues de quantités dont la irrationelles algébriques. C.R. Acad. Sci. Paris 18, 883-885 and 910-911.

J.M. Mack (1977). Simultaneous diophantine approximation. J. Austral. Math. Soc. 24 (Series A), 266-285.

K. Mahler (1932). Zur Approximation der Exponentialfunktion und des Logarithmus I. J. reine ang. Math. 166, 118-136.

_________ (1933). Zur Approximation algebraischer Zahlen (II). Über die Anzahl der Darstellungen ganzer Zahlen durch Binärformen. Math. Ann. 108, 37-55.

_________ (1939). Ein Übertragungsprinzip für konvexe Körper. Časopis Pěst. Mat. Fys. 68, 93-102.

_________ (1955). On compound convex bodies I. Proc. Lon. Math. Soc. (3), 5, 358-379.

_________ (1960). An application of Jensen's Formula to Polynomials. Mathematika 7, 98-100.

________ (1961). Lectures on diophantine approximation. Notre Dame University.

________ (1963). On the approximation of algebraic numbers by algebraic integers. J. Austral. Math. Soc. 3, 408-434.

A.A. Markoff (1879). Sur les formes quadratiques binaires indéfinies. I. Math. Ann. 15, 381-409.

M. Mignotte. Approximation des nombres algébriques par des nombres algébriques de grand degré. Ann. Fac. Sci. Toulouse, 1 (1979), 165-170.

M. Mignotte and M. Payafar (1979). Distance entre les racines d'un polynôme. R.A.I.R.O. Analyse numérique 13, 181-192.

M. Mignotte and M. Waldschmidt (1978). Linear forms and Schneider's method. Math. Ann. 231, 241-267.

H. Minkowski (1896 & 1910). Geometrie der Zahlen. Teubner: Leipzig u. Berlin (The 1910 ed. prepared posthumously by Hilbert and Speiser).

__________ (1907). Diophantische Approximationen Teubner: Leipzig u. Berlin.

H. Montgomery (1975/76). Polynomials in Many Variables. Séminaire Delange-Pisot-Poitou (Théorie des nombres), no 7, 6 p.

L.J. Mordell (1934). On some arithmetical results in the geometry of numbers. Compositio Math. 1, 248-253.

I. Niven (1962). On asymmetric Diophantine approximations. Michigan Math. J. 9, 121-123.

C.F. Osgood (1970). The simultaneous approximation of certain k-th roots. Proc. Camb. Phil. Soc. 67, 75-86.

__________ (1977). Concerning a possible "Thue-Siegel-Roth Theorem" for algebraic differential equations. Number theory and algebra. Collected papers ded. to H.B. Mann, A.E. Ross and O. Taussky-Todd. New York-London: Academic Press.

J.C. Oxtoby (1957). The Banach-Mazur game and Banach category theorem. Contrib. to the theory of games, vol. 3, 159-163. Annals of Math. Studies No. 39, Princeton Univ. Press.

O. Perron (1921a). Über die Approximation irrationaler Zahlen durch rationale I, II. S.B. Heidelberg Akad. W.

________ (1921b). Über diophantische Approximationen. Math. Ann. 83, 77-84.

________ (1954). Die Lehre von den Kettenbrüchen. 3. Aufl. B.G. Teubner: Stuttgart.

M. Ratliff (1978). The Thue-Siegel-Roth-Schmidt Theorem for Algebraic Functions. J. of Number Th. 10, 99-126.

K.F. Roth (1955a). Rational approximations to algebraic numbers. Mathematika 2, 1-20.

________ (1955b). Corrigendum. Ibid, 168.

A. Schinzel (1967). Review of a paper by Hyyrö. Zentralblatt Math. 137, 257-258.

H.P. Schlickewei (1976). Die p - adische Verallgemeinerung des Satzes von Thue-Siegel-Roth-Schmidt. J. reine ang. Math. 288, 86-105.

______________ (1977). On Norm Form Equations. J. of Number Th. 9, 370-380.

W.M. Schmidt (1960). A metrical theorem in diophantine approximation. Canad. J. Math. 12, 619-631.

___________ (1964). Metrical theorems on fractional parts of sequences. Trans. Am. Math. Soc. 110, 493-518.

___________ (1965a). On badly approximable numbers. Mathematika 12, 10-20.

___________ (1965b). Über simultane Approximation algebraischer Zahlen durch rationale. Acta Math. 114, 159-206.

___________ (1966). On badly approximable numbers and certain games. Trans. A.M.S. 123, 178-199.

___________ (1967). On simultaneous approximation of two algebraic numbers by rationals. Acta Math. 119, 27-50.

___________ (1968). T-numbers do exist. Rendiconti convegno di Teoria dei numeri, Roma, 3-26.

___________ (1969). Badly approximable systems of linear forms. J. Number Theory 1, 139-154.

___________ (1970). Simultaneous approximation to algebraic numbers by rationals. Acta Math. 125, 189-201.

___________ (1971a). Mahler's T-numbers. Proc. of Symp. of Pure Math. 20 (1969 Inst. of Number Theory) 275-286.

___________ (1971b). Linear forms with algebraic coefficients. I. J. of Number Th. 3, 253-277.

___________ (1971c). Linearformen mit algebraischen Koeffizienten. II. Math. Ann. 191, 1-20.

___________ (1971d). Approximation to algebraic numbers. L'Enseignement

Math. 17, 187-253.

___________ (1972). Norm form equations. Ann. of Math. 96, 526-551.

___________ (1973). Inequalities for Resultants and for Decomposable Forms. Proc. Conf. Dioph. Approximation and its Applications. (Ed. C.F. Osgood). Academic Press: New York and London.

___________ (1975). Simultaneous Approximation to Algebraic Numbers by Elements of a Number Field. Monatsh. Math. 79, 55-66.

W.M. Schmidt and Wang Y. A note on a transference theorem of linear forms. Scientia Sinica 22, 276-280.

Th. Schneider (1936). Über die Approximation algebraischer Zahlen. J. reine ang. Math. 175, 182-192.

___________ (1957). Einführung in die transzendenten Zahlen. Springer Grundlehren 81. Berlin-Göttingen-Heidelberg.

B. Segre (1945). Lattice points in infinite domains and asymmetric Diophantine approximations. Duke Math. J. 12, 337-365.

J.A. Serret (1878). Handbuch der höheren Algebra, vol. I. (German translation).

C.L. Siegel (1921). Approximation algebraischer Zahlen. Math. Zeitschr. 10, 173-213.

___________ (1970). Einige Erläuterungen zu Thues Untersuchungen über Annäherungswerte algebraischer Zahlen and diophantische Gleichungen. Nachr. Akad. d. Wiss. Göttingen, Math. Phys. Kl., Nr. 8.

V.G. Sprindzuk (1965). A proof of Mahler's conjecture on the measure of the set of S - numbers (Russian). Izv. Akad. Nauk SSSR Ser. Mat. 29, 379-436.

______________ (1977). Metrical Theory of Diophantine Approximation. (Russian). Moscow.

A. Thue (1908). Om en generel i store hele talulφsbar ligning. Kra. Vidensk. Selsk. Skrifter. I. Mat. Nat. Kl. No. 7. Kra.

_______ (1909). Über Annäherungswerte algebraischer Zahlen. J. reine ang. Math. 135, 284-305.

_______ (1977). Selected Mathematical Papers. Edited by T. Nagell, A. Selberg, S. Selberg, K. Thalberg. Universitetsforlaget Oslo-Bergen-Tromsφ.

K. Th. Vahlen (1895). Über Näherungswerte und Kettenbrüche. J. reine angew. Math. 115, 221-233.

B.L. Van der Waerden (1955). Algebra (4th ed.) Springer Grundlehren 33. Berlin-Göttingen-Heidelberg.

L. Vehmanen (1976). On linear forms with algebraic coefficients. Acta Univ. Tamperensis, ser A, 71, 36 p.

Wang Yuan, Yu Kunrui and Zhu Yao-cheng (1979). A note on a transference theorem concerning the linear forms. Acta Math. Sinica 22.

E. Wirsing (1961). Approximation mit algebraischen Zahlen beschränkten Grades. J. reine ang. Math. 206, 67-77.

__________ (1971). On approximation of algebraic numbers by algebraic numbers of bounded degree. Proc. of Symp. in Pure Math. XX (1969 Number Theory Institute) 213-247.

Printing: Weihert-Druck GmbH, Darmstadt
Binding: Theo Gansert Buchbinderei GmbH, Weinheim